Human Genetics

2nd edition

Human

Edward Novitski

University of Oregon, Eugene

Genetics

Macmillan Publishing Co., Inc.
NEW YORK

Collier Macmillan Publishers
LONDON

Macmillan Publishing Co., Inc.
866 Third Avenue, New York, New York,   10022

Collier Macmillan Canada, Ltd.

Library of Congress Cataloging in Publication Data

Novitski, Edward.
  Human genetics.

  Includes bibliographies and index.
  1. Human genetics.   i. Title.
QH431.N68   1982   573.2'1   81-8303
ISBN 0-02-388570-X   AACR2

Printing: 1 2 3 4 5 6 7 8   Year: 1 2 3 4 5 6

To the Memory of Curt Stern

Teacher, Colleague and Friend

Preface

The swift pace of medical and biological advances has necessitated an extensive revision of the original edition of *Human Genetics*. Some sections, such as those on cancer and recombinant DNA, have been extended; others (sex ratio, radiation effects) cut back. At the same time, an effort has been made to integrate more completely the biochemical aspects of genetics with human phenomena.

Following the suggestions of a number of helpful critics, the order of presentation of topics in the first edition has been altered, although most instructors have their own individual preferences and will arrange the sequence of ideas to suit their own style. In any case, the amount of material available is too great for a typical course lasting only one term or one semester and a number of judiciously selected chapters (or segments thereof) must be omitted. For this reason, I have shortened, concentrated, and simplified the material on blood groups, radiation effects, chromosome variants, and several other topics.

The questions at the ends of chapters have been broadened to include more questions of an elementary nature on the one hand and more questions dealing with societal and ethical problems on the other. The Instructor's Manual indicates the author's answers to some of the questions at the ends of chapters, although many questions can be answered in several ways, and those dealing with ethical problems will always elicit several different points of view.

Other differences include a change in the readings presented at the end of each chapter. These are now intended for the benefit of those students who might be interested in pursuing certain areas, or for an instructor interested in making additional assignments. For this reason, the readings are taken, whenever possible, from readily available publications.

The Instructor's Manual includes those references which document the material presented in this text and additional references for the instructor's use, as well as occasional explanatory notes on points which seem to be oversimplified or left incomplete in some respects. It is hoped that the manual, with its supplementary remarks and illustrations, will prove a useful adjunct in the presentation of this material.

Acknowledgments

The author wishes to thank the many students and colleagues who have made helpful suggestions during the preparation of this edition. In particular, Professor H. Eldon Sutton deserves special thanks for the unusual care with which he examined the manuscript. My wife Esther Ellen has improved the text immeasurably by her patient and skillful editing. Any errors and defects that remain in the final work must be attributed entirely to the author.

Contents

xi

Detailed Contents

1

Some Common Human Syndromes

Variability in the Human Species

If all human beings were exactly alike, the science of human genetics would not exist. The variability that exists from one individual to the next, and from one group of people to the next, makes it possible to study the basic mechanisms responsible for the transmission of characteristics from parents to offspring.

The study of genetics demands the use of inherited differences, characteristics determined by the basic cell components; thus we study as wide a variety of human types as possible. All of us carry heritable differences; certainly sex is the most obvious of the differences in the normal range. All persons also carry in a single dose genetic factors that would, if present twice, lead to clear abnormalities, but because these are not usually expressed, most people consider themselves genetically normal. In a smaller number of humans these departures from the normal do express themselves, and it is these persons that we must study for much of our detailed knowledge of human genetics.

The contribution that a specific human type may make to our knowledge may be quite independent of its overall frequency in the population. In many cases we shall find it profitable to consider in detail human characteristics that occur with frequencies of less than one per many thousands. In many cases a single pedigree, or even a single individual, may clarify a puzzling problem. The selection of specific human examples in the study of human genetics is determined neither by their appearance nor by their frequency, but by the extent to which they make a contribution to our knowledge of humans.

Although we must necessarily examine other human beings (and ourselves as well) for evidence of genetic factors responsible for the human makeup, we must never lose sight of the fact that these variations are minor differences superimposed on the common base of biological identity that characterizes all human beings. If the study of human genetics teaches us anything at all, it is that the genetic difference between those within the range of "normality" and those outside that range is exceedingly fine and that it is a matter of chance where in that range a person will be found.

In many cases it will be necessary for us to consider individuals with "defects" of varying degrees of severity. It is a natural human reaction for us to feel discomfort at the sight of another human being who is less perfect than we consider ourselves to be. This uneasiness may be manifested by a refusal to admit that such cases exist, or by a preference that they be kept out of sight so that they do not intrude on our sensibilities. Although at first such an attitude may appear to be sensitive, perhaps a more humane approach is to realize, first, that a person's genetic makeup is never his or her own responsibility and, second, that only by scrutinizing other people's constitutions most minutely and, where possible, relating their characteristics to identifiable biological phenomena can we approach an understanding of the biological nature of humans.

Down Syndrome

Out of every 600 births there appears one mentally defective child with a set of distinctive characteristics which, taken together, form a single medical entity. These characteristics, together referred to as a *syndrome,* include the *epicanthus,* an unusual form of the upper eyelid that superficially gives the affected person an Oriental appearance (Figure 1-1). This syndrome was recognized as a distinct medical entity by Langdon Down, who pointed out in 1866 that if two such affected but unrelated children were placed side by side, it would be difficult to say that they were not of the same parents. In recognition of Down's early observations, this defect is often referred to as the *Down syndrome,* in preference to the common designation "mongolism," which carries unfortunate and quite inaccurate racial implications. About ten years after Down's original description of this syndrome, it was observed that its incidence of occurrence was much higher when the mother was older than average. The differences in the eye folds of an affected person and those of an Oriental and a non-Oriental person are shown in Figure 1-2.

PHYSICAL CHARACTERISTICS. Some of the features of Down syndrome are evident in Figure 1-3. Along with the typical eye fold, affected persons have *hypotonia,* a relaxed condition of the muscles. Because the mouth cavity does not grow to normal size, the tongue tends to protrude, and the mouth may be held slightly open. Often the tongue itself is large and has a furrowed

Figure 1-1 A young girl with the Down syndrome with the characteristic eye folds that superficially resemble those of Orientals. (Courtesy of the Medical Genetics Group, University of Oregon Health Sciences Center.)

surface. In addition, the affected person's stature tends to be short, the nose is broad and somewhat depressed, and the hands are clumsy and stumpy. These individuals very often have a *simian crease,* a single fold across the palm of the hand, instead of the parallel palmar creases found in normal individuals (Figure 1-4). On occasion they also have irregular, abnormal sets of teeth. The growth of pubic hair is sparse, as is the beard in males. The iris of the eye, if blue, may show a speckling sometimes referred to as *Brushfield spots.*

Other physical characteristics include a markedly increased frequency of heart malformations and a low resistance to illness, especially respiratory

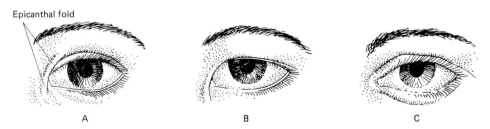

Epicanthal fold

A   B   C

Figure 1-2 A detailed drawing showing the basic difference between the eye fold of a patient with the Down Syndrome *(A)* and the eye fold of an Oriental eye *(B)* and a non-Oriental eye *(C).* (O. Solnitsky, *Georgetown Med. Bull.* **15**:276, 1962.)

infections. In the past, affected persons tended to die fairly young from infections, but with the advent of antibiotics, this problem has largely been removed. The physical features characterizing this syndrome are listed in Table 1-1. No one affected individual shows all of these characteristics, but enough of them appear in any one person with this syndrome that a diagnosis based on physical inspection alone is likely to be correct.

Affected males are sterile; there is no recorded case of any offspring from a Down male. On the other hand, affected females have occasionally pro-

Figure 1-3 Other manifestations of the Down syndrome, including the oversized tongue with partially open mouth and relaxed muscles. (Courtesy of Medical Genetics Group, University of Oregon Health Sciences Center.)

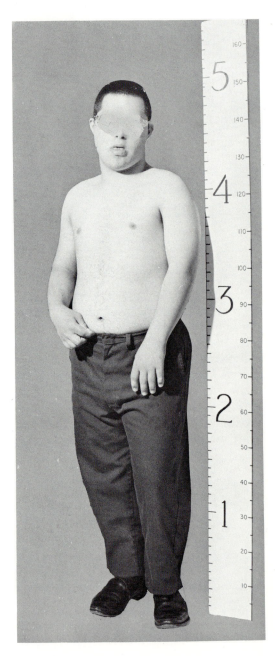

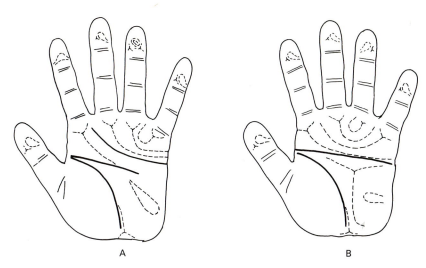

A   B

Figure 1-4   The palmar crease of a normal person *(A)* and of a person with the Down syndrome *(B)*. (From R. Turpin and J. Lejeune, *Les Chromosomes Humains,* Gauthier-Villars, Paris, 1965.)

duced offspring, about half of which were affected like their mother. This tells us the Down syndrome must be determined by something transmitted from one generation to the next, that is, genetic in nature.

PSYCHOLOGICAL CHARACTERISTICS.   Persons with the Down syndrome have low intelligence. At young ages their mental ability is generally that of a child about half their age; that is, their intelligence quotient, or IQ, ranges around 50. Their capabilities level off early in life so that even in middle age they appear childish. The distribution of IQs in Down patients runs from 25 to 50 in 60–70 percent of the cases, classifying them as "severely" retarded, and from 0 to 24 in 25–40 percent of the cases, putting them in the category of the "profoundly" retarded.

Description of Symptom	Number Showing Symptom/ Number Checked	Percent Showing Symptom
Abundant neck skin	48/51	94
Mouth corners turned downward	42/50	84
General hypotonia	28/34	82
Flat face	39/49	80
At least one dysplastic ear	39/50	78
Epicanthus in at least one eye	38/50	76
Gap between toes one and two	35/52	67
Tongue protruding	32/51	63
Head circumference at birth not exceeding 32 cm	20/47	43
Simian crease in at least one hand	22/52	42

Table 1-1   A list of the most accurate clinical symptoms for diagnosing the Down syndrome. (J. Wahrman and K. Fried, The Jerusalem Prospective Newborn Survey of Mongolism, *Ann. N.Y. Acad. Sci.,* **171**:341–60, 1970.)

Deafness has been reported as a variable characteristic of this syndrome; recent studies show that more than three quarters of all individual ears suffer some hearing impairment and about two-thirds of Down patients have subnormal hearing in both ears. It seems likely that in some cases the apparent intelligence of the Down child must be somewhat underestimated (from 10 to 15 IQ points) for this reason.

About one in twenty-five can read with some understanding and one in fifty can write. In general, affected children can learn simple tasks and can speak with a limited vocabulary. With proper guidance and early training, their range of capabilities can be extended considerably but will not usually extend into the normal range. They are good mimics and may have a well-developed sense of rhythm. They enjoy singing and dancing and are particularly partial to watching television. Although they have a reputation for being good-natured and sociable in general, some affected individuals are reported to be as unpleasant as normal children can be.

Among the inmates of institutions for the mentally retarded, the Down syndrome is found in about one case in ten and is the most frequent single diagnosis of the severely mentally retarded. In the general population, it characterizes about 30 percent of all severely retarded children in the United States and Western Europe.

The Human Chromosomes

To see the basic difference between the genetic makeup of a person with the Down syndrome and that of a normal person, we must look at the body cells of each. If white blood cells are taken from an individual, grown in culture for several days in a special medium, and then stained with a suitable dye, some of the cells arrive at the proper stage of cell division to show distinct bodies called *chromosomes*. A typical set of chromosomes from a normal human cell as seen under the microscope is found in Figure 1-5.

Figure 1-5 The chromosomes of a normal human female. In this case the chromosomes have been stained with an ordinary dye that simply darkens the chromosome. (Courtesy of R. E. Magenis, University of Oregon Health Sciences Center.)

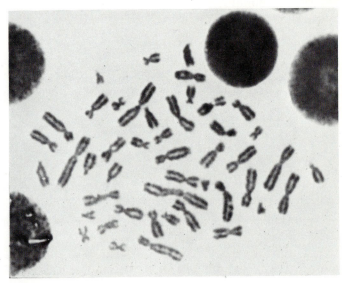

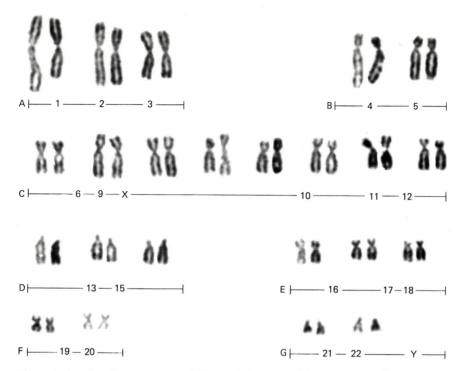

Figure 1-6   The chromosomes of Figure 1-5 arranged by pairs according to size and shape in a standard karyotype. (Courtesy of R. E. Magenis, University of Oregon Health Sciences Center.)

THE KARYOTYPE.   In order to examine the chromosomes more closely, it is convenient to photograph the cell, cut out the chromosomes individually, and place them in order according to size and staining pattern. When this is done for normal individuals, we find that there are forty-six chromosomes; of these, forty-four can be assigned to seven major size groups (labeled A to G), and in each group similarly stained chromosomes can be arranged in pairs. This standard arrangement is called a *karyotype* (Figure 1-6). Using special staining techniques, it is possible to bring out variation in staining density along the length of the chromosome. This variation is distinctly different for each pair; as a result the chromosomes can be unambiguously assigned a specific number from 1 to 22.

However, twenty-two pairs of chromosomes make a total of forty-four, not forty-six. What about the other two chromosomes? In the cells of just about half of all humans, each chromosome has a duplicate of approximately the same size and shape, making a total of twenty-three pairs. Such cells come from females (Figure 1-6). In the other half of the cases, cells from males, the chromosomes do not occur as twenty-three similar pairs; one of the large chromosomes found in the female cell is replaced in the male by a smaller one, the *Y chromosome* (Figure 1-7). The chromosome found twice in the female but only once in the male is the *X chromosome*. The X and the

Y chromosomes are referred to as the *sex chromosomes;* the other twenty-two pairs are *autosomes.*

When we examine the karyotypes of individuals with the Down syndrome, we discover that in about ninety-five percent of the cases, instead of the normal forty-six chromosomes (twenty-two pairs of autosomes plus two sex chromosomes), the total count comes to forty-seven, the extra chromosome being an additional autosome number 21. Because there are three chromosomes numbered 21 characteristic of the Down syndrome, it is often referred to as *trisomy-21* (Figure 1-8). In approximately 3 percent of the Down syndrome cases, the additional chromosome 21 material is present but may be less obvious, being attached to one of the other chromosomes in a way to be described in Chapter 8. In the remaining several percent, the Down patient is a *mosaic* and has body cells some of which have an extra chromosome 21 and others which are normal.

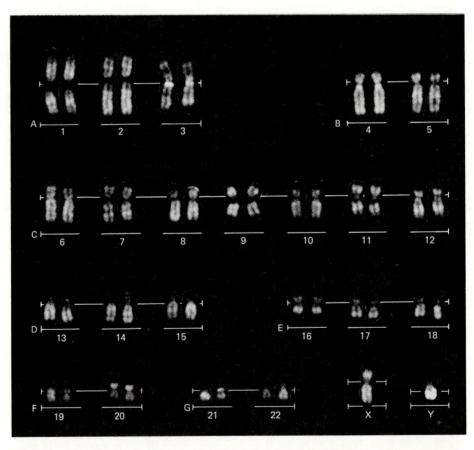

Figure 1-7 A karyotype of a normal male cell, with one Y chromosome and one X chromosome. In this case the chromosomes have been stained with a special dye called *quinacrine* that causes different parts of each chromosome to fluoresce differently when exposed to ultraviolet light. (Courtesy of F. Hecht, The Genetics Center of the Southwest Biomedical Research Institute.)

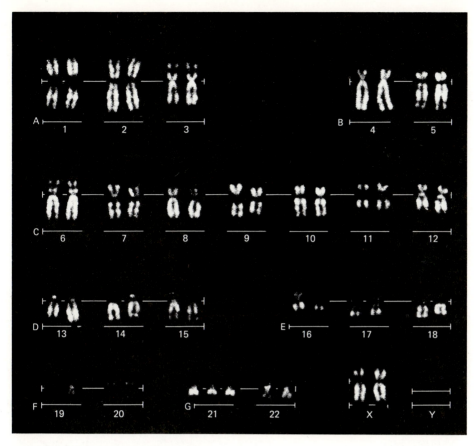

Figure 1-8 The karyotype of a person with the Down syndrome. The presence of three number 21 chromosomes (in the G group), instead of the normal two, has led to the alternative name *trisomy-21*. (Courtesy of Medical Genetics Group, University of Oregon Health Sciences Center.)

Turner Syndrome

On rare occasions a young girl at the normal time for puberty fails to commence menstruation, and at the same time her *secondary sexual characteristics* (breasts, pubic hair) fail to develop normally. When the anxious parents take the girl to a physician, her difficulty may be identified as the *Turner syndrome,* which afflicts about 1 in 10,000 female births.

PHYSICAL CHARACTERISTICS.   Affected women of this type are almost normal in appearance. In addition to the failure of the secondary sexual characteristics to develop, such women tend to have short stature, widely spaced nipples, and, when young, a characteristic shape of the chest called a "shield" chest (Figure 1-9). At birth there may be folds of skin at the neck. This problem is sometimes relieved by surgery. Older children usually manifest a flaring of the neck as it joins the shoulders. In addition, the hairline at the nape of the neck may be unusually low (Figure 1-10).

In most cases the slowness of growth is noticed from birth. Administration of female hormones (*estrogens*) may cause the secondary sexual characteristics to develop; male sex hormones may promote growth. On very rare occasions Turner females are fertile; their usual infertility cannot be corrected, having been irrevocably impaired prior to birth.

PSYCHOLOGICAL CHARACTERISTICS. Women affected with the Turner syndrome do not do quite as well on IQ tests as their sisters. When they have any academic weakness, it is likely to be in mathematics, probably because,

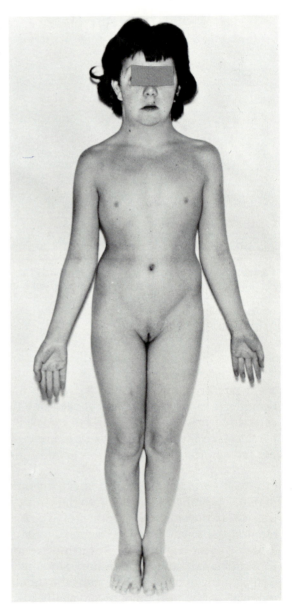

Figure 1-9 An eighteen-year-old with infantile secondary sexual characteristics and other manifestations of the Turner syndrome noted in the text. (Courtesy of P. Pearson, University of Leiden, The Netherlands.)

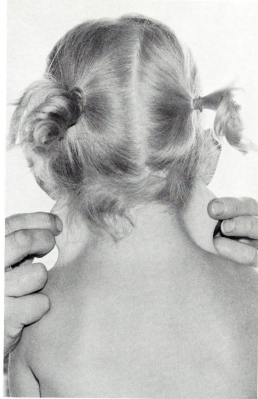

Figure 1-10 The low hairline at the base of the neck and the looseness of the skin characteristic of the Turner syndrome.

as a group, they do appear to have one interesting psychological peculiarity: faulty spatial perception. This is illustrated in Figure 1-11.

A comparison of more than a dozen Turner girls with normal controls suggests that their feminine characteristics were somewhat more extreme than those of the controls. On the average they showed less interest in athletics and childhood fighting, and a greater interest in dress and adornment.

KARYOTYPE. A karyotype of the chromosomes of an affected woman usually reveals a conspicuous deficiency: instead of forty-six chromosomes, she has only forty-five. The missing chromosome is an X. (Sometimes the chromosome is not missing but is abnormal.) When instead of having two X chromosomes she has only one, her condition may be referred to as *XO* to distinguish it from that of the normal female, *XX*. Corresponding to the word *trisomy*, used when there are three instead of the usual two chromosomes of a kind present, the word *monosomy* is used to indicate that only one of the usual two is present (Figure 1-12).

Figure 1-11 Two figures from a visual retention test, showing how a ten-year-old girl with the Turner syndrome reproduced the figures at the left after viewing them for ten seconds. (J. Money, in *Endocrinology and Human Behavior,* ed. by R. P. Michael, Oxford University Press, New York, 1968.)

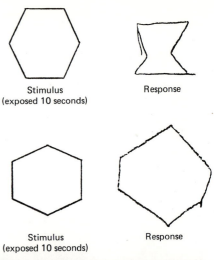

Stimulus
(exposed 10 seconds)

Response

Stimulus
(exposed 10 seconds)

Response

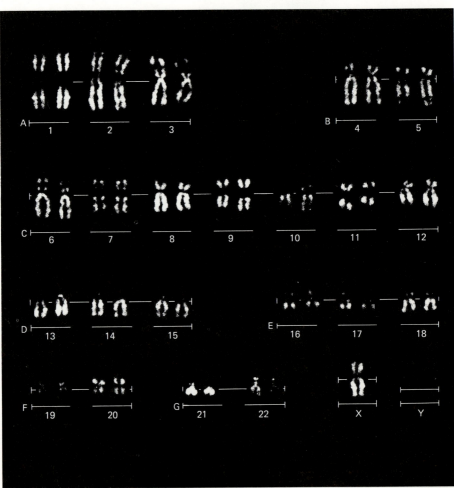

Figure 1-12 The karyotype of a Turner syndrome patient. Note that there is only one X chromosome and no Y. (Courtesy of Medical Genetics Group, University of Oregon Health Sciences Center.)

The Chromosomal Basis for Sex Determination

We know from our examination of karyotypes of normal males and females that the female has two X chromosomes, whereas the male has only one. But the male has a Y chromosome, which the female lacks. Is a woman female because she has two Xs and a man male because he has only one X? Or is a man male because he has a Y and a woman female because she does not have one? The genetic constitution of the female with Turner syndrome gives us a partial answer to this question. As she has only one X chromosome, like a normal male, it would appear that it is the additional presence of the Y chromosome in the male that is ordinarily responsible for the difference between the sexes, not the number of X chromosomes.

The XXX Female

Females with three X chromosomes occur in the population with a frequency of 1 per 1,000 female births. They are indistinguishable from XX (normal) females physically; for this reason they are usually discovered by accident when chromosome studies are made in population surveys or are made in families for other reasons. On the average, they show a slightly depressed mental ability. When they produce children, the children are usually normal.

Klinefelter Syndrome

PHYSICAL CHARACTERISTICS. In about 1 of every 1,000 male births, there appears a boy who seems quite normal until the onset of puberty. At this time, he may show partial breast development and other less obvious female characteristics, a slightly altered set of bodily proportions with somewhat narrow shoulders and larger hips than are typical of the normal male (Figure 1-13). Such persons with the *Klinefelter syndrome* may be taller than average. Generally there is a deficiency of the male hormone level, resulting in sparse and female distribution of pubic hair, small testicles, and a somewhat high-pitched voice. This person may appear to be a perfectly normal male and may marry, but is infertile. About 5 percent of all men who appear at fertility clinics prove to have this syndrome.

PSYCHOLOGICAL CHARACTERISTICS. In common with most cases of abnormal numbers of X chromosomes (XO and XXX females being good examples), these individuals tend to show a slightly depressed intelligence, with an average IQ of the order of 90. About 1 person in 100 institutionalized for mental retardation has the Klinefelter syndrome. These patients have been described as passive and withdrawn and tend to have other psychological problems, such as delayed speech and emotional development. Their motor coordination tends to be poor.

KARYOTYPE.   In Figure 1-14 we see the karyotype typical of a male with the Klinefelter syndrome. These men have two X chromosomes as well as a Y; that is, they are *XXY*. This information confirms our earlier conclusion based on the Turner syndrome: it is the presence of the Y chromosome that is characteristic of maleness, not the single X chromosome. More than one X does not necessarily direct development toward femaleness, as the presence of a Y along with two Xs causes the individual to develop into a male.

The XYY Male

Figure 1-15 shows an interesting set of twins. These two boys, nonidentical twins, are grossly different in physical characteristics, one being excessively tall. Chromosome analyses show that the taller twin has two Y chromosomes. In general, about two thirds of all *XYY* males are over six feet tall. In a sample of twenty tested, half had an IQ of less than eighty. In most instances, males with this chromosome anomaly appear to be quite normal. Males with an XYY condition are born with a frequency of about 1 per 1,000, and they seem to occur more frequently in whites than in nonwhites.

Figure 1-13   A male with the Klinefelter syndrome, showing partial breast development. (Courtesy of G. Prescott, University of Oregon Health Sciences Center.)

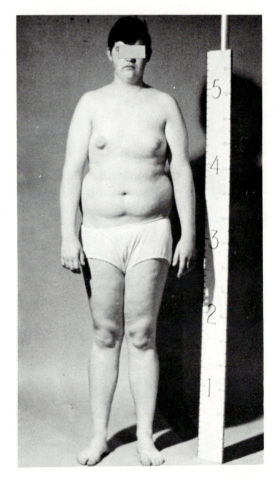

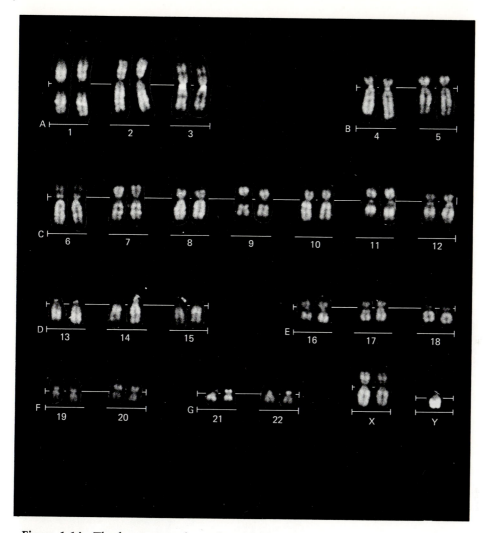

Figure 1-14 The karyotype of a male with the Klinefelter syndrome showing the presence of two X chromosomes as well as a Y chromosome. (Courtesy of B. Kaiser-McCaw, The Genetics Center of The Southwest Biomedical Research Institute.)

BEHAVIORAL TRAITS. It was pointed out by Patricia Jacobs and her co-workers in 1965 that an unduly high proportion of males found in penal institutions are XYY. This study in England showed that in a few prisons the frequency went as high as one in twenty-five, indicating that such individuals might be innately more aggressive than normal and therefore more likely to run afoul of the law. In contrast to the possibility that aggressive tendencies may be inherently part of this chromosome makeup, it has been suggested that the effects of tallness and a slower mentality might combine in early childhood to promote the development of an aggressive personality, which would then lead secondarily to behavioral problems. Further, because many cases are known of XYY males who are of normal height, intelligence, and behavior, the proposal that persons with this chromosome constitution are characterized by behavioral disturbances that lead to antisocial acts has sometimes been considered an unwarranted supposition.

On the other hand, many workers in the field have produced evidence that the XYY condition has psychological manifestations (other than a common reduction in intelligence). One group was able to identify without error seven XYY males from twenty-eight controls solely on the basis of test results that covered maturity level, emotionality, use of defense mechanisms, and intellectual level. They noted that XYY males appear to have a lower threshold for the control of aggression in frustrating or provocative situations. Another study uncovered three XYY boys, two of whom, at age seven, were already having behavioral problems at home and school. One was under the care of a psychiatrist, even at this early age. These observations were made prior to the time that they were karyotyped.

This condition has been of great interest from the legal standpoint, because in a few cases (the first occurring in Australia and subsequent ones in

Figure 1-15 Two nonidentical twin brothers, one XY and the other XYY, showing the increased height often associated with the presence of an extra Y chromosome. (Courtesy of G. Prescott, University of Oregon Health Services Center.)

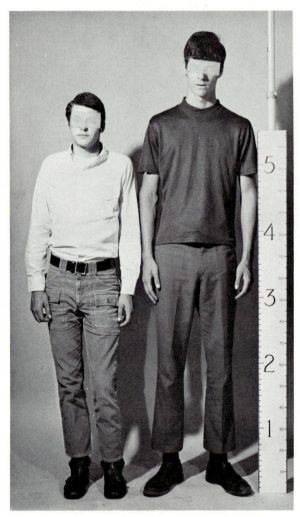

France and the U.S.) men accused of murder who proved to be XYY used their abnormal chromosome constitution in their defense, claiming that this abnormality had led them to commit crimes over which they, as individuals, had little control. In the Australian case the XYY male was acquitted, and in several other cases such persons have had their sentences reduced on the grounds that their chromosome complement, for which they could hardly be held responsible, was in part the cause of their antisocial behavior.

A large-scale study made in Denmark in 1978 led to the conclusion that although the frequency of XYY men in prisons was higher than expected, the crimes for which they were convicted were mostly nonviolent in nature, and the most reasonable explanation for their high frequency in prisons was simply their depressed intelligence. The report goes on to note, however, that, compared with their siblings, the XYY males were more immature, were more impulsive, had greater contact difficulties, were less reflective, were emotionally unstable, and had weak defense mechanisms.

In one case in 1975, a previously convicted murderer was indicted for a subsequent murder, and as part of an insanity defense, the attorney asked that the defendant be karyotyped. The judge denied the request with the comment that "the genetic inbalance theory of crime causation has not been satisfactorily established and accepted in either the scientific or legal communities to warrant its admission in criminal trials." This is probably a fair statement of the most prevalent view of workers in this field at the present time.

The data are still fairly limited, but the evidence for a higher-than-expected proportion of XYY males in penal institutions comes from studies of incarcerated whites and does not seem to be true of blacks or Orientals. Perhaps the basic frequency of the XYY condition is different in the different ethnic groups. Studies made to determine whether institutionalized XYY men have been predominantly in the lower socioeconomic classes show that, in fact, they come from a wide spectrum of backgrounds. Moreover XYY males with a criminal record usually come from relatively stable, law-abiding families, whereas the average convicted offender comes more often from unstable families in which other members have also been incarcerated. This finding strongly suggests that the chromosome composition of these males is in some way correlated with their antisocial behavior. The exact psychology of the XYY male is still a matter of controversy, and it continues to be one of the more interesting abnormalities because of the legal and ethical issues involved.

Questions Raised by These Unusual Chromosomal Types

A study of the abnormal chromosome complements associated with the syndromes described in this chapter points up the importance of the normal chromosome set in determining normal development. Abnormalities of chromosome constitution can be responsible for deviations from normality in physical, mental, and behavioral traits; thus an understanding of the normal human makeup depends on a greater appreciation of the role our chromosomes play.

It would be possible to expand our list of genetically informative syndromes to include some caused by simple genetic factors, rather than chromosome additions or subtractions, or to describe some syndromes that appear as birth defects after exposure of the mother to harmful chemicals or to diseases. However, such discussion must be deferred; our description of these four chromosomal syndromes has raised other questions that must be answered first. What about anomalies involving other chromosomes? What happens when a developing individual has, say, an extra chromosome two or nine? What about those cases in which a person might have only one autosome instead of the usual autosomal pair?

The X chromosome seems to behave somewhat differently from the autosomes: an extra chromosome 21 has a profound effect on physical and mental development, whereas the much larger X chromosome may be present as an extra chromosome in XXX females with very little effect. Does this mean that there is some essential difference between the X chromosome and the autosomes?

There are even more fundamental and more interesting questions we can ask. What is the basic mechanism by which these syndromes are produced? Is there anything that can be done to decrease the frequency with which they occur in the population? Is there any way that a family can safeguard against having such an abnormal child? Will an affected person, if fertile, produce similarly abnormal children?

As a first step in attempting to answer these questions, we will consider in some detail the mechanism by which the chromosomes are distributed during the process of cell division, or *mitosis*.

Additional Reading

For reading in the area of ethical issues, as well as elaboration of some of the ethical questions raised in these sections, the *Bibliography* (1979–1980) published by the Hastings Center, Hastings-on-Hudson, New York 10706, is an invaluable guide.

BECKWITH, J., and J. KING. 1974. The XYY syndrome: A dangerous myth. *New Scientist*, **64**:474–476.

HAMERTON, JOHN L. 1976. Human population cytogenetics: Dilemmas and problems. *American Journal of Human Genetics*, **28**:107–122.

HOOK, E. B. 1973. Behavioral implications of the human XYY genotype. *Science*, **179**:139–150.

JACOBS, P. A., and J. A. STRONG. 1959. A case of human intersexuality having a possible XXY sex-determining mechanism. *Nature*, **183**:302.

SMITH, G. F., and J. M. BERG. 1976. *Down's Anomaly*, 2nd ed. New York: Churchill Livingstone.

WITKIN, H. A., et al. 1976. Criminality in XYY and XXY men. *Science*, **193**:547–555.

The XYY Controversy: Researching Violence and Genetics. *Hastings Center Special Supplement*, August 1950.

Questions

1. Define *syndrome*. Would a person who is only near-sighted be properly described as having a syndrome? Why not?
2. Can you draw a typical human chromosome? How many are there in a normal human cell?
3. Some chromosomes are found in both males and females, whereas others are distributed unequally in the two sexes. What terminology do we use to distinguish the two types?
4. What is the general term for the condition in which there is an extra chromosome beyond the normal number? For the condition with one too few chromosomes?
5. How frequent is Down syndrome among the newborn? What proportion of inmates of institutions for the mentally retarded have this syndrome?
6. What are the major differences in the manifestations of Turner and Down syndromes? Do you think that all children should be karyotyped and their parents informed of any chromosome abnormality? Defend your point of view.
7. How many chromosomes would be found in a karyotype of the very rare combination, in the same person, of Turner and Down syndrome?
8. What does Turner syndrome tell us about the chromosomal basis for sex? How is this confirmed by Klinefelter syndrome?
9. If a male has an extra Y chromosome, how is he likely to differ from the average? Why is this condition of legal interest?
10. How are the twenty-three pairs of chromosomes arranged in a karyotype? Is the chromosome pair numbered two larger or smaller than the pair numbered nine?
11. In theory, how many different monosomic types of humans could occur? How many trisomic types? Can you suggest any reason that, out of all of these possibilities, the Down, Turner, and Klinefelter syndromes were selected for the introductory discussion?
12. Phillip B., an institutionalized Down patient in California, had a heart defect that could be repaired by surgery, thereby prolonging his life by twenty or thirty years. His parents, whose contact with him was that they visited him a couple of times a year, refused to give permission for the operation. The courts agreed that the parents had this right. Do you think that a Down child is less entitled to medical care than a normal child?
13. Should a person who is XYY be allowed to use this fact as a defense after committing a serious crime?

Chromosome Structure and Cell Division

The Cell: The Basic Unit of Structure

All living organisms, with the exception of the minute viruses, are made up of units called *cells,* and all of the components of the organism are made of combinations of cells or of their products. The cells of a mature animal are specialized in their work, and their form enables them to carry out their function. A nerve cell that transmits information from the tip of a limb to the central nervous system may be several feet long, whereas a red blood corpuscle, which must carry oxygen through the smallest capillaries of the body to every permanently fixed cell, has a microscopic ovoid structure that allows it free passage.

All cells come from the division of one cell into two daughter cells. Every mature person started initially as a single cell, a fertilized egg, which, by mitosis, became two, and the two became four, and, as the total number grew, the cells gradually became different in structure in different parts of the developing organism until the mass of cells took on the appearance of a small child. The human body, as it is finally constituted, seems so complex that it is sometimes difficult for the nonbiologist to keep in mind that the entire body is made up only of minute cells and, to a small extent, their immediate products (hair, nails, bone).

IMPORTANCE OF THE GENETIC MATERIAL.   The internal machinery of the cell is a complicated assembly of chemical systems (Figure 2-1) for carrying on a certain number of basic functions. At the same time the cell performs some special task, depending on the tissue in which it is found (muscular, nervous, dermal, or digestive). Whatever the assigned job, the directions are given by the *genes* that are located in the chromosomes of the cell. In this book we are concerned primarily with the manner of transmission of the genes and chromosomes from one generation to the next—in other words, with inheritance—and secondarily with the functioning of the genes as they determine the characteristics of the final product, the fully developed organism. In many cases our attention will be directed to the chromosomes, which, as aggregates of genes, affect many human characteristics; in other cases it will be more appropriate to look at the specific gene responsible for the production of a particular characteristic.

Research over the past few decades has produced information of fundamental importance regarding chemical interactions within the human cell, but it is important to realize that present knowledge of the details of cellular mechanisms is quite incomplete. Despite newspaper reports on the great biological breakthroughs of the last several years, and the imminent possi-

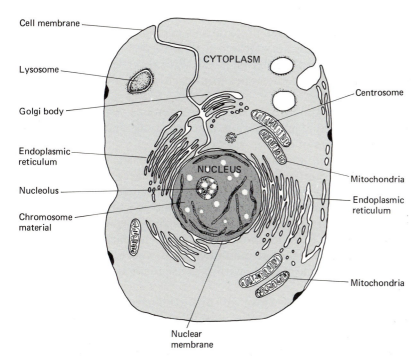

Figure 2-1   An idealized representation of a "typical" cell. The synthetic activities of the cell are carried out in an intricate structural complex, including specialized bodies indicated as mitochondria, endoplasmic reticulum, Golgi bodies, and lysosomes located in the cytoplasm. However, our attention will be focused primarily on the nucleus and on the genetic contents carried in the chromosomes that provide the instructions for the operation of the cell. (From J. Brachet, *The Living Cell*, © September 1961 by Scientific American, Inc. All rights reserved.)

229113

bility of "creating life," that possibility is far from being realized, and it will be many years, if ever, before even the most primitive kind of cell can be synthesized. However, this fact does not prevent us from making observations on the relationship between the final product—the completely formed individual—and the genes that are fundamentally responsible for the final form of the organism.

Techniques of Human Chromosome Studies

The art of chromosome study in humans is of fairly recent origin; as late as 1956 it was generally thought that the diploid chromosome number was forty-eight. The early techniques involved laborious sectioning of material (usually testicular) with a tedious fixing, staining, and sectioning procedure that gave relatively few good preparations. At best, the chromosomes were small, easily misinterpreted, and generally uninformative, even in the hands of the most skilled workers. Researchers preferred to work with other organisms such as lilies or salamanders, with large, clearly staining, and more manageable chromosomes.

In the 1950s, a number of technical advances combined to transform this difficult and unrewarding area into an exciting and highly profitable one:

1. It was shown that metaphase preparations could be vastly improved if the cells were placed in a *hypotonic solution*—one in which the salt concentration is lower than that of a cell. With this treatment the cells swell, separating the chromosomes from each other and making them easier to identify.
2. At about the same time, *tissue culture* techniques were being developed that allowed rapid proliferation of human cells, making it possible to get large numbers of cells in active division.
3. The classical sectioning technique (in which the tissue is cut into very thin slices to be examined individually) was replaced by a *squashing* technique in which whole cells were simply flattened on a microscope slide. This procedure not only reduced the time and care necessary for the preparations but also ensured that all of the chromosomes of one cell would be found together more frequently. The hazard of the early sectioning technique was that individual cells were sliced at random, sometimes through the nuclei, with a consequent inaccuracy in the chromosome counts.
4. *Lymphocytes,* one type of white cell in the bloodstream, proved to be an easily available source of cells capable of dividing.
5. *Phytohemagglutinin,* a plant extract, agglutinates the red blood cells to facilitate their removal from the blood sample, at the same time inducing growth and division in the lymphocytes, which otherwise never divide in culture.
6. For many years a drug called *colchicine,* from the autumn crocus, had been known to inhibit cell division. (Colchicine is, incidentally, one of the oldest prescriptions for the treatment of gout.) This drug made it possible to obtain much larger numbers of dividing cells than are found

normally. Colchicine also contracts the chromosomes, making them appear more distinctive.

7. The use of many different kinds of chromosome-specific chemicals and stains (such as *quinacrine*, Giemsa, and acridine orange), combined with a variety of recipes for pretreating the chromosomes, made it possible to stain the chromosomes in distinctive ways with a minimum of fuss. Now, by a combination of some or all of these innovations, a small quantity of blood can be extracted and cultured for several days, and a slide of the cells can then be made fairly quickly to give a clear, unambiguous karyotype.

EARLY OBSERVATIONS.   In 1956, J. Tjio and A. Levan in Sweden and C. E. Ford and J. L. Hamerton in England made an important discovery. These workers established the correct number of chromosomes as forty-six. This was followed by the demonstration three years later that persons with the Down syndrome had an extra chromosome, now numbered twenty-one. Almost simultaneously it was shown that females with the Turner syndrome usually had only one X chromosome and that males with Klinefelter syndrome had two X chromosomes as well as a Y. The XYY constitution was identified in 1961 and four years later data from penal institutions was published suggesting that such males might express unusual behavior patterns.

A significant advance was made in the late 1960's when it was found that different regions of plant chromosomes stained differently, after having been treated with a fluorescent dye called *quinacrine mustard*. They subsequently showed that with this substance each human chromosome displays its own unique banding pattern, making it possible to identify without question every chromosome of the complement. These details of chromosome structure have now been incorporated (by a conference in Paris in 1971) into a numbering system specifying the regions of chromosomes involved (Figure 2-2).

Basic Structure of DNA

HISTORICAL BACKGROUND.   In 1913 Alfred H. Sturtevant showed that *Drosophila* genes were arranged in a linear sequence on the chromosome, and for years there was speculation about the kind of molecule that might make up the genes in that linear order. For about thirty years, the most popular idea was that of all the large molecules found in the cell, only proteins were complex enough to store the wide variety of genetic information that would be necessary to produce a complicated organism. In 1944 Oswald Avery and his colleagues devised an important experiment involving what is now known as transformation. They isolated the *DNA* (*d*eoxyribo*n*ucleic *a*cid) from a strain of pneumococcus which ordinarily manufactures a capsule surrounding each cell. This DNA was then added to a strain which had lost the ability to make the capsule. Some of the cells of the second strain then regained the ability they had lost, and these cells transmitted their

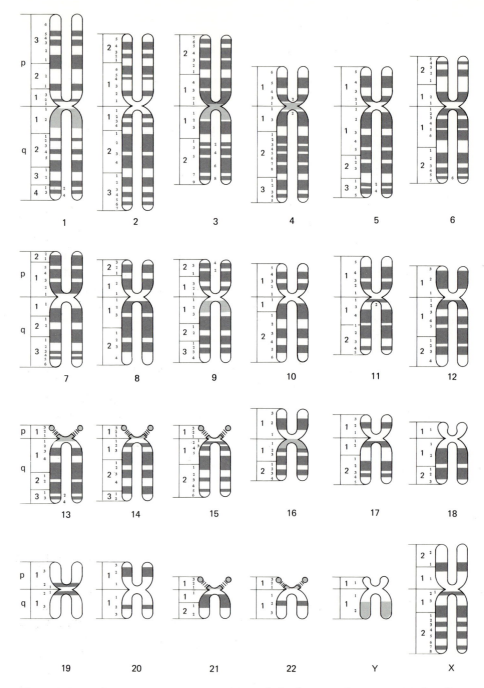

Figure 2-2 A diagrammatic representation of the human chromosome's banding patterns revealed by the different kinds of staining techniques. The letter *p* represents the short arm, *q* the long arm. Two-digit numbers specify the bands. Thus, the light band (36) at the tip of the short arm of chromosome 1 is designated as 1p36. (*Birth Defects: Orig. Art. Ser.,* ed. D. Bergsma. Published by Williams & Wilkins Co., Baltimore, for The March of Dimes Birth Defects Foundation, Vol. VIII [7], 1972.)

newly acquired capacity to all their descendants. This showed that the genetic information necessary for capsule formation was contained in the purified DNA.

An understanding of how DNA could contain genetic information did not come until 1953, when, at Cambridge University, James D. Watson and Francis H. C. Crick discovered the structure of DNA, using new high-resolution X-ray diffraction patterns they borrowed from Wilkins, a research worker at King's College, University of London.

THE DNA HELIX. Because DNA contains all of the information on how to construct an entire organism, it might be expected to be too complex a molecule to be understood. On the contrary, its simplicity is striking.

A single strand of a DNA molecule is a long chain of just four different subunits occurring repeatedly in what, at first sight, might appear to be a random order. These four subunits are commonly referred to as *adenine, thymine, guanine* and *cytosine**. We will use the abbreviations A, T, G, and C (Figure 2-3A). In nature, however, DNA is found as two long strands held tightly together by many very weak hydrogen bonds according to the following simple rule. An A on one strand binds weakly to a T on the other, and a G on either one of the two strands binds with a C on the other (Figure 2-3B). Any variations that occur can be considered mistakes. In addition, the strands can be considered directional, and we will use the number 5′ at the beginning and 3′ at the end of a sequence to indicate one direction, 3′ at the beginning and 5′ at the end for the reverse.

* For the student with a better-than-average background in chemistry, however, the subunits of DNA are deoxyadenylic acid (A), deoxyguanylic acid (G), deoxythymidylic acid (T), and deoxycytidylic acid (C). They are covalently linked through phosphodiester bonds between the 3′-carbon of one subunit and the 5′-carbon of the next.

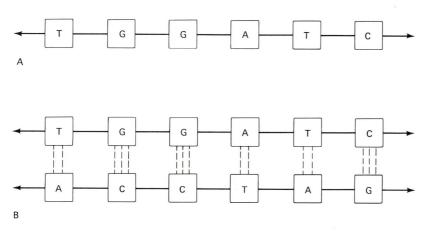

Figure 2-3 *A.* A small section of a single strand of DNA showing the linear attachment of molecules of the four subunits A, T, G, and C, apparently in no special order.

B. Diagrammatic representation of the rule that an A subunit on one strand will pair only with a T on an adjacent strand (and vice versa) and that a G on one strand will pair only with a C on the other (and vice versa).

Figure 2-4 Replication of a section of a DNA strand. When two complementary strands *(A)* separate *(B)*, each can then form a new complementary strand *(C)*. The end result consists of two double strands with exactly the same composition as the original.

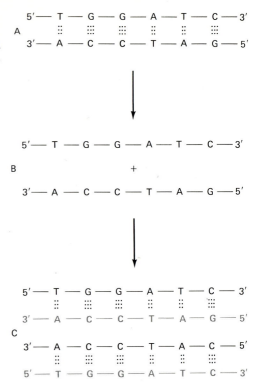

It is clear, then, that if we know the sequence on one strand as 5' - - - AATGCTAC - - - 3', we can state categorically the sequence in the parallel strand, in the case given, as 3' - - - TTACGATG - - - 5'. Because the sequence of the subunits on one strand predetermines the sequence on the other, they are said to be *complementary* to each other.

REPLICATION. Double-stranded DNA found in chromosomes in the nucleus can direct its own *replication*. If the two strands of a double-stranded DNA are separated and a new complementary strand is manufactured parallel to each, the result is two identical double-stranded DNAs (Figure 2-4). The description of the specific enzymes and the exact chemical steps involved in DNA replication is currently a subject of intense research.

Watson and Crick showed that the two strands of double-stranded DNA are not straight like a ladder but form a *double helix* resembling a spiral staircase (Figure 2-5). The problem of exactly how the cell unwinds its long double-stranded DNA during replication is still a puzzling one.

AMOUNT OF GENETIC MATERIAL IN THE CELL. The length of threadlike DNA in each human cell cannot be measured directly, but techniques exist for determining the total quantity of DNA present in an individual nucleus.

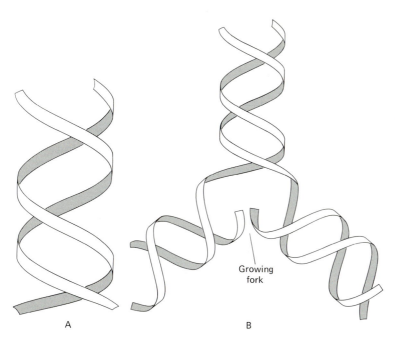

Figure 2-5 The basic spiral structure of the DNA molecule *(A)* showing how the spiral unwinds and replicates at the growing fork *(B)*. (Redrawn from D. Fraser, *Viruses and Molecular Biology*, Macmillan, 1967. By permission.)

Growing
fork

A   B

This quantity turns out to be approximately 20,000 times more than the amount of DNA in a virus particle. By electron microscopy, the virus DNA has been shown to have a length of 60 μm (60 millionths of a meter). As virus and human DNA are similar in basic structure, we can calculate the length of DNA in the nucleus of a human cell as 60 μm times 20,000, or 120 cm, which amounts to about 4 feet.

It is the precise order of the base pairs, the A–T or the G–C combinations that hold the two parallel strands together, that is responsible for the information that the DNA molecule possesses. The apparently random sequence of the A–T and G–C base pairs along the length of the DNA molecule is an illusion; as we shall see later, this order is very precisely determined, and on it depends the specific chemical activity performed by the machinery of the cell. To give an example of the amount of information in the DNA in one set of human chromosomes, if it were to take one second to count one base pair, a person would finish counting the total number at about his ninety-fifth birthday, provided that he neither slept nor took time off for vacation.

This length must be packed into the cell nucleus, which is possible only because the thread is so thin that the total volume the thread occupies is much less than the size of the nucleus. When a cell divides to produce two new cells, the genetic material must reproduce itself exactly and partition itself precisely into the two daughter cells so that each of them receives exactly the same genetic material present in the original cell.

This is an incredibly difficult problem. Suppose that you were present at the Creation and were asked to help design a system with the following properties: a cell of microscopic size must contain about fifty thin threads with a total length of several feet; each thread must be able to replicate itself exactly; and when this minute cell divides into two, each of the products must receive precisely one half of the forty-six threads. This problem would baffle the most skillful and imaginative engineer. Yet this feat is performed thousands of times a second in every human, with such accuracy that only infrequently can we detect any kind of error. This problem is solved by the cell during the process of nuclear division, or *mitosis.*

The problem of handling the total length of genetic material within the cell nucleus is helped by its being divided into a number of individual pieces, the chromosomes. As we have seen, there are forty-six of these in the body cells of humans.

One might imagine that the long lengths of DNA are converted into the short thick chromosomes by being stuffed into an inert outer chromosome sheath, much as a sausage is manufactured. However, this cannot be the case because the individual chromosomes of the complement have charac- teristic shapes and, after staining, show unique banding patterns along their lengths. Furthermore, when one chromosome is accidently broken and a piece of it becomes attached to another, that misplaced piece retains its original chromosomal banding pattern indefinitely. This must mean that the DNA thread converts itself into a chromosome in a manner that is highly dependent on the constitution of the DNA (and any associated mate- rial) in that particular chromosome.

Actually, the packaging of the long thin threads of DNA into the much shorter and thicker arrays that we see as chromosomes is accomplished by the association of the DNA thread with small bodies made up of proteins called *histones.* The DNA is wrapped around this body to form a *nucleosome;* the effective length of the DNA is decreased by a factor of six (Figure 2-6). The nucleosomes then align themselves in a closely packed configuration that reduces their length by another factor of from five to ten. It is hypothe- sized that these shorter threads then coil themselves up into the much shorter, thicker bodies that we see as chromosomes.

In this chapter we shall see how this large amount of genetic material is moved from one cell into two daughter cells and how identical sets of ge- netic material are delivered to the two daughter cells.

Figure 2-6 The way which the DNA is wrapped around nucleosomes to achieve a much shorter length.

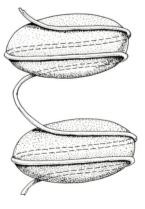

Rate of Cell Division

INTERPHASE: THE "RESTING STAGE." Between cycles of division, the cell goes into the stage of *interphase,* in which the chromosomes are contained within a nucleus as quite diffuse and inconspicuous bodies. The chromosomes are uncoiled into long threads and are engaged in synthetic activity. The name *resting stage* has historically been used for this period to emphasize the absence of cell division, but it is a complete misnomer from the standpoint of cell activity, because this is the time when the cell does its main work, whatever that may be.

TIME SPENT IN INTERPHASE. The cells of some tissues, such as those of the blood system and the epithelium of the intestinal tract, undergo rapid turnover, with an accompanying high rate of division. Then there are other cells—for example, nerve cells—that do not divide at all after embryonic development, remaining in a permanent interphase condition. Within those cells in the process of division, an important event takes place during each interphase: the replication of the genetic material. In actively dividing human white blood cells grown in culture, the whole process of cell division may average about twenty-four hours, of which interphase takes up twenty-two. The chromosome replication (or synthetic) period, the S period, occupies about seven hours, being preceded by a nonreplicating period of interphase (G_1) of eleven hours and followed by a period of completed replication (G_2) of about four hours. Mitosis, or nuclear division, itself actually occupies only a very small part of the total cycle, in this case only about two hours (Figure 2-7).

Details of Mitosis

SELECTION OF APPROPRIATE MATERIAL. As our main interest in this book is the human being, it might naturally be assumed that the best material for a demonstration of mitosis would come from humans. The fact is, however,

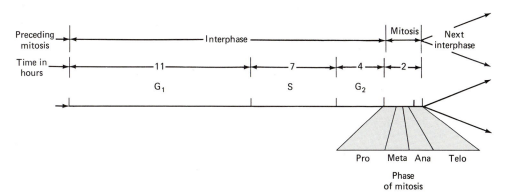

Figure 2-7 The time scale of events in hours showing the relatively long duration of the G_1, S, and G_2 stages of interphase, and the speed of the other mitotic stages, indicated as *Pro, Meta, Ana,* and *Telo* for prophase, metaphase, anaphase, and telophase.

that the phenomenon of mitosis is found in all plants and animals, with only minor variations, so that we are free to choose an organism that shows the chromosomes clearly as they go through cell division, with the understanding that the essential features of mitosis are the same for human beings.

An unusually favorable species for demonstrating mitosis is a plant with very large and clear chromosomes, the blood lily from South Africa, *Haemanthus*. The cells being observed are those from tissue, called the *endosperm*, found in the developing seed. These cells can be removed and, with a special optical system, can be observed in the living state as they go through the complete cycle of mitosis. The first visible evidence that a cell is about to undergo mitosis and produce two identical daughter cells comes about when certain changes take place in and around the interphase nucleus. Starting with this first stage, called *prophase*, we will follow the details of mitosis.

PROPHASE. At prophase the nucleus is clearly visible (Figure 2-8). The chromosomes, closely packed together within the nucleus, are so intimately associated that they are indistinguishable as separate entities. Previously, in interphase, the chromosomes existed as elongated, exceedingly thin strands, and now these strands, in the early prophase stage, are arranged into compact coils so that they appear thicker and shorter than at interphase. Outside the nuclear envelope, clear spaces are present. These indicate the formation of a special cellular mechanism, the *spindle*, which will eventually be responsible for moving the chromosomes to the daughter cells. A later prophase stage is shown in Figure 2-9. The nucleus is still present, but the clear zone is becoming more elongate, taking on the shape of a spindle. In *Haemanthus* endosperm, prophase lasts about three or four hours.

METAPHASE. In Figure 2-10 the clear zone is becoming increasingly spindle-shaped, and the envelope around the nucleus no longer encases the nuclear contents. Studies with the electron microscope have shown that the nuclear envelope at this stage of mitosis, called *metaphase*, breaks into small pieces, which then mostly disintegrate, a process taking only about two minutes. With the breakdown of the nuclear envelope, the chromosomes begin movements that will take them to the cell's equator. Those movements are going on in Figure 2-11.

The fact that each chromosome consists of two strands is clearly evident in Figure 2-11. Chromosome duplication occurred back in the interphase stage preceding this mitosis, but in living *Haemanthus* endosperm, it is only now that the two-strand structure of each chromosome can be easily seen. The two strands making up a chromosome are called *chromatids;* thus each of the chromosomes that are present in Figures 2-8, 2-9, 2-10, and 2-11 consists of two identical chromatids, as a consequence of interphase duplication.

In Figure 2-12 the chromosomes have almost achieved their final metaphase positions at the equator of the cell. In this photomicrograph, as in Figures 2-11 and 2-13, the spindle is developed, its fibers radiating out from the poles at either end of the cell. The spindle consists of long organic mole-

Figure 2-8 Early prophase, showing the nucleoli within the nucleus. (Figures 2-8 through 2-19 are reproduced courtesy of Andrew S. Bajer, Department of Biology, University of Oregon.)

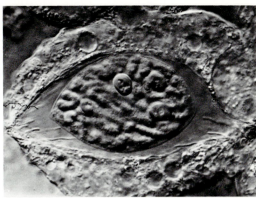

Figure 2-9 Late prophase stage with the appearance of the spindle.

Figure 2-10 Beginning of metaphase, with the breakdown of the nuclear membrane.

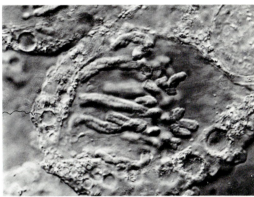

Figure 2-11 Metaphase movement of chromosomes toward the cell's equator with the duplicated structure of the chromosomes clearly visible.

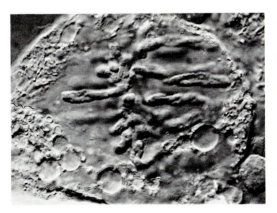

Figure 2-12 Metaphase positioning of chromosomes.

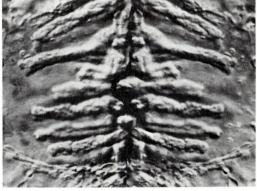

Figure 2-13 Late metaphase stage.

cules that have formed filamentous tubular elements called *microtubules,* which individually are beneath the resolution of the light microscope. The spindle fibers that are seen at relatively low magnifications consist of bundles of microtubules that have reached a dimension great enough for observation with light microscopy. To execute the complex movements of mitosis, each chromatid must make contact with the spindle microtubules at a particular specialized region of its axis, the *centromere.* It is the centromeres of the chromosomes that move to the equatorial position within the cell, and at that position they oscillate to and fro, suggesting pulling forces that are acting first toward one pole of the spindle and then toward the opposite pole. These movements are believed to be associated with dynamic changes in the spindle microtubules.

ANAPHASE. In Figure 2-14 the two chromatids of which each chromosome is composed have been separated from each other, producing two identical sets of chromosomes. This is the *anaphase* of mitosis. About two hours are required by the cell from the beginning of metaphase until the start of the anaphase movements, which are completed in thirty-five to forty-five minutes. Each chromosome now consists of only a single chromatid. Figures 2-14, 2-15, and 2-16 are a sequence showing a cell advancing from the middle of anaphase in Figure 2-14 to its completion in Figure 2-16. The centromeres lead in the movements of anaphase, apparently in response to pulling forces administered by way of the spindle microtubules. It is a curious fact that such a simple and universal phenomenon as the anaphase movement of the chromosomes during mitosis still defies complete explanation. Much research effort is spent in this area and many rather advanced theories have been proposed. For our purpose it is sufficient to regard the system as one like a rope and pulley, with the rope (the microtubules) pulling the chromosomes toward one end of the cell or the other.

TELOPHASE. Anaphase has been completed in the cell illustrated in Figure 2-17, and the events of *telophase* are now beginning. The spindle fibers are disintegrating and the chromosomes are clumping together. In Figure 2-18 the telophase clumping of chromosomes has progressed to the point at which individual chromosomes can no longer be clearly distinguished. The single cell is forming into two new ones. In the late telophase stage of Figure 2-19, the tightly packed gyres of the chromosome strands have relaxed, producing long, greatly extended chromosomes. The two new nuclei are now surrounded by their nuclear envelopes and the division into two cells is complete. At this point the two daughter cells go into interphase and the mitotic cycle has been completed.

From the foregoing we can readily visualize the process of mitosis: At interphase, the long chromosomes replicate themselves exactly. At prophase, they shorten and thicken, and move to the center of the cell, where they are found at metaphase. At anaphase, the replicated chromosomes separate from each other so that each of the two products of mitosis has precisely the same genetic makeup.

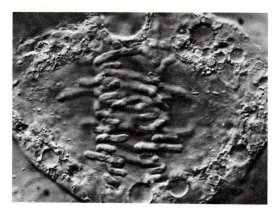

Figure 2-14 Anaphase separation of the two chromatids making up each chromosome.

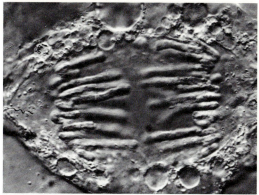

Figure 2-15 Further movements of chromosomes during anaphase.

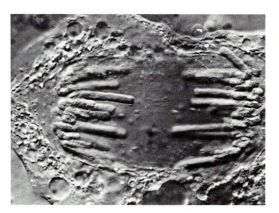

Figure 2-16 Completion of anaphase.

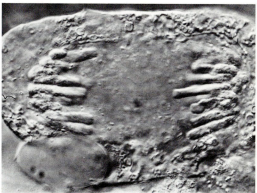

Figure 2-17 Beginning of telophase.

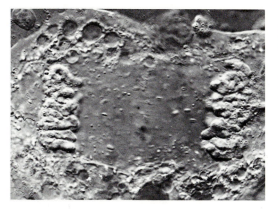

Figure 2-18 Midtelophase, with the chromosomes becoming more tightly packed.

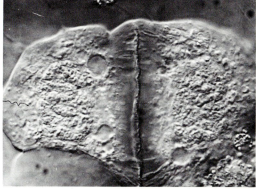

Figure 2-19 Later telophase and the appearance of the characteristic features of the new interphase.

The Form of the Chromosomes

THE FOUR-ARMED STRUCTURE. In the karyotypes in Chapter 1, most of the chromosomes appeared to have four arms. This is so because these are metaphase chromosomes, with each of the two chromosome arms split into two chromatids, but with the centromere still joining them together. Chromosomes may be obtained in this condition by treating dividing cells with the drug *colchicine*, which prevents the separation of the centromere into two and thus effectively halts the mitotic process at metaphase. This treatment is almost universally used in chromosome studies because it yields a very large number of cells with the visible chromosomes of this stage; without the treatment, the number of cells at the metaphase stage would be a small fraction of the total since most cells at any given time will be in interphase, at which stage the chromosomes are virtually invisible. Chromosomes in this divided condition have several advantages. The position of the centromere is clearly marked, making it possible to distinguish chromosomes of equal total length, but with different centromere positions; the short and the long arms of the chromosomes can be identified and, since there are two identical chromatids still attached, if one of the two chromatids for some reason is not clear in a given cell, the other presents a second opportunity.

TERMINOLOGY BASED ON CENTROMERE POSITION. Chromosomes are given different names according to the position of their centromere (Figure 2-20). When the centromere has a central (medial) position, the chromosome is a *metacentric*; if it is slightly off center, it is *submetacentric*; if near the end of the chromosome, *acrocentric*; if at the extreme end of the chromosome (as far as can be determined), *telocentric*. Telocentric chromosomes do not normally occur in humans. The acrocentric chromosomes of humans may have stalklike appendages, called *satellites*, attached. The chromosomes of

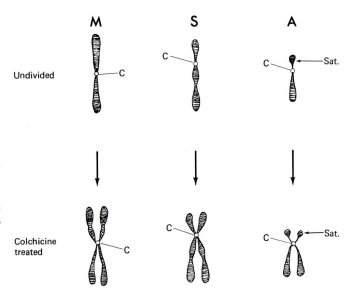

Figure 2-20 A comparison of the metaphase configurations of a metacentric *(M)*, a submetacentric *(S)*, and an acrocentric *(A)* chromosome. When the centromere splitting is suppressed, as it may be when chromosomes are treated by the application of colchicine, the two chromatids are held together at the centromere region, marking the location of the centromere *(C)*, or primary constriction, unambiguously. The chromosome depicted in column *(A)* is like that of the D or G group, with obvious satellites *(Sat)*.

groups A and F are metacentric; those of B, C, and E are submetacentric; and groups D and G are acrocentric, the latter also showing satellites in favorable preparations.

Because each chromosome has a centromere which is not located precisely at its midpoint, the two arms will be of unequal length. The shorter arm is given the designation p and the longer arm q.

DETAILS OF CHROMOSOME STRUCTURE. Since so much information about a person's genetic make-up depends on a close look at the individual chromosomes, it might seem profitable to magnify them many thousands of times for greater detail. This can be done with an electron microscope (Figure 2-21); the results are disappointing, at least as far as viewing genetic defects is concerned.

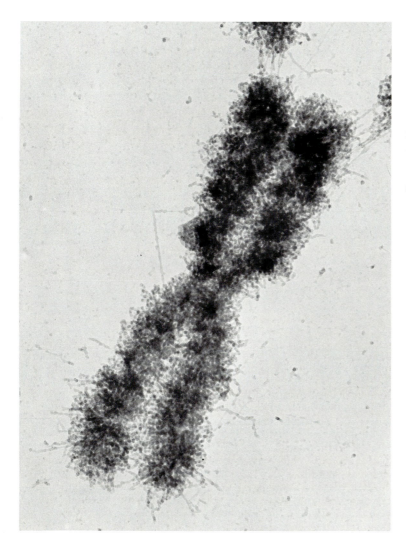

Figure 2-21 An electron-microscopic photograph of a metaphase X chromosome, showing its highly fibrous structure. Although some differentiation along the fibers is evident, unfortunately not as much detail can be seen as might have been hoped. (G. Bahr, *Human Chromosomes and Chromatin,* Tutorial Proceedings of the International Academy of Cytology, 1972.)

Analysis of Abnormal Chromosome Behavior

SHORTCUTS IN CHROMOSOME REPRESENTATION. Because many of the analyses in this book depend on a visualization of chromosome form or behavior during cell division, drawings of chromosomes are indispensable in illustrating important points. It is almost never necessary to draw, laboriously, all twenty-three pairs of chromosomes found in the normal cell; only as many chromosomes are depicted as are necessary to illustrate the particular point being made. Very often only a few or just one pair of chromosomes are shown in a cell and the student must imagine the presence of the rest of the complement, behaving normally. Similarly there is usually no point in drawing a chromosome very precisely; a single line generally does quite well. A centromere may, or may not, be indicated on the chromosome, depending on its relevance to the issue under consideration. Two homologues may be included in a figure if the relationship of the two is important; otherwise only one need be shown.

ABNORMAL CHROMOSOME BEHAVIOR IN MITOSIS. Although we must marvel at the accuracy with which the chromosomes are reproduced at mitosis and partitioned to the daughter cells at each division, it is not surprising that on occasion a serious mistake occurs. The most common error in the regular distribution of chromosomes during mitosis is referred to as *mitotic nondisjunction.* Figure 2-22 shows one of several ways by which mitotic

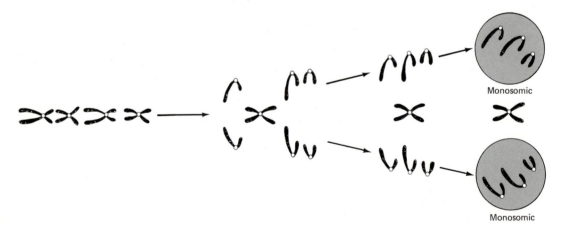

Monosomic

Monosomic

Figure 2-22 Sequence of events showing how a single chromosome may be lost from a diploid cell during cell division. The number of chromosomes depicted here is greatly reduced from the normal human diploid number for the sake of clarity. At the left are four chromosomes of the complement, each already split into two chromatids at metaphase. During early anaphase the individual chromatids separate, becoming new chromosomes. As they proceed to opposite poles, one set accidentally becomes stalled on the spindle. At late anaphase the chromosomes have moved to opposite poles, except that the immobile pair have remained on the spindle. After the completion of the division, when the new cell membranes form, each of the two parts of the division lacks one of the members of a pair and the two products are therefore monosomic for that particlar chromosome.

nondisjunction can occur. It is possible for a set of two chromatids, each of which would ordinarily go to each of the poles, to lag behind the rest of the chromosome complement at anaphase and to remain on the metaphase plate, with their subsequent loss. The two daughter cells, then, would each have one chromosome missing.

Another possibility, shown in Figure 2-23, is that one of the two chromatids might lag at metaphase. If this accident occurs very early in development, then an embryo would be formed that, theoretically, would on the average have half normal cells and half lacking one chromosome; it would therefore be a *mosaic* with respect to chromosome number.

Yet a third possibility is that the two chromatids of one replicated chromosome might go the same pole (Figure 2-24). This event gives rise to two cell lines, one of which is trisomic, the other monosomic. The fractions of the body made up of each type depend on the stage of development at which the mitotic nondisjunctional event takes place. All individuals are, without doubt, to some extent mosaic, because among the billions of somatic cells, mitotic nondisjunction must have altered at least some of them. In an individual of otherwise normal constitution there are always abnormal cells present; this frequency increases as the individual ages. Fortunately the majority of cells with normal chromosome numbers are sufficient to carry on all of the necessary functions of life.

Individuals of both sexes produce *germ cells*, or *gametes* (sperm in the male, eggs in the female), highly specialized cells that combine to produce the fertilized egg and, eventually, the newborn. Nondisjunction can also occur during gamete formation, and when it does, the resulting chromosome unbalance may seriously affect development. It is, in fact, during the formation of the germ cells that the nondisjunction most often occurs that gives rise to the trisomies and monosomies that we have already discussed

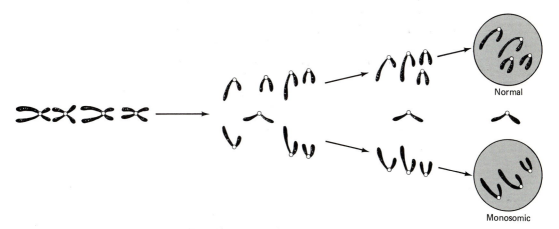

Figure 2-23 A second way in which mitotic nondisjunction can produce cells of abnormal constitution. In this instance only one of the two chromatids has lagged, with the result that of the two products, one has a normal chromosome number, whereas the second is monosomic.

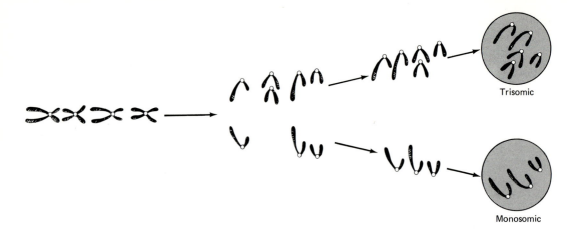

Figure 2-24 Still a third way in which mitotic nondisjunction may produce cells of altered chromosome constitution. In this case the two chromatids of a given chromosome, instead of going to opposite poles, both go to the same pole. As a result, of the two products of the division, one lacks the chromosome completely, whereas the other one has an extra copy. The result is that one cell is monosomic and the other trisomic.

in Chapter 1. We shall therefore examine still other ways by which nondisjunction can produce abnormal chromosome numbers when we take up germ cell formation.

Additional Reading

BRACHET, J. 1961. The living cell. *Scientific American,* **205**:3–13.
CRICK, F. H. C. 1954. The structure of the hereditary material. *Scientific American,* **191**:54–61.
MAZIA, D. 1974. The cell cycle. *Scientific American,* **230**:54–64.
OLINS, D. E., and A. L. OLINS. 1978. Nucleosomes: The structural quantum in chromosomes. *American Scientist,* **66**:704–711.
ROBERTSON, M. 1979. Beads of life. *New Scientist,* **83**:8–10.
ROGERS, J. 1977. How DNA folds into the living cell. *New Scientist,* **73**:699.
VAN'T HOF, J., and C. A. BJERKNES. 1979. Chromosomal DNA replication in higher plants. *BioScience,* **29**:18–22.
Why DNA doesn't get its knickers in a twist. 1980. *New Scientist,* **85**:699.

Questions

1. What improvements in technique have been made in the last few decades in the study of human chromosomes?
2. Which of the karyotypes presented in Chapter 1 are stained with ordinary dyes and which are stained with quinacrine, which reveals the fluorescing bands?

 3. Who discovered the basic structure of DNA? Can you briefly describe this structure?
 4. What is meant by the *double helix*?
 5. How much DNA is there in a human cell? Express your answer in terms of (a) the total length and (b) the number of AT and GC base pairs.
 6. Illustrate with simple drawings the way in which a length of DNA replicates to produce two identical products. Use a dozen or so base pairs on the original segment of DNA.
 7. What is a nucleosome? Is it smaller or larger than a chromosome?
 8. What is the basic problem that is solved by the process of mitosis?
 9. Describe the changes in form that the chromosomes undergo during a mitotic division. Name the different stages of a single division.
10. What is nondisjunction? Why does mitotic nondisjunction make every individual a chromosomal mosaic?
11. Why do the chromosomes in a typical karyotype appear to have four arms? In what way does this configuration change in different chromosome types (i.e., the acrocentric and metacentric types)?

3

Formation of the Germ Cells

August Weismann was one of the many brilliant biologists of the last half of the last century. His name is commonly associated with experiments designed to demonstrate the lack of inheritance of acquired characters by the amputation of the tails of mice for a number of generations to show that newborn mice with amputated ancestors still had tails of normal length. Among his more substantial contributions was his deduction that there must exist, as a matter of simple logic, a stage in the formation of the germ cells when the genetic material is halved. The reasoning went as follows:

It is a matter of common observation that two individuals of different races produce children intermediate in their racial characteristics. This must mean that both parents make a contribution to the next generation, and because the children are pretty much the same regardless of which parent is of which race—that is, the same in *reciprocal crosses*—the contribution of each parent must be about equal. However, the simple conclusion that both parents contribute all of their characteristic genetic material to their progeny leads to a logical absurdity, for then each offspring would have twice as much as either parent. Thus the amount of DNA would double generation after generation until the human body would be too small to hold all of the DNA it had received from its ancestors.

To resolve this dilemma, Weismann postulated that the genetic material must be precisely halved at some point in the formation of the gametes, so that when two gametes join to form the zygote, each progeny will have the same amount as each parent. This halving of the genetic material is accomplished during two distinctive cell divisions that occur only in those cells that are immediate ancestors of the gametes. These divisions are designated *meiosis* and consist of two divisions, the first and second meiotic divisions. Because the cellular events of meiosis determine what the parents transmit to their offspring, an understanding of the behavior of the chromosomes during the two meiotic divisions is basic to an understanding of genetics.

Occurrence of Chromosomes in Homologous Pairs

Because the *zygote* from which all the cells of the mature individual are derived is formed from the fusion of the egg and the sperm, it must contain two sets of chromosomes, one set from the male and the other from the female parent. For each chromosome brought in by the sperm, there is a "partner" chromosome in the set contributed by the egg. In humans there are forty-six chromosomes in the zygote, a *haploid* set of twenty-three brought in by the sperm and another haploid set of twenty-three by the egg, resulting in the *diploid* set of twenty-three pairs of chromosomes.

Gametes are formed by certain specialized cells called *gonial* cells, found in the testes of the male and the ovaries of the female. These divide by mitosis, producing millions of other cells like themselves. After a certain number of mitotic divisions in the male testis, a *spermatogonium* becomes a *spermatocyte,* which will then go through two meiotic divisions and become a mature *sperm.* Similarly, in the female ovary, the oogonial cells produced by mitosis differentiate into *oocytes* and, after the two meiotic divisions, become mature eggs, or *ova.* Although a number of events of genetic importance occur during meiosis, the basic accomplishment of the meiotic divisions is that the homologous chromosomes are separated from each other so that the resulting gametes carry a haploid set of twenty-three chromosomes.

Outline of Meiosis

Before considering the two meiotic divisions in detail, we shall take a quick overview of the process. In the upper left of Figure 3-1, a simplified cell with six chromosomes is shown. Note that the six chromosomes occur in pairs: chromosomes A and A' are homologues, chromosomes B and B' are homologues, as are C and C'.

In the interphase prior to this stage, replication of the chromosomes occurred, so that each chromosome entering the first meiotic prophase consists of two identical strands. Note that the two strands (or *chromatids*) of each chromosome are still held together at the centromere region.

PROPHASE OF THE FIRST MEIOTIC DIVISION. Early in the prophase of the first meiotic division, each chromosome moves to a position alongside its homologue in a process known as *synapsis.* This is a most remarkable phenomenon. At this stage the chromosomes are still long threads, with forty-six

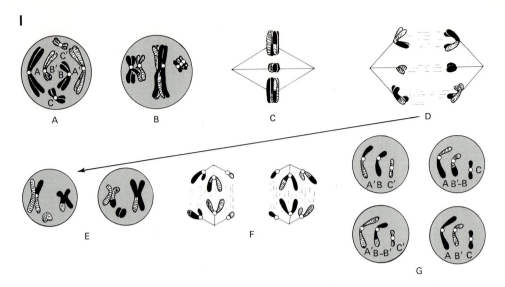

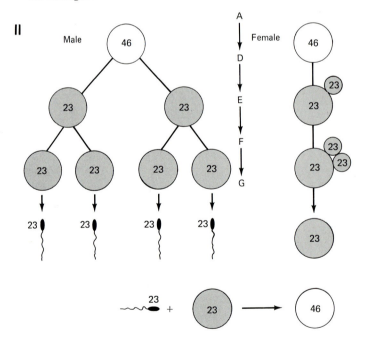

Figure 3-1 *I. (Above)* A diagram illustrating the important aspects of chromosome behavior during the two meiotic divisions. For simplicity, only three pairs of homologues are shown. The detailed explanation is given in the text. *II. (Below)* The sequence of events showing the reduction of the chromosome number and the formation of four sperm in the male, or an egg in the female, from the final products. The letters *A* to *G* indicate where the corresponding stages of chromosome behavior in Figure 3-1*I* are found.

individual chromosomes seemingly scattered more or less at random throughout the nucleus in a fashion resembling a tangled web of string. Yet somehow the homologues of each of the twenty-three pairs manage to synapse without becoming hopelessly entangled with the others.

Each of the chromosomes making up a synapsed pair consists of two identical strands, making a total of four strands. This group of four is called a *bivalent.* By synapsis, the six chromosomes become three bivalents (Figure 3-1,*IB*).

METAPHASE OF THE FIRST MEIOTIC DIVISION.   The bivalents now line up along one plane (Figure 3-1,*IC*). The nuclear membrane disappears and spindle microtubules can be seen connecting the centromere regions of the chromosomes to two opposite poles.

ANAPHASE OF THE FIRST MEIOTIC DIVISION.   At anaphase of the first meiotic division, the spindle fibers appear to pull the homologous centromeres of each bivalent to opposite poles (Figure 3-1,*ID*), thus separating the four strands of the metaphase bivalents into two groups of half bivalents. The two strands of a half bivalent are closely associated at their centromere regions (Figure 3-1,*IE*).

THE SECOND MEIOTIC DIVISION.   The interphase between the first and second divisions is atypical, usually of short duration without the return of the chromosome to its active extended state. Replication of DNA does not occur during the interphase between the first and second meiotic divisions. At metaphase of the second division, the half bivalents move to the equatorial position within the nucleus and at anaphase, the sister centromeres of these half bivalents are pulled by the spindle fibers to opposite poles (Figure 3-1, *IF*).

The two meiotic divisions result in the separation and the segregation into separate nuclei of all homologous material that was present in the cell that entered the first meiotic prophase. The two divisions produce four products, each of which has precisely an accurate haploid set of chromosomes from the diploid gonial cell (Figure 3-1, *IG*).

These haploid products now go on to differentiate into the mature gametes. In the male, each of the four products differentiates into a mature sperm. In the female, however, the four products of meiosis, although equal in chromosome distribution, are grossly unequal in size. At the first division, one of the two products retains the mass of the original cell, while the other is pinched off as a tiny *polar body.* At the second division, the process is repeated for the larger cell, and the polar body may (or may not) divide again, with the result that one large cell, the functional egg, and two or three smaller nonfunctional polar bodies are produced (Figure 3-1,*II*).

Production of New Gene Combinations

SHUFFLING OF WHOLE CHROMOSOMES.   Meiosis separates the homologues so that the gametes have only one member of each of the homologous pairs. Just which member of a given pair of homologues gets into a particular gamete depends on the orientation of the bivalent relative to the poles of the meiotic apparatus at the first meiotic metaphase (Figure 3-1, *IC*). This orientation is entirely random from one pair of homologues to the next, so that the various gametes may contain any possible combination of the homologous chromosomes. If we take the simplified case in which a zygote is formed from the combination of three chromosomes, ABC from the mother and a homologous (but not genetically identical) set A′B′C′ from the father, the diploid condition of that progeny may be represented by AA′BB′CC′. A gamete from such a person could, by chance, happen to have the three chromosomes that originally came from the mother (ABC) or from the father (A′B′C′). With equal frequency, however, the gamete could have one of the other combinations ABC′, AB′C, A′BC, AB′C′, A′BC′, or A′B′C. With three sets of homologues, there are 2^3 or 8 different possible products. In the case of humans, with twenty-three pairs of homologous chromosomes, there would be 2^{23} (or more than eight million) possible different combinations of maternal and paternal chromosomes in the gametes.

REDISTRIBUTION OF CHROMOSOME SEGMENTS. In addition to the random distribution into the gamete of either the maternally or paternally derived member of each homologous set, another process shuffles the genetic material even further. In the prophase of the first meiotic division, homologous strands within a particular bivalent exchange segments with each other. This process, called *crossing over*, is shown in the second column of the page of diagrams. In Figure 3-1 (*IB*), two strands have crossed over, with a resulting exchange of material between them. The effect of this crossover on the composition of the chromosomes of the four nuclei that result from the meiosis is shown in Figure 3-1 (*IG*). Because the homologues usually differ in the alleles they carry, crossing over produces strands that contain new combinations of genetic material.

EVOLUTIONARY SIGNIFICANCE OF RECOMBINATION. By these mechanisms, genes found in certain combinations in the parents are *recombined* during the meiosis of each of the gametes produced by their offspring. This recombination happens when different combinations of maternal and paternal chromosomes are produced at the metaphase of the first division and also when segments of maternal and paternal chromosomes exchange segments by crossing over during the prophase of the first meiotic division. An additional randomization occurs when two such new gametes fuse at the time of fertilization to form an even more distinctive zygote. In this way, each new birth represents a unique combination of genes, unlike any ever to have appeared earlier in human history, or ever to appear again.

When two or more advantageous genes appear in different persons, *recombination* provides the opportunity for them to be combined in a single individual, with obvious beneficial results. For this reason, recombination is considered one of the fundamental processes of evolution.

Details of Meiosis

Meiosis, like mitosis, is a universal phenomenon, found in both plants and animals, at all levels of complexity. It has been the subject of research for almost a century, with certain organisms such as grasshoppers, salamanders, and corn among the preferred objects of study because of the clear pictures provided by their large, and relatively few, chromosomes. It is only within the past decade that reasonably good preparations of meiosis have been made using human cells, but these have not shown anything that had not been seen previously in other organisms. Because we have a choice in the selection of organisms for the demonstration of fundamental properties of the cell, in this case we choose to make use of the spermatocytes of salamanders, in which meiosis is unusually clear. The salamanders from which the illustrations have been obtained are members of a large family of lungless salamanders found in North America and in Latin America.

First Meiotic Division

PROPHASE. The prophase of the first meiotic division is long and complicated, differing in a number of important respects from the prophase of mitosis. Figure 3-2 shows an unfixed and unstained cell, in a tissue culture chamber, just entering the first meiotic division. Although chromosome strands cannot be seen, the parts of the chromosomes associated with the centromeres are visible as small dark granules scattered throughout the nucleus. The conspicuous spherical structure in the cytoplasm is a *centrosome*. A pair of *centrioles*, not visible in the photomicrograph, is embedded in the centrosome. During the first meiotic division, this centrosome divides, giving rise to the poles from which the spindle fibers radiate.

Figure 3-3 illustrates the appearance of chromosomes that are just entering the prophase of the first meiotic division. At this early stage of meiosis, the chromosomes are resolved by the microscope as elongate single strands, all tangled together within the nucleus. Although the microscope cannot reveal this phenomenon, the chromosomes consist of two identical strands because of the duplication that took place during the immediately preceding interphase. In Figure 3-3 it is possible to see that these elongate strands do not have an even surface contour; rather they seem to be constructed of small granules and lumps strung together by thinner regions. The small granules are called *chromomeres,* and this chromomeric structure is characteristic of meiotic chromosomes. It is also apparent that the chromosome

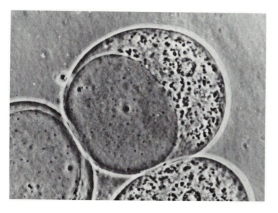

Figure 3-2 Cell just entering first meiotic division. (Figures 3-2 through 3-21 are reproduced courtesy J. Kezer, Department of Biology, University of Oregon.)

Figure 3-3 Chromosomes just entering prophase of first meiotic division.

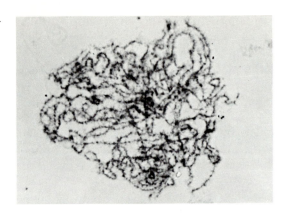

strands have a "fuzzy" appearance. In favorable preparations it can be seen that this "fuzz" consists of pairs of fine, threadlike loops that originate from the chromomeres.

In salamander spermatocytes, synapsis begins simultaneously at both ends of a pair of homologues and proceeds inward, bringing homologous chromomeres together. In Figure 3-4 synapsis has been completed in the thicker, deeply staining strands but has yet to occur in the thinner, more lightly staining material of this nucleus.

Figure 3-5 illustrates a spermatocyte nucleus in which synapsis has been completed. The chromosomal structures present in this nucleus are now complete four-strand bivalents, but the two members of a homologous pair have become so closely associated by synapsis that they appear as a single strand. The pairs of fine loops of DNA coming from the chromomeres are packed in among the bivalent strands. The relatively large, deeply staining lumps of material that are conspicuous in this nucleus mark the location of the centromeres. This centromere-associated material consists of condensed DNA called *constitutive* (or centromeric) *heterochromatin.*

The structure of the bivalents at this stage of meiosis is more clearly shown in the severely squashed nucleus of Figure 3-6. In this nucleus the

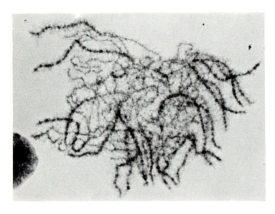

Figure 3-4 Synapsis is about half completed.

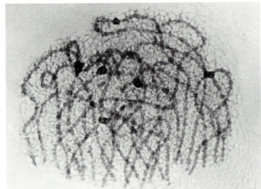

Figure 3-5 Spermatocyte nucleus in which synapsis has been completed.

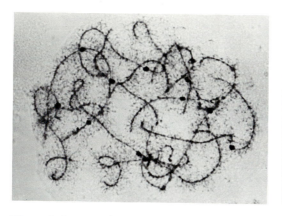

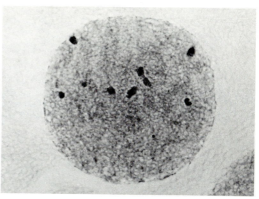

Figure 3-6 Severely squashed nucleus showing structures of the completely synapsed bivalents.

Figure 3-7 DNA spun out into elongated loops, producing a diffuse condition within the nucleus.

bivalents have been separated and pushed out of their normal position. The centromeres of the bivalents are marked by deeply staining constitutive heterochromatin. The chromomeres and their associated loops are clearly visible.

In the next stage, Figure 3-7, the DNA of the chromomeres of the preceding stage spins out into such greatly elongated loops that the entire nucleus is packed with the greatly extended filamentous threads of the loops. The material associated with the bivalent centromeres, the centromeric heterochromatin, remains condensed and is conspicuous as deeply staining granule-like particles. At the earliest stage of the first meiotic prophase, threadlike loops of DNA springing from the chromomeres were present. It is at the stage illustrated in Figure 3-7 that they reach their maximum extension, producing a diffuse nucleus lacking clearly visible bivalent strands. It seems that at this diffuse stage of meiosis the DNA that had been packed in the chromomeres is spun out so that it is placed in intimate contact with the metabolic pool of the nucleus.

To produce the nuclei of the next stage in the prophase of the first meiotic division, the greatly extended loops of genetic material of the diffuse stage must again become folded into the chromomeres of the bivalents. The appearance of nuclei just emerging from the diffuse stage is illustrated in Figure 3-8. As the bivalents become more clearly visible (Figure 3-9), it can be seen that the component halves, so closely associated earlier, are separated except at certain positions where they are engaged in intimate contact. These places of intimate association of the strands are known as *chiasmata* (singular, *chiasma*). It is believed that at these locations the breaks and exchanges of crossing over occur. The detailed structure of a bivalent at this stage is illustrated in Figure 3-10. Chiasmata are present and there is a terminal contact of strands at each end of the bivalent. The centromeres appear as more deeply staining short regions on each of the four strands.

Figure 3-11 shows the bivalents at this stage of meiosis with their four strands involved in the intricate twists and turns caused by the occurrence of chiasmata between the homologous members of the bivalents.

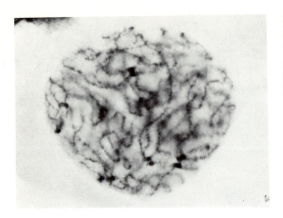

Figure 3-8 Nuclei just emerging from the diffuse stage.

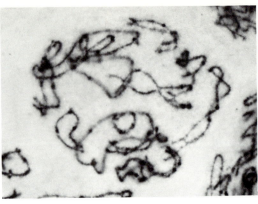

Figure 3-9 Bivalents becoming visible in prophase, showing chiasmata.

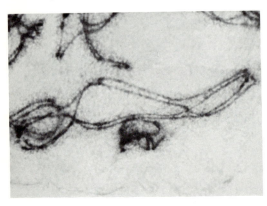

Figure 3-10 Detailed structure of a bivalent.

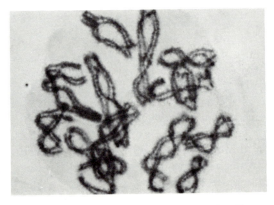

Figure 3-11 Twists and turns in strands of bivalents when chiasmata are present.

METAPHASE. In Figure 3-12 we see the chromosomes as they approach the first meiotic metaphase. The bivalent axes possess the same kinds of filamentous loops that were present in the earlier meiotic stages.

The metaphase of the first meiotic division is illustrated in Figure 3-13. The bivalents have moved to the equatorial position within the cell, but in this illustration they have been pushed out of their normal equatorial arrangement during preparation. The homologous centromeres of the bivalent halves have become more widely separated, suggested the pulling forces of the spindle microtubules that are operating at this stage. The arms on either side of the centromeres of the bivalent halves remain locked in chiasmata. The first meiotic metaphase bivalents are shown in Figure 3-14, the equatorial arrangement again having been disturbed by the squashing.

ANAPHASE. The early anaphase of the first meiotic division is illustrated in Figure 3-15. Homologous centromeres of the bivalent halves are moving in opposite directions so that the bivalent halves are separated to give two sets of two-stranded structures. In Figure 3-16, the first anaphase has been completed and the resulting chromosome sets have been separated.

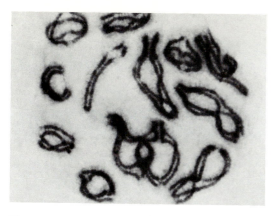

Figure 3-12 Beginning of first meiotic meta-phase.

Figure 3-13 Metaphase of first meiotic division.

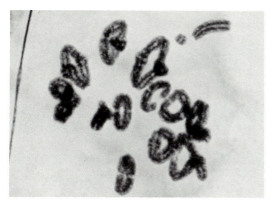

Figure 3-14 First meiotic metaphase bivalents.

Figure 3-15 Early anaphase of first meiotic division.

Figure 3-16 Completion of first anaphase.

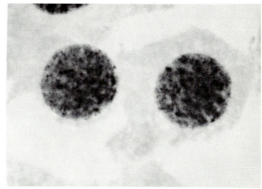

Figure 3-17 Two nuclei in interphase between first and second meiotic divisions.

Second Meiotic Division

INTERPHASE AND PROPHASE.   The two nuclei presented in Figure 3-17 are in the interphase between the first and second meiotic divisions. Chromosome duplication does not occur during this interphase. The interphase nuclei enlarge and enter the prophase of the second meiotic division, indicated by the gradual appearance of the half bivalents. In the early prophase nucleus, the location of the centromeres is marked by the more deeply staining centromeric heterochromatin from which the faintly visible strands extend (Figure 3-18). Somewhat later, the half bivalents become completely distinguishable, with the two strands of each half bivalent closely associated at their centromere regions (Figure 3-19).

METAPHASE AND ANAPHASE.   The second meiotic metaphase is shown in Figure 3-20, with the half bivalents somewhat squashed out of their normal equatorial position. The centromeres of a centrally located half bivalent are stretched so that they have the shape of a V, suggesting the pulling forces of the spindle fibers. The anaphase of the second meiotic division (Figure 3-21) separates the sister centromeres of the metaphase half bivalents. Thus the

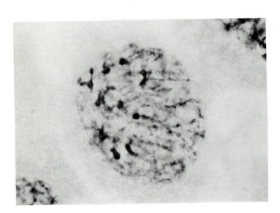

Figure 3-18   Nucleus in early prophase of second meiotic division.

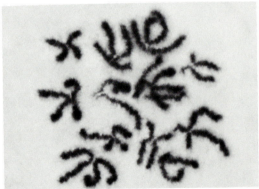

Figure 3-19   Half bivalents at early metaphase of the second meiotic division.

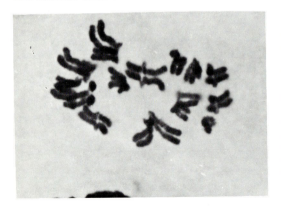

Figure 3-20   Second meiotic metaphase.

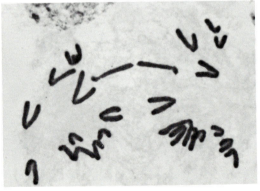

Figure 3-21   Second meiotic anaphase.

two meiotic divisions are completed with the production of four products, each with a precisely halved (haploid) set of chromosomes.

To summarize the major features:

1. Chromosomal duplication takes place during the interphase prior to the prophase of the first meiotic division. This doubleness of the chromosome strands does not become visible until later.
2. Synapsis occurs early in the first meiotic prophase, bringing the homologous strands into intimate contact.
3. The synapsed homologues, or bivalents, are held together by chiasmata, which are the positions at which the breaks and exchanges of crossing over are believed to occur.
4. The four-stranded bivalents are now separated by the two meiotic divisions; at anaphase of the first division the two chromatids attached to the same centromere go as a set to each of the two poles, and at the second division those two chromatids separate.
5. The four cells that result from the two meiotic divisions contain only half the number of chromosomes of the cell from which they have been derived, as there was no duplication of the genetic material in the interphase between the first and second divisions.

The basic differences between the two meiotic divisions and a set of two successive mitotic divisions are shown in Figure 3-22.

Thus the two members of each homologous pair of chromosomes, genetically modified by crossing over, are separated from each other by meiosis and distributed into different nuclei so that each of the four products of a meiosis has only one member of a given homologous pair. These four haploid products of meiosis differentiate into four sperm in the male and into one egg and three nonfunctional polar bodies in the female. The diploid number is then restored for the next generation when the haploid egg and the haploid sperm unite to produce a diploid zygote.

Continuity of the Germ Line

During the very early stages of embryonic development, a small number of cells are set aside to become the germ cells for the forthcoming generation. Other cells are destined to go through mitosis, to differentiate, and to form specific tissues of the developing embryo. These somatic cells will be represented by their mitotic descendants in the mature animal and will cease to exist when the animal dies. The germ cells, however, will form gametes, and it is the combination of two gametes that will form a new individual. Hence the germ cells are characterized by potential immortality; these are the cells that concern us when we consider the propagation of genetic changes from one generation to the next.

Timing of Meiosis

The course of sperm cell formation in the human male parallels the natural maturation of the sexually mature male. During infancy and childhood the primitive germ cells (spermatogonia) remain relatively quiescent and do

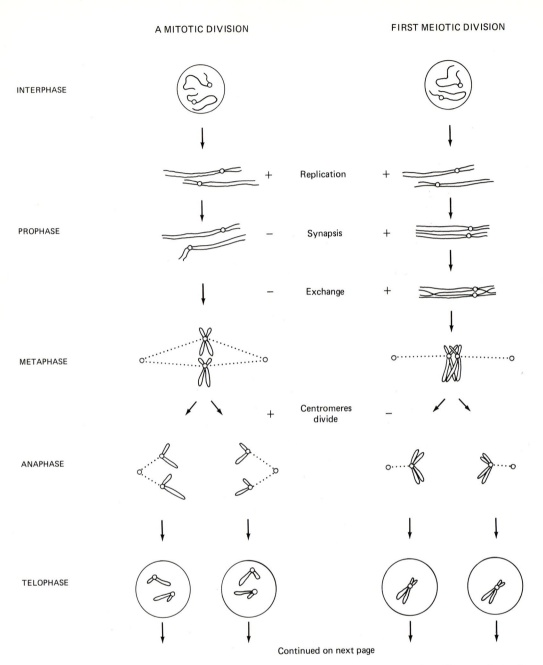

A MITOTIC DIVISION   FIRST MEIOTIC DIVISION

INTERPHASE

PROPHASE   Replication

Synapsis

Exchange

METAPHASE

Centromeres
divide

ANAPHASE

TELOPHASE

Continued on next page

Figure 3-22 Comparison of chromosome behavior when a body cell goes through
two mitotic divisions with the behavior when a germ cell goes through the two
meiotic divisions.

not develop into spermatocytes and mature sperm until the onset of puberty
in the early teens (Figure 3-23). After this time the male produces sperm
until the end of the reproductive lifetime, many decades later.

One might therefore imagine that the course of events in the human fe-

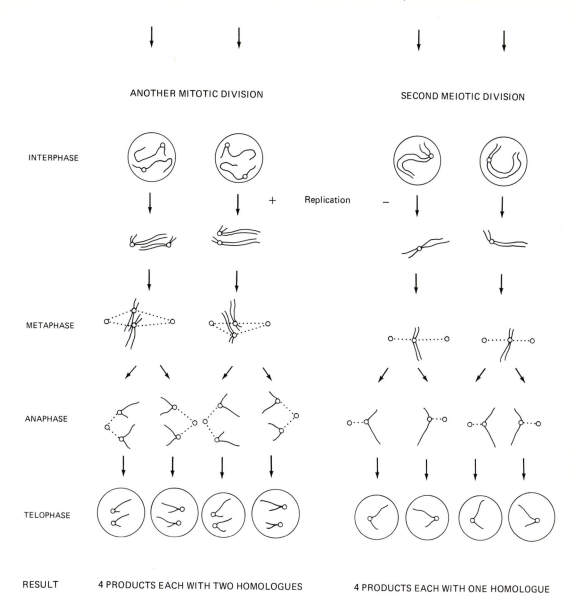

Figure 3-22 (continued)

male would parallel that of the male. This is not the case. The course of oogonial proliferation is at its maximum during the middle of the fetal life, well before birth, at which time the total number of potential egg cells runs into the millions (Table 3-1). During the latter part of this period the cells enter the late prophase of the first meiotic division. The chromosomes then assume a fuzzy appearance, and the cell goes into a prolonged prophase stage called *dictyotene* (or dictyate). At the same time, large numbers of these oocytes disintegrate so that the total number of potential egg cells at the time of birth has decreased to several hundred thousand. This number con-

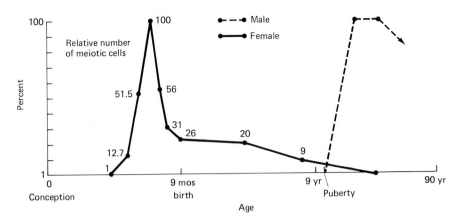

Figure 3-23 The relative number of germ cells in meiotic division in the male and female at different ages. In order to expand the prenatal period on the graph, the ages are plotted logarithmically so that the first third of the graph represents conception to nine months (birth), the middle third represents birth to nine years, and the last third represents nine to ninety years. (Data for female meiosis from T. G. Baker, *Proc. R. Soc. Lond.* [*Biol.*] **158**:427, 1963.)

tinues to decrease throughout the reproductive lifetime of the woman. At the time of puberty the egg cells are released individually at intervals of about once a month. At the time of their release from the ovary, at ovulation, the cells resume meiosis and continue as far as the metaphase of the second meiotic division, at which point the process again halts. It continues when the egg is fertilized by a sperm, after which time the meiotic products, one mature egg and several polar bodies, are present (Figure 3-24).

Thus it can be seen that the egg that gives rise to an embryo during the middle of the reproductive life of a woman has remained in the dictyotene stage for two or three decades.

Age After Conception	Total Number of Germ Cells
2 mo	596,800
3 mo	1,421,200
4 mo	3,577,200
5 mo	6,831,600
6 mo	3,609,400
7 mo	2,277,600
At Birth	**2,023,600**
Age After Birth	
6 mo	383,000
10 mo	288,600
2 yr	402,000
7 yr	188,000

Table 3-1 Calculated populations of germ cells in the human female. Note that in the upper half of the table, age is given in months after conception of the female embryo. (T. G. Baker, *Proc. R. Soc. Lond.* [*Biol.*] **158**:417–33, 1963.)

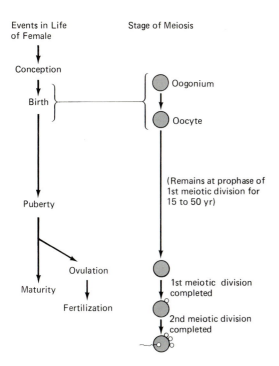

Events in Life of Female

Stage of Meiosis

Conception

Birth

Oogonium

Oocyte

(Remains at prophase of 1st meiotic division for 15 to 50 yr)

Puberty

Ovulation

Maturity

Fertilization

1st meiotic division completed

2nd meiotic division completed

Figure 3-24 Time sequence of events of meiosis in the female, showing the prolonged prophase stage of the first division, the resumption of meiosis after ovulation of the egg up to the second division, and the completion of meiosis after fertilization.

MEIOTIC NONDISJUNCTION. Just as chromosomes may be lost or may go to the wrong pole during anaphase of mitosis, they may behave similarly during meiosis, giving rise to *meiotic nondisjunction* (Figure 3-25). It can occur at either the first or the second division of meiosis. If it occurs at the first meiotic division, and if the second meiotic division separates the sister strands normally, the gamete may contain two homologous chromosomes that, when uniting with a gamete from the opposite sex, can produce a trisomic embryo. The complementary class lacks a chromosome and can give rise to an embryo monosomic for the chromosome involved.

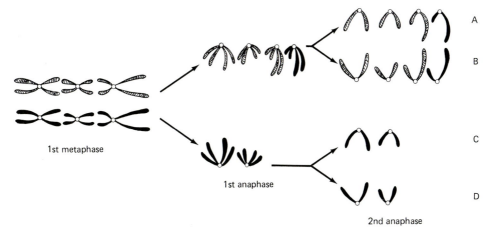

A

B

1st metaphase

C

1st anaphase

D

2nd anaphase

Figure 3-25 Nondisjunction of two homologous chromosomes at the first meiotic division. At the conclusion of meiosis, two of the four products (*A* and *B*) will have an extra chromosome beyond the haploid set, and two (*C* and *D*) will lack one.

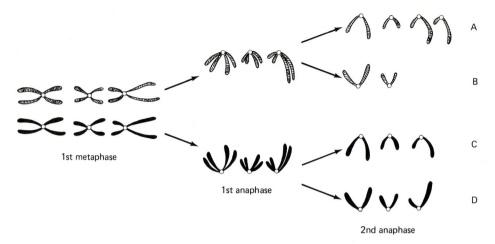

Figure 3-26 Nondisjunction of chromosomes occurring at the second anaphase division, resulting in an extra chromosome in one of the four products *(A)* and a missing one in another *(B)*.

Nondisjunction at the second meiotic division may yield a gamete that carries two sister chromatids (Figure 3-26) and that is generally indistinguishable from the gamete resulting from a first meiotic nondisjunction. There are several cases, however, in which they may be distinguished . For instance, the YY sperm that produce XYY males must come from second meiotic nondisjunction, because first meiotic nondisjunction would give only XY sperm (Figure 3-27).

TRISOMY-21. One of the interesting characteristics of *trisomy-21* is that its frequency is higher in offspring born to older mothers than in those born to younger ones (Figure 3-28). This is known to be an effect primarily of the mother's age, rather than the father's, because if the ages of both parents are checked the increased incidence of trisomy-21 is found predominantly when the mother is older. It seems reasonable to suppose that this maternal-age-dependent event is related to some alteration of normal meiosis in the older mothers.

One common explanation is that the oocytes, which have been stalled at the dictyotene stage since about the time of birth, deteriorate with age, and some 40 years later do not go through meiosis properly, but instead the chromosomes show a tendency to nondisjoin. This feature of female meiosis provides a good explanation for the increased frequency of trisomy-21 in older women—the prolonged stay at the dictyotene stage may result in an increased inability of the chromosomes 21 to proceed through the meiotic stages without experiencing a nondisjunctional accident.

Abnormalities of Chromosome Number

The condition of duplicate chromosomes, or *diploidy,* is characteristic of the vast majority of existing species on this planet. One might suspect, therefore, that the presence of two of each kind of chromosome is of some

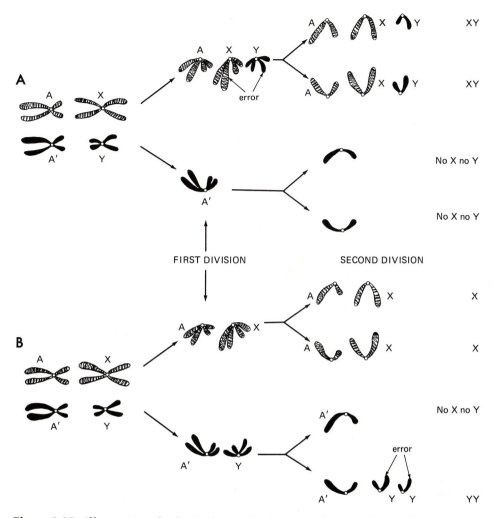

Figure 3-27 Illustration of a basic distinction between first- and second-division nondisjunction. When the homologues are different, as they are in the case of the X and Y chromosomes, first-division nondisjunction *(A)* can produce only XY or no-X no-Y sperm. Second-division nondisjunction *(B)*, however, can produce YY sperm but not XY sperm. The unlabeled chromosome is an autosome with normal behavior.

advantage in the survival of most forms of life. The argument that such an advantage exists is very simple: each chromosome is made up of a large number, probably several thousand, of different genes, and most of these genes are essential to the life of the organism. Diploidy provides a backup system in the event that some of these essential genes are destroyed. During the long life span of a multicellular organism, it can be anticipated that some of the genes carried by the chromosomes will be lost or inactivated by normal molecular agitation, by factors from the external environment such as radiation, or by poisonous chemicals. In that event, the unaffected genes on the other chromosome of the homologous pair are usually able to take over the functions of the damaged ones. It is easy to understand that this

Figure 3-28 The greatly increased frequency of trisomy-21 in two widely separated populations, with increasing age of the mother. (J. Wahrman and K. Fried, *Ann. N.Y. Acad. Sci.* **171**:341–60, 1970.)

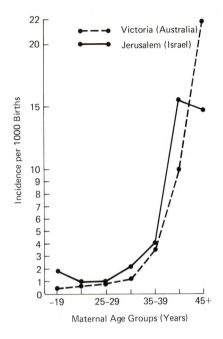

backup system would be much more important to a long-lived organism than to a short-lived one; most of the organisms that are haploid (i.e., with only one set of chromosomes), such as bacteria and molds, have relatively short life spans.

Because the diploid organism has become adjusted to the presence of two chromosome sets (and therefore to two complete sets of genes) over countless generations of natural selection, the addition of an extra chromosome throws the cellular mechanism off balance; the genetic factors no longer interact in the proper proportions. When this occurs, an abnormal individual is produced. As a simple analogy, let us suppose that an expert mechanic has designed an excellent two-cylinder gasoline engine. If one were to try to add to this machine one additional spark plug or cylinder or piston, the net result would be chaotic; the engine, after such a piecemeal addition, could never work as well as the original optimum design. Quite likely, it would not work at all.

TRISOMY-18 AND TRISOMY-13.   Autosomal trisomies other than trisomy-21 may be even more disastrous to the newborn. Trisomy-18 is one of the more common: generally embryos with such trisomies succumb prior to birth; when affected infants do survive to term, they usually do not live beyond a few months (although in a few rare cases survival has been recorded beyond several years). Some surveys have suggested frequencies of incidence as high as, or even slightly higher than, 1 per 20,000.

Like trisomy-18, trisomy-13 is usually lethal; affected fetuses are most often aborted spontaneously, or, when viable as a newborn, the infant rarely survives beyond the age of three months. The incidence of this abnormality among newborns is about 1 per 7,000 births. A list of physical characteris-

	Trisomy-13 (percentage)	Trisomy-18 (percentage)
Mental retardation	51	85
Low-set ears	51	82
Malformed ears	77	75
Congenital heart disease	83	82
Flexed fingers	40	75
Overlapping fingers	14	50
Rocker-bottom feet	—	40
Microcephaly	54	—
Receding chin	—	87
Defects of eyeball	88	—
Polydactyly	63	—
Cleft lip or palate	71	—

Table 3-2 Frequencies of developmental anomalies associated with trisomy-13 and trisomy-18. (Condensed from J. Nusbacher and K. Hirschhorn, *Adv. Teratol.* **3**:11–53, 1968.)

tics in these two syndromes is given in Table 3-2. Although we can expect occasional medical reports of survival of trisomies other than 13, 18, or 21 (and, indeed, trisomy-22 and trisomy-8, although rare, are now established as specific syndromes), it is a general rule that the other autosomal trisomies are lethal during early prenatal development.

MATERNAL AGE AND OTHER CHROMOSOME ANOMALIES. Because of the lower frequency of birth of infants affected with other trisomies, it is more difficult to make a categorical statement about an age effect. However, what information is available suggests that trisomy-13 and trisomy-18 births may also appear at advanced maternal ages. This, however, is not true for the births of infants with Turner syndrome, suggesting that the syndrome does not arise primarily from nondisjunction in the oocyte as it proceeds through meiosis. This fits in very well with the observation that a large proportion (up to 30 percent) of all Turner syndrome individuals are known to be mosaics, and this method of origin would require a mitotic nondisjunctional event occurring during the early cleavage divisions shortly after fertilization of the egg. Furthermore, as we will see later, many XO females carry the X chromosome of the mother, indicating that in such cases the nondisjunctional event occurred in the father.

Spontaneous Abortion

Spontaneous abortion is generally defined by national and state laws, and therefore its definition is subject to considerable variation. For our purpose, we shall define spontaneous abortion as the natural termination of pregnancy before the fetus has reached an age of 22 weeks or a weight of 500 gm (slightly over 1 lb). The spontaneous loss of the fetus after this period is a *stillbirth*.

The estimate of the frequency of spontaneous abortions in the population depends on the source of the data. Estimates based on interviews with women suggest a figure close to 15 percent, but this is undoubtedly an underestimate, because most women may not be aware of early embryonic losses. Studies of the early stages of pregnancy in women normal in fertility show that approximately 30 percent of all fertilized eggs are clearly physically abnormal and incapable of further development. This additional 30 percent represents fertilized eggs that would have been lost prior to three weeks of pregnancy, that is, probably before the women themselves would have been aware of their existence. This gives a total of 45 percent for spontaneous loss of very early embryos.

Nearly half of the embryos and fetuses spontaneously aborted are morphologically abnormal. It seems quite likely that an additional percentage are abnormal to some degree not obvious by ordinary inspection. From this point of view, then, it would appear that in most cases early spontaneous abortions have a beneficial effect in preventing grossly abnormal fetuses from coming to term.

CHROMOSOME ANOMALIES IN SPONTANEOUS ABORTIONS. About 1 newborn in 200 has a chromosome abnormality of some sort. On the other hand, in a large series of spontaneous abortions examined for abnormal chromosome numbers, about a third proved to have some deviation from diploidy, or *aneuploidy*. Of the aneuploid specimens, about 50 percent had simple trisomy. The frequency of chromosomally abnormal abortuses depends on the mother's age; older females show higher incidences than women aged twenty to thirty years. Furthermore, the actual frequency may depend on the age of the embryo or fetus at the time it was aborted. Those that are aborted at less than three and one half months of age are chromosomally abnormal almost half the time, whereas those aborted after four and one half months have as few as 5 percent abnormalities. Table 3-3 shows the relative frequencies of trisomy for the different chromosome groups out of a total of 690 trisomic spontaneous abortions. Chromosome 16 is for some reason unusually frequently involved in trisomies (Figure 3-29). Even those trisomies that appear at times in liveborn infants—13, 18, and 21—result in abortion far more often than they do in live births.

Table 3-3 The frequencies of trisomies for the seven chromosome groups in a series of abortions. The "expected" percentage is based simply on the number of chromosomes belonging to each group. A comparison of the observed percentages with the expected suggests that, except for the F group, the smaller chromosomes are more likely (and the larger ones less likely) to be found in an inviable trisomy recoverable from a spontaneous abortion. (Courtesy of C. E. Ford, Oxford University.)

| | Chromosome Group | | | | | | | Total Trisomies |
	A	B	C	D	E	F	G	
Number	20	9	112	150	252	22	125	690
Percentage observed	3	1	16	22	37	3	18	
Percentage expected	14	9	31	14	14	9	9	

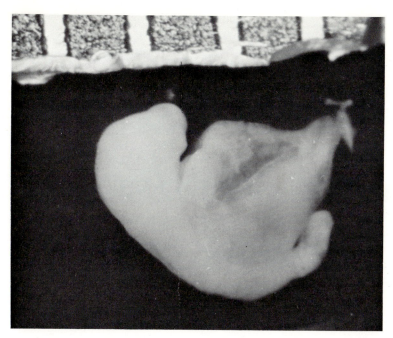

Figure 3-29 A grossly abnormal early embryo, identified as having trisomy-16. (D. H. Carr, Chromosomes and Abortion, in *Advances in Human Genetics,* copyright © Plenum, New York, 1971.)

MONOSOMY. From our earlier discussions of nondisjunction, we have seen that trisomy and monosomy may be complementary products of one event (Figure 2-24), but that monosomy may arise in ways that do not also produce a complementary trisomic product. (Figures 2-22 and 2-23). We would therefore predict that the frequency of embryos monosomic for one of the chromosomes would be greater than the frequency of trisomies for that chromosome. In fact, the frequency of observed monosomies is very much less, the reason being that autosomal monosomies are quite inviable and are presumably lost prior to the time of birth—in fact, even before they can be detected as miscarriages. There appear to be a few well-authenticated cases of monosomy-21 and monosomy-22 in viable children who are characterized by mental retardation and developmental anomalies. As a general rule, however, it is probably true that the rare occasional cases of monosomy reported in the medical literature represent mosaic individuals with some normal cells present that allow them to survive, or individuals with complicated chromosome anomalies who appear to be monosomics but, in fact, are not.

An exception to this rule is monosomy for the X chromosome. XO individuals may survive to term and, in fact, are relatively normal except for disorders of fertility (Turner syndrome, Chapter 1). The curious fact is that although XO newborns are relatively uncommon (1 in about 4,000 newborn females), they comprise about one fifth of all abortuses with abnormal chromosome numbers. It can be calculated that less than 5 percent of all fertilized eggs that start out as XO actually come to term.

POLYPLOIDY. In about 20 percent of the abortions with chromosome abnormalities, three sets of chromosomes (*triploidy,* Figure 3-30) are found instead of two (sixty-nine chromosomes instead of forty-six). The possession

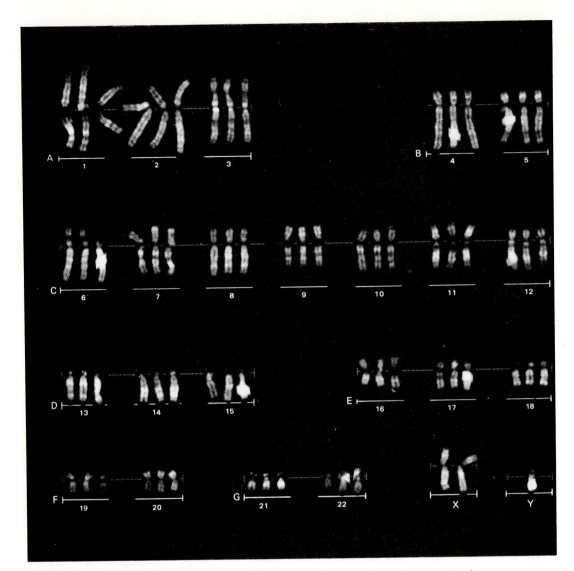

Figure 3-30 Triploidy in a human abortus, one of the most common types of inviable chromosome anomalies. Note that there are three chromosomes present in each set, except for the sex chromosomes, which are represented by two X chromosomes and a Y chromosome. (Courtesy of R. E. Magenis, University of Oregon Health Sciences Center.)

of more than two chromosome sets (*polyploidy*) is lethal in humans. Why this is so is not known; not only is this condition quite viable in plants and in some animals other than mammals, but in many cases polyploids are difficult to distinguish from diploids.

In one study of more than 300 triploids, about 20 survived to birth but lived for only a few hours or days. In all cases where apparent triploids have survived beyond a few days, detailed chromosome analyses have shown

Type	Number of Observations	Percentage
Monosomics (mostly XO)	262	19
Trisomics	703	51
Triploids	239	18
Tetraploids	76	6
Mosaics and rearrangements	79	6
Total	1,359	100

Table 3-4 Types of chromosome abnormalities identified in human spontaneous abortions and their relative frequencies. (Courtesy of C. E. Ford, Oxford University.)

them to be mosaic, with most of their cells diploid and only some of their cells triploid.

The relative frequencies of the types of chromosome abnormalities in spontaneous abortions are given in Table 3-4.

Additional Reading

CREASY, M. R., J. A. CROLLA, and E. D. ALBERMAN. 1975. A cytogenetic study of human spontaneous abortions using banding techniques. *Human Genetics,* **31**:177–196.

HOOK, ERNEST B. and IAN H. PORTER, eds. 1977. *Population Cytogenetics: Studies in Humans.* New York, N. Y.: Academic Press.

JACOBS, P. A., M. MELVILLE, and S. RATCLIFFE. 1974. A cytogenetic survey of 11,680 newborn infants. *Annals Human Genetics,* **37**:359–376.

JAGIELLO, G. M., and P. E. POLANI. 1969. Mammalian meiosis with special reference to man (a pictorial presentation). *Guys Hospital Report,* **118**:413–431.

WARBURTON, D., CHIH-YU YU, J. KLINE, and Z. STEIN. 1978. Mosaic autosomal trisomy in cultures from spontaneous abortions. *American Journal Human Genetics,* **30**:609–617.

Questions

1. What is meant by a reciprocal cross? What important biological fact comes from similarity of offspring from reciprocal crosses between two different races?
2. Can you present any argument in favor of a halving of the genetic material in each generation?
3. List the most important ways in which the two meiotic divisions differ from two successive mitotic divisions.
4. In what two ways are genes shuffled during meiosis?
5. Describe the difference between meiosis in the male and in the female, with respect to (a) the different kinds of cells produced during the meiotic divisions and (b) the timing of the different meiotic stages during the lifetime of the two sexes.

6. What is the difference between meiotic nondisjunction and mitotic nondisjunction?
7. How can nondisjunction be responsible for an XYY male?
8. Where do the following terms fit into a discussion of meiosis: *gonial cells; spermatogonia; oogonia; spermatocyte; oocyte; polar body; synapsis; chiasma; crossing over; recombination?*
9. How many years might an oocyte stay in the dictyotene stage?
10. How many chromosomes are found in the sperm? In the ovum? In the zygote?
11. What are the main differences between trisomy-21 and the other two autosomal trisomies, 13 and 18?
12. What is the approximate rate of loss of very early embryos? How can such loss sometimes be considered beneficial?
13. Discuss the relative frequencies, among spontaneous abortions, of (a) aneuploids; (b) trisomics; (c) monosomics (including XO); (d) triploids and tetraploids.
14. Does life cease at any point between successive generations? What is your view about the continuity of *human* life? Is there some period during which human life may be considered nonexistent?

4

Simple Inheritance

From even the most casual observations of parents and children, it is clear that there exist many characteristic differences inherited from one generation to the next. A more detailed examination of families by physicians, human geneticists, biochemists, and similar inquisitive people yields a total of several thousand inheritable traits, but the vast majority of them are of interest primarily to specialists in the field.

Two common characteristics with a simple inherited basis are *albinism*, the absence or reduction of dark pigment in the hair and skin, and "dwarfism" of a type known as *achondroplasia*. For certain common traits, such as eye and hair color, minor complications in the inheritance patterns prevent us from using them as examples. For other characteristics, such as allergies, there is almost certainly some elementary genetic mechanism involved, but it has not yet been precisely defined. Other traits, such as interracial differences in pigmentation, have a definite genetic basis but are inherited in a more complex fashion. Some characteristics that we believe to have an inherited basis, such as cleft lip, depend to some extent on the person's genetic composition, but developmental variables may be more important. The height and weight of an individual also fall into this category. Finally, at the other extreme, there are certain traits for which the extent of the genetic component is still under investigation, and in which environmental influences must play a large part, such as intelligence as defined by performance on standard IQ tests. For now, however, we shall consider only a few traits that are particularly useful in illustrating the nature of simple genetic inheritance in humans.

Dominant Inheritance

Let us take as our first example the inheritance of achondroplasia. In this type of dwarfism, the long bones do not develop properly, with the result that the individual is usually quite short, although the trunk and head may approach normal size (Figure 4-1). The intelligence of these persons is unaffected and they mature physically and mentally. They are, of course, well aware of their own decreased stature and may suffer from some psychological problems for this reason.

Figure 4-1 An achondroplastic boy whose well-being is clearly evident. (Courtesy of Judith Hall, University of Washington Medical School.)

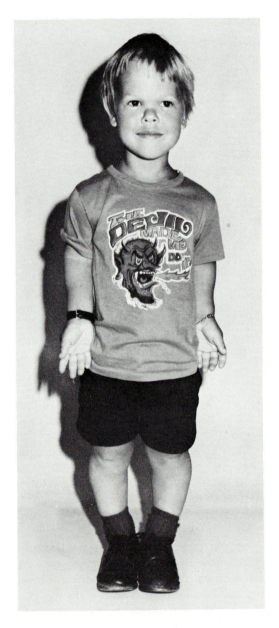

THE PEDIGREE. The relationships in a family with this characteristic are shown in Figure 4-2. In a standard *pedigree* chart, the squares refer to males and the circles to females. A horizontal line connecting the two represents a mating, and a vertical line connects them to their progeny, who, as a group, are referred to as a *sibship*. The first child is given on the left, and successive progeny are given to the right. Individuals who are affected by the given characteristic are designated by a solid symbol. In order to make it possible to specify certain individuals in the pedigree, the rows of individuals (generations) are numbered with Roman numerals, *I, II, III,* and so on, and the individuals in a given row are numbered consecutively with Arabic numerals.

In this pedigree, an affected male (*I1*) has married a normal female (*I2*), and they have seven progeny, four who are affected and three who are not. The marriage of one affected child (*II1*) has once again produced affected children (*III1* and *III3*), but the marriage of the normal progeny (*II7*) has produced only normal grandchildren.

Twins are represented by two lines that have a common point of origin, suggesting their shared birth (as in *II3* and *II4*, and *II6* and *II7*). A distinction is made between identical and fraternal twins by a small line projecting vertically from the sibship line to indicate the common origin of the identical twins. If the sex of an individual is unknown, the symbol is indicated as a diamond (*IV9*); if there are several similar individuals in a sibship, the pedigree may be abbreviated with the number put inside the symbol, as in *IV7* and *IV9*, the first representing two female progeny and the latter two children of unspecified sex. A stillbirth is indicated by a smaller symbol than usual. At any point in the pedigree, a small question mark can be inserted within a symbol to indicate some doubt as to the reliability of the information indicated by that symbol.

THE PROPOSITUS. Except in those cases where complete populations are being examined for specific diseases, attention is ordinarily drawn to a fam-

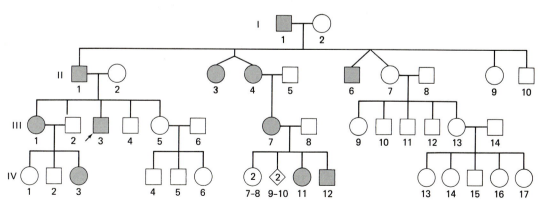

Figure 4-2 A pedigree showing the pattern of simple dominant inheritance, along with the more common symbols used in pedigrees. Detailed explanation of symbols in text.

ily by the existence of a particular genetic trait in one individual. Such an affected individual is called either the *propositus,* the *proband,* or the *index case* and is indicated on a pedigree by an arrow (as in *III3*).

This pedigree of a family affected with achondroplasia illustrates three special properties of the inheritance of certain defects: (1) all affected individuals have at least one parent who is also affected; (2) about half of the progeny of an affected individual are also affected; and (3) both sexes appear to be affected with equal frequencies. These rules characterize *simple dominant inheritance.*

THE RELATION TO MEIOSIS. It is reasonable to ask why the progeny show a characteristic to the same extent as the parents: Would it not make more sense if each child were a blend, intermediate between the two parents? Furthermore, why are only half the children of an affected parent also affected, and not all of them?

The answers to these questions can easily be found if we consider the consequences of the meiotic divisions. The end result of meiosis is the production of gametes, half of which carry one of the two chromosomes of each homologous pair, the other half the other chromosome. Let us suppose that the achondroplastic dwarf parent has a genetic factor, which we will call D, that is responsible for his condition. This factor is present in one of the two homologous chromosomes present in the cells. Which chromosome carries this particular trait, whether number 3, number 17 or some other, is unknown at present, but it is unimportant to us now. On the other homologue of the chromosome carrying D is the normal factor, represented by the letter d. This person's genetic composition with respect to the achondroplastic factor is therefore Dd; we refer to this as his *genotype.* After meiosis, half of his gametes will carry D and half d.

Because it is the presence of D on one of the chromosomes that makes an individual achondroplastic, all normal persons have a genotype of dd and all of their gametes carry d. Thus it is easy to see that when an achondroplastic individual mates with a normal person, the chance combination of their gametes will result in half of the progeny with achondroplasia and the other half normal (Figure 4-3). Because the allele D manifests its characteristic *phenotype,* achondroplasia, in the Dd combination, and d does not, D is said to be *dominant* over d, and d is *recessive* to D.

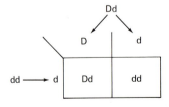

Figure 4-3 A simple checkerboard, or Punnett square, illustrating the standard method for predicting the kinds of progeny from a mating. In this case, one parent carries two dd factors so that all of the gametes are d. In the other parent, one chromosome carries D and the other d so that half of the gametes carry one and half the other. These symbols on the margins are then combined in the squares, to give the expected proportions of progeny, $^1/_2$ Dd and $^1/_2$ dd.

Linear Order of the Genes

In certain tissues of the fruit fly *Drosophila melangaster,* we find a unique situation in which the chromosomes enlarge by successive replication without cell division, so that they attain a size a thousand times greater than normal, with resulting visible detail that is far greater than can be obtained from human chromosomes at the present time. Thus they are ideal for genetic research and experimentation. Because each chromosome is composed of so many strands joined together, these chromosomes are referred to as *polytene chromosomes* (*poly* = "many," *tene* = "thread"). Figure 4-4*A* shows a set of polytene chromosomes from a cell of the salivary glands of *Drosophila.* They form an unusual configuration because (1) the two homologues are completely fused together so that each arm is actually two homologues, and (2) all of the chromosomes are joined together at their centromere regions because of an apparent "affinity" for the chromosome material (*heterochromatin*) that surrounds the centromere.

Figure 4-4*B* is a line drawing showing how the four polytene chromosomes of *Drosophila,* including an X chromosome and a "dot" chromosome, make up this configuration. At higher magnification it can be seen that each chromosome arm is crossed by bands, approximately a thousand on each arm. These bands may be considered the genes, those factors located on the chromosomes that are responsible for the development of specific traits in the individuals carrying them. Figure 4-5 shows the end of the X chromosome of *Drosophila,* along with a detailed picture of the locations, or *loci* (plural of *locus*), of some genes. In *Drosophila* it is possible to find the exact locus on each chromosome arm of a large number of genes whose functions are known. The limitations of the chromosome detail in humans have prevented any comparably exact assignment of genes within a chromosome, although great strides are currently being made, by means of the newer staining techniques, in assigning various types of genes to specific chromosome regions.

Allelic Relationships

In the example of achondroplasia, the gene *D* is responsible for dwarfism and *d* is responsible for normal development in the absence of *D*. These two different forms of a gene are called *alleles.* Thus *D* is an allele of *d* and vice versa; but *D*, responsible for achondroplasia, would not be an allele of the gene for, say, eye color, because the eye-color gene would be found at a completely different locus, probably even on a different chromosome. If the two alleles on the two homologous chromosomes carried by an individual are the same (*DD* or *dd*), that person is said to be *homozygous* for those alleles; if they are different (*Dd* or *dD*), the person is *heterozygous.* It is natural to wonder what the appearance, or phenotype, of the homozygote for the gene *D* for achondroplasia would be. Such homozygotes are rare, because the *D* allele would have to be contributed by each parent; in other words, each parent would have to be affected. In the case of achondroplasia, such chil-

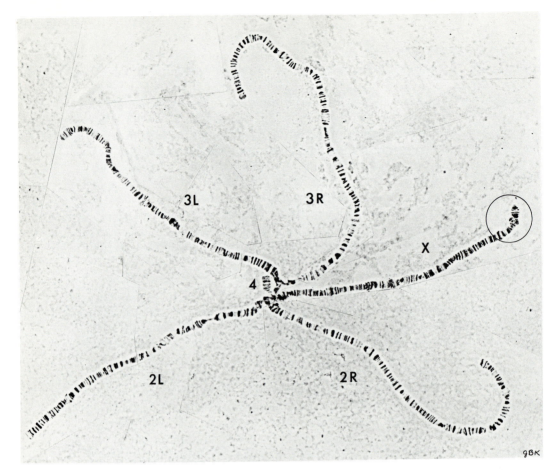

A

Figure 4-4 *(A)* The polytene chromosomes found in the salivary glands of the fruitfly *Drosophila,* much larger than usual chromosomes because they consist of thousands of chromosome strands side by side. The individual bands may be considered the location of specific genetic factors. *(B)* A line drawing showing how the *Drosophila* chromosomes form this typical polytene configuration. All the chromosomes come together at the locations of their centromeres, and the two homologues of each chromosome of the diploid are so tightly synapsed that the two together appear to be a single strand. (Courtesy of G. Lefevre, California State University at Northridge.)

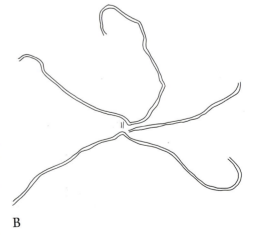

B

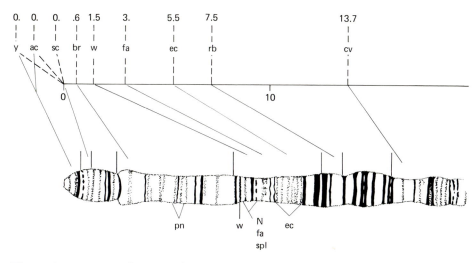

Figure 4-5 A magnification of the tip of the X chromosome of *Drosophila*, the section encircled in Figure 4-4*A*. The letters in the top row represent symbols for genetic factors (*y* = yellow body color, *w* = white eyes, *rb* = ruby-colored eyes, etc.), and the lines to the section of the polytene X chromosome show the location of that particular gene.

dren are probably very severely affected (Figure 4-6) and die young, if they survive to birth. For many relatively rare dominant genes, the phenotype of the homozygote is unknown.

In adopting symbols for the designation of alleles, the student has wide freedom. Following the practice initiated by Gregor Mendel in his pioneering work with the garden pea, however, it is customary to use a letter of the alphabet to designate a locus, and to use the capital letter to denote the

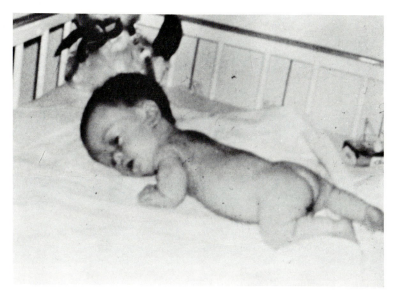

Figure 4-6 A severely affected child, the progeny of two parents with achondroplasia. She probably received the *D* allele from each parent and represents the very rare case of the homozygote for this gene. (Courtesy J. G. Hall, J. R. Dorst, H. Taybi, et al.: Two Probable Cases of Homozygosity for the Achondroplasia Gene. In *Birth Defects: Orig. Art. Ser.*, ed. D. Bergsma. Part IV. Skeletal Dysplasias. The National Foundation— March of Dimes, White Plains, N.Y. Vol. IV [4] pp. 25, 26, and 28, 1969.)

dominant allele and the lower case letter for the recessive allele. Some confusion can be avoided if different loci are given different letters, and if the letter chosen is the initial letter of a word describing the phenotype, such as *D* for dwarfism or *a* for albinism.

Recessive Inheritance

Dominance requires the presence of only one different allele, of the two present in the cell, to produce a distinctive appearance, or phenotype. The alternate, recessiveness, requires the simultaneous presence of two like alleles for the new phenotype to appear.

In the common forms of albinism, the biochemical reactions that lead to the formation of *melanin*, a dark skin pigment, are blocked so that albinos have extremely light skin and hair, with little or no coloration in the iris of the eye, giving their eyes a pale blue or pink appearance (Figure 4-7). Although this is a relatively rare defect, with only about 1 in 20,000 people showing it, it is so striking that most adults have at one time or another noticed an albino human. In any case, it is a defect well known to children from popular white-colored varieties of pets, such as mice, rats, guinea pigs, and rabbits.

In order to manifest albinism, an individual must have two *a* alleles, that

Figure 4-7 A group of Shimopaui Indians including three albinos. (Courtesy of The Smithsonian Institution.)

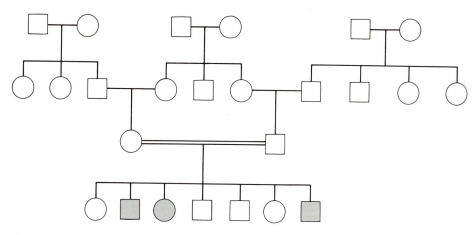

Figure 4-8 A pedigree showing the inheritance of albinism. In this particular case, the two parents of the albinos are first cousins.

is, must be homozygous *aa*. Those who have one *a* allele and one *A* allele (i.e., are heterozygous *Aa*) are normal in appearance, as of course are those who are homozygous *AA*. In many cases, the allele defined as the recessive one may show some slight effect in the heterozygote, but the relationship is conveniently referred to as a recessive one nevertheless.

A typical pedigree showing the inheritance pattern of albinism is given in Figure 4-8. Here we can see that an affected individual appears in a family without affecting either of the two parents. More than one child in a sibship may be affected, depending on the family size; however, the overall expectation in a large number of offspring is not one half, as it is in dominant inheritance, but one fourth. Figure 4-9 shows why this is so. The homozygous recessive type appears because both parents are heterozygous, and with two heterozygous (*Aa*) parents, we can see that if half the gametes from each of the parents carry *a*, then the probability that any one zygote produced will be homozygous *aa* is one half of one half, or one quarter. This checkerboard diagram, introduced in the early days of genetics by the British geneticist Reginald C. Punnett and therefore referred to as a *Punnet square,* is a simple and accurate way of determining the types and frequencies of genotypes expected from parents of given genetic constitutions. Even the most professional geneticist makes use of this system to explain or predict the results of genetic crosses.

With respect to the locus for albinism under consideration here, most

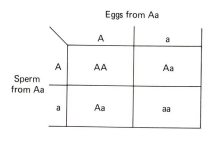

Figure 4-9 A checkerboard showing how two heterozygotes, *Aa*, will produce, on the average, one quarter homozygous *aa* children.

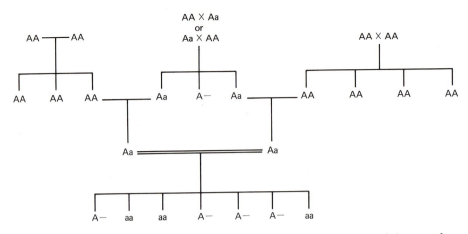

Figure 4-10 The pedigree shown in Figure 4-8 with the genotypes of the members replacing the usual pedigree symbols. Because the recessive allele for albinism is relatively rare, we are justified in assuming that any normal person is homozygous *AA*, unless the person is either a parent of a homozygous *aa* child or one of a sibship in which either a homozygote or a heterozygote has appeared.

people are homozygous for the dominant allele (*AA*), and it is therefore possible for us to designate the genetic constitutions or genotypes of the individuals in a pedigree with some degree of assurance, as in Figure 4-10. Where there is a good chance that a phenotypically normal person may be heterozygous *Aa* (as would be the case, for instance, if one of his parents were known to be *Aa*), but it cannot be stated for certain whether he is *AA* or *Aa*, then that person's genotype may be given as *A—*. Another convention for expressing genotypes is to separate the alleles on the two chromosomes by a slash (/). Thus a homozygous dominant is A/A, a heterozygote is A/a and an individual with at least one dominant allele is A/—.

CONSANGUINITY. Under ordinary circumstances, the occasional rare individual homozygous for a deleterious recessive allele is produced because each of his parents happened by chance to be heterozygous for the same recessive. However, in those cases where the parents are related (i.e., have a recent ancestor in common), that ancestor might have carried a recessive, which, by chance, was transmitted to the two parents. In this case, the probability that a homozygote will appear is very greatly increased.

The increased frequency of homozygotes from marriages of related persons, or *consanguineous* matings, is of great importance in human genetics studies, and such unions, where known, are represented in a pedigree by a double line connecting the partners. The appearance of *consanguinity* in a pedigree is itself a strong reason for suspecting recessive inheritance, and, equally, pedigrees that involve recessive traits often prove to have consanguinity in the ancestry of the affected members.

A quick way of recognizing a pedigree involving consanguinity is to note that it is characterized by an area that is completely enclosed by pedigree lines. Figure 4-8 is that of a consanguineous marriage: the two parents giving rise to the sibship with the albino children are first cousins, as they have a pair of grandparents in common, and it was from one of these grandparents that both received the recessive allele.

The Work of Gregor Mendel

What we have described earlier in this chapter is an illustration of the rules of inheritance that were first derived in the middle of the nineteenth century by Gregor Mendel, an Austrian monk (Figure 4-11). His experiments with the garden pea *Pisum sativum,* published in 1866 in a scientific paper entitled "Experiments in Plant Hybridization," can truly be considered one of the great landmarks of modern biology, and if not the first careful and thoughtful attempt to analyze breeding experiments, then certainly the first to achieve success.

Gregor Mendel was born in 1822 in northern Moravia, which was then a part of Austria, now in Czechoslovakia. At the age of twenty-one, he was admitted to the Augustinian monastery at Brünn (now Brno, Czechoslovakia), where he spent the rest of his life. After self-training in the natural sciences, he took an examination for a teaching certificate in 1850 and failed. His poor performance resulted more from his lack of a formal university education than from poor aptitude; his natural intelligence was well appreciated by those around him.

He then spent several years at the University of Vienna, where he studied

A

B

Figure 4-11 Portraits of Gregor Mendel, as a young man (*A*) and late in life (*B*). The miter and scepter shown in *B* represent his final position as an abbot. (Courtesy of V. Orel, Moravian Museum, Brno, Czechoslovakia.)

all the basic sciences, including paleontology. His emphasis was on physics, and his physics course was taught by Professor Christian Doppler (for whom the Doppler effect is named), and his courses in mathematics, chemistry, and botany were also taught by well-known scientists of the time. A second attempt to pass the examination for a teaching certificate was thwarted by an indisposition at examination time. (This phenomenon has been observed as well among present-day university students.) Although classroom teaching, as a substitute teacher, occupied a good part of his time at the monastery, he nevertheless managed to devote considerable time to his research interests. We may surmise that if his talents had been directed toward full-time classroom teaching, as they are by many teachers today, his scientific experimentation would have been greatly curtailed and he might never have started his experiments in plant breeding.

In any case, instead of a simple-minded monk puttering about his small garden and happening on some basic scientific principles by accident in the best Hollywood tradition, we have the picture of a trained scientist setting up his experiments with knowledge and purpose. In fact, in addition to his work with the garden pea, Mendel carried out breeding experiments as well with the pea weevils, with honeybees, and with mice. He was, in addition, an amateur meterologist and made daily records of rainfall, temperature, humidity, and barometric pressure until his death. He kept records of sunspots and published a detailed description of a tornado that passed over the monastery. He requested, and obtained, permission to be present at an autopsy performed by local authorities. In 1868, after his experimental work with the garden pea, Mendel became abbot of his monastery. This office led him to excessive involvement in administrative duties, including a bitter controversy with the state on problems of taxation. The harassments of his administrative chores contributed to the phenomenal consumption of more than twenty cigars a day. Nevertheless, he continued his scientific experimentation to the end of his life. As he grew older, he suffered from a long progressive disease of the kidney with dropsy, uremia, and hypertrophy of the heart. He died in 1884 at the age of sixty-one.

MENDEL'S RESEARCH. The best account of Mendel's work comes from his own paper. Although Charles Darwin's epoch-making work on evolution entitled *On the Origin of Species by Means of Natural Selection* did not appear in print until 1859, Mendel stated as one of his main reasons for starting the breeding experiments with the garden pea in 1854 that "It requires indeed some courage to undertake a labor of such far-reaching extent; this appears, however, to be the only right way by which we can finally reach the solution to a question the importance of which cannot be overestimated in connection with the history of the evolution of organic forms." If Darwin had been aware of Mendel's work and had appreciated its significance, he would have surmounted one of the major obstacles to his theory of natural selection: how it might be possible for desirable characteristics to be transmitted in full form from one generation to the next without being diluted in successive generations. In fact, Darwin himself had observed the segregation of simple characteristics in the second generation of crosses of the snapdragon and had made counts that approximated a 3:1 ratio. But not being sensitive to the significance of this fragmentary observation, Darwin proposed a theory of heredity he called "pangenesis," according to which the germ cells somehow received and incorporated information that came to them from the extremes of the body and then passed this information on to the next generation.

Darwin's theory of pangenesis was never really accepted by the scientific world, which would have to wait until 1900 for the rediscovery of Mendel's paper and the elucidation of the simplest rules of inheritance. In this paper, Mendel had pointed out the necessary criteria for setting up a proper genetic analysis: (1) making specific crosses with well-defined characteristics; (2) keeping track of the results for each generation; and (3) making exact counts to, as he put it, "ascertain their statistical relations." The organism to be chosen for the experiment, according to Mendel, must possess a variety of constant characteristics from one strain to the next, must be easily crossed experimentally and readily protected from foreign pollen, and must not suffer from problems of infertility of the hybrids. The garden pea had additional advantages: it could be grown easily and had a relatively short period of growth.

Mendel obtained a large number of different strains of peas from nurseries throughout Europe and grew as many as thirty-four strains in the summer of 1854; the following year he checked to see which of the strains had constant characters, and the following year he selected a total of seven different characters for his final experiments.

A typical result from crossing different varieties of plants is shown in Figure 4-12. In the parental generation, a tall plant (genotype TT) is crossed with a short plant (genotype tt) by the placing of the pollen from one (A) strain on the stigma of the flower of the other (B) strain. The plant that has been pollinated (B) then matures, and its flowers become pods of seeds (C).

The next year, the seeds, which are heterozygous Tt, are planted and produce all tall plants (D), because tallness is dominant. If these first-generation heterozygotes are mated to each other or are allowed to self-fertilize, they will produce seeds of the genotypes TT, Tt, and tt in the ratio of $1:2:1$ (Figure 4-13). However, the seeds must be planted in the following year and allowed to grow if one is to determine their phenotypes, which, in this case, is simply $3:1$ (E).

Thus, if we started the experiments in one year, we might expect to get out the first generation (the F_1) in the following year and the second generation (the F_2) one year later. Further, we must be prepared to raise many plants in that last year in order to get a reasonable agreement with some genetic ratio.

Mendel was faced with problems of space and time, and so he took advantage of the fact that some characters express themselves in the seed just a few weeks after pollination, as illustrated in Figure 4-14. Round (R) and wrinkled (r) are two such seed characters. If pollen from a round-seed plant (A) is placed on the stigma of a wrinkled-seed plant (B), the seeds that are produced on that parental wrinkled plant will be round (C). This happens because the Rr seeds express their own genotype, not that of the plant on which they are growing.

Next year, these heterozygous Rr seeds are planted, and the resulting

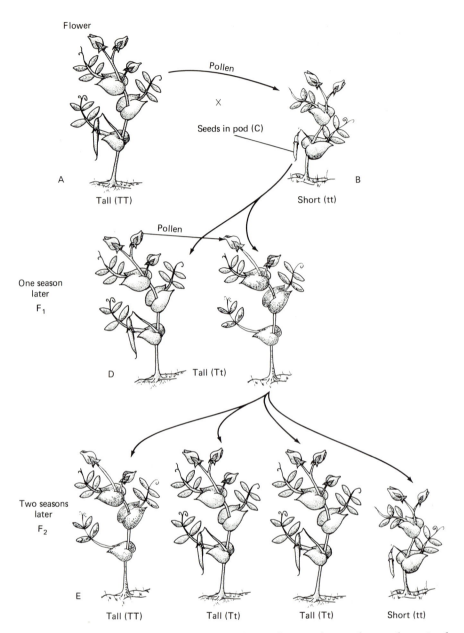

Figure 4-12 The results after crossing a tall pea plant with one that is short. In the parental generation, pollen from the tall plant *(A)* fertilizes the eggs of the short plant *(B)*. The seeds that develop in the pod *(C)* are then planted.

The seeds produce the first-generation plants *(D)*, which are all tall, because tallness is dominant. If the first-generation plants are intercrossed (or allowed to self-fertilize) and the seeds that are produced are then planted, the plants appearing in the second generation, two seasons after the start of the experiment, will appear in the ratio of three tall plants to one short plant.

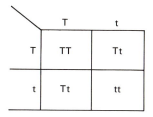

	T	t
T	TT	Tt
t	Tt	tt

Figure 4-13 A Punnet square showing how the ratio of 3 tall to 1 dwarf comes from the random combination of eggs and pollen, both with equal numbers of dominant *T* and recessive *t* alleles.

plants are mated to each other or are allowed to self-fertilize. Because this is a mating of two heterozygotes, *Rr* x *Rr*, the resulting phenotypic ratio should be 3:1. When the seeds eventually mature and express their genotype, they are indeed found in a 3:1 ratio. But this 3:1 ratio, characteristic of the F₂, is found on the seeds of the F₁ plant (*D*).

In this way Mendel was able to shorten his experiments from parent to F₂ by one generation time. In addition, because each seed (rather than plant) was the individual counted, and each plant could bear about thirty seeds, he

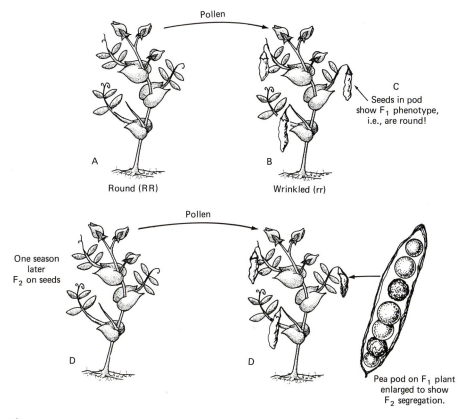

Figure 4-14 The use of seed characters to increase the efficiency of plant-breeding experiments. The parental mating of round *(A)* with wrinkled *(B)* produces seeds, on the female parent plant, that already show the characteristics of the first generation. As round is dominant, these heterozygous *Rr* seeds *(C)* also appear round.

After the seeds are planted, the first-generation plants *(D)*, which are heterozygous, are mated to each other or are allowed to self-fertilize. The seeds that appear on those plants show the second-generation segregation ratio of 3 round to 1 wrinkled.

Figure 4-15 A view of the area in which Mendel performed his experiments. It was his inspired use of seed characters that enabled him to accumulate large quantities of data in such a small space over a short period of time. (Courtesy of V. Orel, Moravian Museum, Brno, Czechoslovakia.)

was able to get the same numerical results in one thirtieth of the space it would have taken for the usual experiments involving plant characters.

This procedure explains how he was able to collect such very large numbers for his ratios, on such a small plot of land and in such a short period of time (Figure 4-15). These advantages become apparent if his original figures are examined for a comparison of the number of experimental individuals involving the two seed characters with the number involving the five plant characters he worked with. The two seed characters were the basis for progeny counts more than three quarters of the time; the five plant characters only one quarter.

MENDEL'S RESULTS. From the crosses of lines with two different contrasting characteristics, such as a tall plant crossed with a short plant, Mendel observed that one of the two characters manifested itself in the first generation. He introduced the terms *dominant* (to describe the character that appeared in the F_1 heterozygote) and *recessive* (to describe the one that did not) and suggested the convention of using a capital letter for the dominant allele and the corresponding lowercase letter for the recessive. From matings of the two first-generation hybrids (F_1), from which he got the $3:1$ ratio of phenotypes, he concluded that the factors responsible for the parental characteristics were not lost or changed in the hybrid but were segregated as individual factors in equal proportions in the germ cells of the hybrid. This is the *principle of segregation*, or Mendel's first law. We have already applied that principle to explain the proportions of affected persons in a sibship for simple allelic differences such as achondroplasia.

In experiments involving not just one pair of characters but two, Mendel showed that the characters behaved independently of each other. For instance, consider that pair of characters affecting the seed color: either yellow or green, with yellow being dominant to green. The dominant allele yellow is symbolized by Y and the recessive green by y. If a strain homozygous for yellow (YY) is mated to a strain homozygous for green (yy), the F_1 seeds will be genotypically Yy, and phenotypically yellow. If two plants from these F_1 seeds are crossed, the F_2 seeds on those F_1 plants will be approximately three-quarters yellow and one-quarter green.

Round (R) and wrinkled (r) are also seed characters, with the second generation yielding roughly three-quarters round and one-quarter wrinkled. If a plant taken from a strain characterized by round green seeds is mated to one with wrinkled yellow seeds, all the F_1 seeds will be round and yellow. However, if these F_1 seeds are germinated and a cross is made between two similar plants (or if one such plant is allowed to self-fertilize), then the seeds produced in the F_2 generation will combine the two ¾ to ¼ expectations into a 9/16 : 3/16 : 3/16 : 1/16 distribution. Figure 4-16 shows a simple system making this combination of two independent crosses. The resulting ratio is generally expressed as a $9:3:3:1$ ratio. The independence of different pairs of contrasting characters is embodied in Mendel's second law: the *principle of independent assortment.*

In the century since Mendel did his research, we have come to understand that his law of independent assortment is based on the independent segregation of different chromosomes at meiosis. The locus for the $Y–y$ alleles is on one chromosome, and the locus for the $R–r$ alleles is on another. In the F_1

Proportions of Round vs Wrinkled		
	3/4 Round	1/4 Wrinkled
3/4 Yellow	9/16 Round Yellow	3/16 Wrinkled Yellow
1/4 Green	3/16 Round Green	1/16 Wrinkled Green

Proportions of Yellow vs Green

Figure 4-16 One method for deriving the 9:3:3:1 ratio found when two independent pairs of characters segregate independently. One set of the contrasting pairs, round and wrinkled, are entered on the top along with their expected proportions in the F_2, ¾ and ¼. The other contrasting pair, yellow and green, are entered at the side margin, along with their expectations. Each class is then combined with each other class, to give four phenotypic classes, and the respective marginal frequencies are multiplied to give the proportions expected for each class. From this it can be seen that in an F_2 segregating for both the round–wrinkled and yellow–green pairs of characters, the expected proportions are ⁹⁄₁₆ round–yellow, ³⁄₁₆ round–green, ³⁄₁₆ wrinkled–yellow, and ¹⁄₁₆ wrinkled–green.

double heterozygote *YyRr*, the alleles separate from each other, following the behavior of the homologues carrying them, so that four different types of gametes are produced (*YR, Yr, yR,* and *yr*) with equal frequencies (Figure 4-17).

In retrospect, it appears that Mendel's numerical results were much closer to theoretical expectations than one might have anticipated on a chance basis. In fact, it has been calculated that results as close to the exact ratios as he got should occur by chance only 1 time in about 40,000. These too-good results have worried a number of investigators, some suggesting that perhaps Mendel had an overzealous laboratory assistant who miscounted the various phenotypes to make them conform more closely with his superior's expectations. Still another explanation is that, considering Mendel's calling, he may have had some divine assistance not ordinarily available to the average scientist! Because Mendel could not have profited from any manipulation of his data and, in any case, such a performance on his part would have been completely out of character judging from his well-substantiated willingness to accept contrary data, biologists have made a number of suggestions about how such excellent results might have been obtained. Possibly he extended his experiments when the ratios appeared to depart from the expected to make sure that an unusual result was not a bonafide departure from his simple rules. Such an innocent procedure—perfectly rational and legitimate—would tend to eliminate the exceptional cases and make the rest of the data "look too good."

So it was that the foundations of genetics were laid. Mendel was able to repeat his results with the garden pea using several other plant species (such as corn and four-o'clocks). However, when he attempted to obtain hybrids in the hawkweed (*Hieracium*), he failed completely. It would not be known for

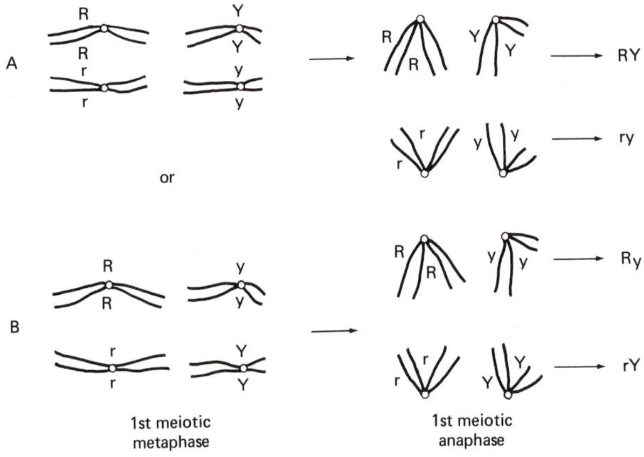

Figure 4-17 The arrangement of chromosomes in a plant heterozygous for two pairs of contrasting characters. Only two pairs of chromosomes are shown—those that are hypothesized to carry the loci of the independent factors. The orientation of the chromosomes at metaphase of the first meiotic division can be such that *(A)* the chromosomes with the dominant alleles of each heterozygous pair *(R* and *Y)* go to the same pole at anaphase, with the recessive alleles *(r* and *y)* going to the other. Equally frequently, however, *(B)* the chance orientation of these independent chromosome pairs will be such that a dominant allele of one pair will go along with the recessive of the other, to give the combinations *Ry* and *rY*. The final result, at the end of the second meiotic division, will consist of four possibilities, *RY, Ry, rY,* and *ry,* each expected equally frequently.

more than thirty years that the reason for this failure was that seeds of that species can be purely maternal in origin and may be produced without either a normal meiosis or fertilization.

We remain fascinated with Mendel's short paper, not just because he explained the simplest rules of inheritance, but also because he implied the basic nature of the genetic mechanism in higher plants and animals. Without any knowledge of the role that chromosomes play in heredity—their significance would not be appreciated until almost a half century later— Mendel implied constancy of genes, diploidy of the mature organism, the reduction division and resultant haploidy of the gametes, and the segregation of independent units (the chromosomes) at meiosis. He explained mathematically the consequences of large numbers of generations of matings of closely related individuals, or inbreeding, and the ways in which genetic factors might interact to produce unusual ratios and unusual phenotypes. Possibly it was just this elegance and abstractness in his approach and presentation, stemming from his training as a physicist, that confused the botanists of his time and led to their failure to recognize the value of his work until it was "rediscovered" in 1900. If any one scientific work were to be described as being ahead of its time, surely Mendel's "Experiments in Plant Hybridization" stands as first choice.

Other Examples of Simple Inheritance

It is sometimes felt by the beginning student that the discipline of human genetics must consist primarily of a listing of all known inherited human characteristics, along with their mode of inheritance. A recent catalog of inherited human traits* lists 1,736 defects as "good" (i.e., unambigous) autosomal dominants, another 1,973 as suspected or probable dominants, 1,521 as "good" autosomal recessives, and another 1,596 as probable recessives. Clearly even a partial listing in an elementary text would be impossible. However, there are so many important principles to be discussed that, in fact, the discussion of specific inherited traits is largely limited to their usefulness in exemplifying basic principles. Furthermore the vast majority of cases of simple inheritance known are of interest only to specialists in medicine, biochemistry, or population studies. It is also true that for many physical characters of general interest—handedness, pigmentation variants such as freckles and moles, height, weight, and so on—the mode of inheritance is more complicated, probably involving numbers of interacting loci. Even for eye and hair color, although it is true that the darker pigmentation types are usually dominant over the lighter (brown eyes dominant over blue, brunette hair dominant over blonde), there exist not only well-established contradictions to these rules in some pedigrees but also an array of intermediate types (red hair, green and hazel eyes) for which no simple explanation suffices.

---

* McKusick V. A., 1978, *Mendelian Inheritance in Man: Catalog of Autosomal Dominant, Autosomal Recessive and X-linked Phenotypes.* Baltimore, MD: The Johns Hopkins University Press.

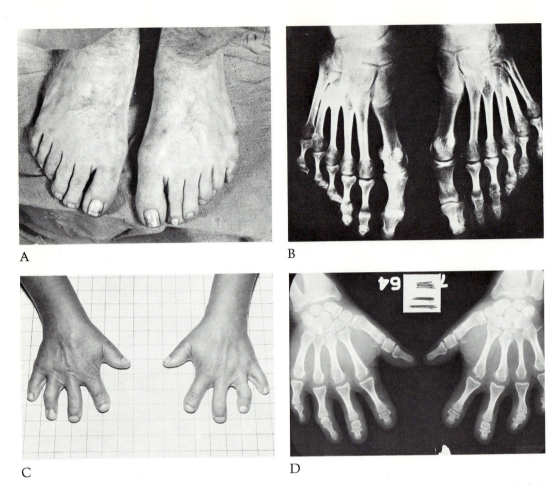

A

B

C

D

Figure 4-18 Two common striking abnormalities of the hands and feet. (*A*) A photograph of a polydactylous pair of feet, along with an X-ray photograph of the same (*B*). (S. B. Pipkin and A. C. Pipkin, *J. Hered.*, **37**:93–96, 1946.) (*C*) A pair of brachydactylous hands, with the X-ray photograph (*D*) showing the shortening of the terminal joints. (D. Hofnagel and P. S. Gerald, Hereditary Brachydactyly, *Ann. Hum. Genet.* **29**:377–82, 1966. Used by permission of Cambridge University Press.)

MALFORMATIONS OF THE LIMBS. Anomalies of the hands and feet are not uncommon, and when they appear they often show dominant inheritance. In Figure 4-18*A* is a clear example of *polydactyly*, in this case six digits instead of five, along with an X-ray photograph (*B*) of the feet revealing the bone structure. A shortening of the fingers (Figure 4-18*C*) known as *brachydactyly* is caused by the shortening and fusing of the terminal bones (Figure 4-18*D*). Brachydactyly is of historical interest, because it was clearly described as a dominant trait in the eighteenth century, and an extensive pedigree was published as early as 1905 (Figure 4-19). In *syndactyly,* two or more of the digits may be joined together.

The preceding abnormalities are dominant in the sense that they produce a characteristic phenotype in the heterozygote. With the exception of the case of the achondroplastic homozygote described earlier, homozygotes for

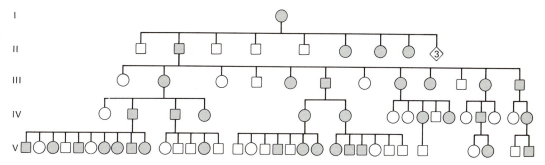

Figure 4-19 The first human pedigree of a simple Mendelian dominant, showing the inheritance of brachydactyly. (*Farabee Papers,* Peabody Museum, Harvard University, 3, 1905.)

these alleles are unknown, because the defects are sufficiently rare so that heterozygotes have not married to produce homozygous offspring. Another exceptional case is shown in Figure 4-20. In this instance two first cousins with minor hand and feet abnormalities produced an offspring with extreme defects, undoubtedly caused by homozygosity for the allele carried by both parents.

CYSTIC FIBROSIS. In the United States, *cystic fibrosis* affects more children than any other serious disease determined by defective genes. It affects about 5 out of every 10,000 newborns in the white population. Its incidence in nonwhites (Orientals and blacks) is very much less. The major medical problems are caused by abnormal secretions that obstruct the ducts of internal organs. The lungs, pancreas, intestine, and liver may be affected; the

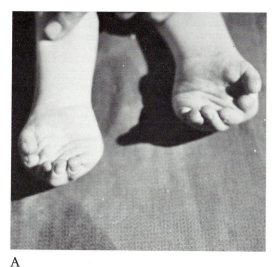
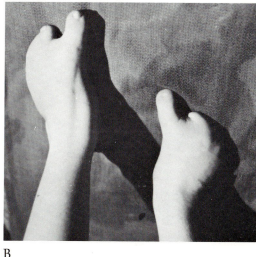

A   B

Figure 4-20 The extremely malformed hands *(A)* and feet *(B)* of an offspring of first cousins, both of whom showed minor foot defects. (J. A. Edwards and R. P. Gale, Camptobrachydactyly: A New Autosomal Dominant Trait with Two Probable Homozygotes. *Am. J. Hum. Genet.* **24**:464–74, 1972. © 1972 American Society of Human Genetics. Published by The University of Chicago Press.)

clinical manifestations depend on the organ, or combination of organs, involved. In addition, the salt concentration of the sweat and the saliva is almost always increased. About three quarters of affected persons who get beyond the first year of life live to fifteen or twenty. Affected females may be fertile, but the males are almost always sterile. This disease is caused by a simple recessive; both parents of an affected child must therefore be heterozygous. It would be helpful in many cases if heterozygotes who are at risk of producing affected children could be identified. This problem may be close to a solution. It has been found that the amount of sodium that a skin cell absorbs after treatment with a drug called *ouabain* is depressed in heterozygous cells, compared to homozygous normal cells.

TAY-SACHS DISEASE. *Tay-Sachs disease,* a recessive condition involving degeneration of the nervous system, appears at about the sixth month after birth. The degeneration of the optic neurones leads to blindness, followed by loss of intellectual capabilities, progressive muscular weakness, and paralysis. It is usually lethal by the age of two or three. Although the biochemistry of this condition is now fairly well understood, there is no known cure. Heterozygotes are quite normal but may be detected by a reduced concentration of a specific enzyme.

The disease occurs with a high frequency in Ashkenasic Jews (those from Central Europe). In the Jewish population of New York City, it is present in about 1 out of 4,000 to 6,000 births. Among non-Jewish births, it has a much lower frequency, about 1 per 500,000. About 1 out of 35 of the U.S. Jewish population is heterozygous for this recessive.

SICKLE-CELL ANEMIA. In 1910 a young West Indian black complained of fever and muscular pain. On medical examination he proved to have a previously unrecognized form of anemia, one in which his red blood cells took on an unusual elongated (or sickle) shape (Figure 4-21). In 1928 *sickle-cell anemia* was recognized as an inherited condition, and in 1949 it was shown

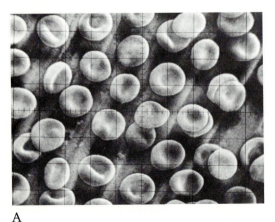
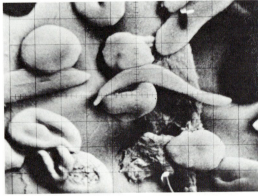

A   B

Figure 4-21 *(A)* Normal red blood cells, magnified 2,000×. *(B)* Sickled red blood cells, magnified 5,000×. Both photographs were taken by a scanning electron microscope. (Photographs by Irene Piscopo, Philips Electronic Instruments, Inc., Mount Vernon, N.Y.)

that the red cells sickled because they contain an abnormal *hemoglobin,* the protein present in red blood cells and responsible for the transport of oxygen from the lungs to the cells of the body.

More than two million people in the world, including 50,000–100,000 blacks in the United States, are homozygous for an allele, *S,* that causes sickle-cell anemia. About 1 in 10 of all U.S. blacks is a heterozygote (sometimes called "sickle cell trait" and usually with no ill effects whatsoever), but about 1 in 500 is a homozygote, with sickle-cell anemia. This disease affects the homozygote by producing sickling crises, during which some cells, with their irregular spindle shape, clog small blood vessels, adhere to the lining of the blood vessels, and cut off the oxygen supply to certain tissues. Under these conditions of low oxygen pressure, the cells sickle even more, aggravating the situation and causing extreme pain and tissue damage. Many affected persons die during childhood, but some live to old age.

Sickling occurs because of an abnormality in the hemoglobin molecules, which attach to each other end to end to form long fibers that distort the cell shape. Whether a cell will sickle or not may depend, in addition, on the protein constitution of the cell membranes. Some chemical treatments (urea, amino acid derivatives, sodium cyanate, aspirin, and others) seem to offer some hope for relief by diminishing the capacity of the abnormal hemoglobin to form long fibers. It has been suggested that there may be a lesser incidence of severe sickling in homozygotes in African groups because of the increased (fortyfold) cyanate content of the common foods in their diets: yams, sorghum, millet, and cassava.

The reason for this unusually high frequency in certain populations seems to be that the heterozygote, who is ordinarily unaffected, has an increased resistance to *falciparum malaria.* Certainly there is a high correlation in Africa between regions with a high incidence of the *S* gene for the abnormal hemoglobin and those with high incidence of malaria (Figure 4-22). In addition, it has been shown that sicklers (i.e., homozygotes) have a lower frequency of infection by the malarial parasite than nonsicklers, and that the heterozygotes, when infected, have lighter infections than do those without the *S* gene.

Thus is appears that this allele performs a useful function in malaria-ridden areas by providing the heterozygotes with some protection against the disease. The price paid by the population for this protection is the general debilitation and high mortality of the homozygotes produced by two heterozygotes. At the present time, in nonmalarial regions such as in the United States and around the Mediterranean Sea, where some Italian and Greek populations carry the allele, the presence of the *S* allele serves no useful purpose; and except for some cataclysmic event like the breakdown of civilized society, it is not likely to in the future.

THALASSEMIA. Abnormalities of hemoglobin in which the normal amount is reduced are found throughout the world. One group, the α-thalassemias, are found primarily in Asia; another group, the β-thalassemias, are found around the Mediterranean Sea, through the Middle East, India, and Southeast Asia. It has been estimated that as many as 100,000 deaths per year may be

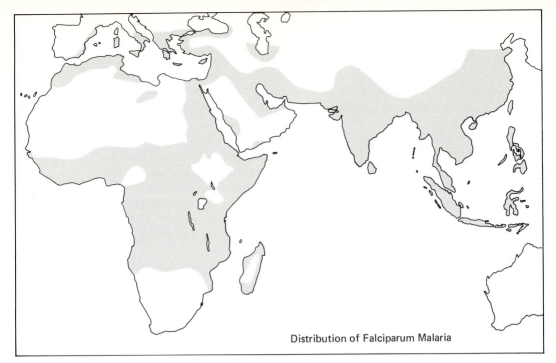

A

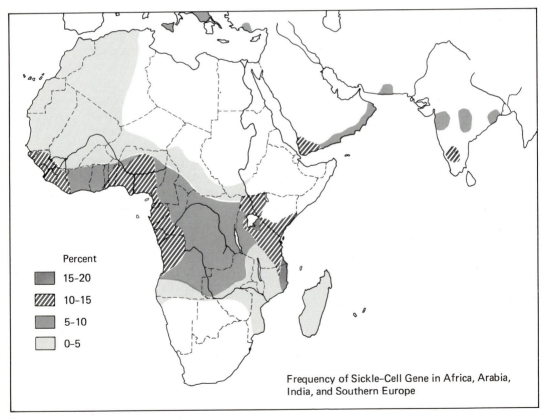

Distribution of Falciparum Malaria

Percent
- 15–20
- 10–15
- 5–10
- 0–5

Frequency of Sickle–Cell Gene in Africa, Arabia, India, and Southern Europe

B

Figure 4-22 *(A)* The distribution of malaria in Africa, southern Asia, and southern Europe. *(B)* The frequency of the sickle-cell allele in the same area. (A. C. Allison, in *Genetical Variation in Human Populations,* Pergamon Press, Inc., out of print.)

caused by homozygosity for β-thalassemia. It is thought that, like sickle-cell anemia, β-thalassemia confers resistance to malaria in the heterozygotes.

The molecular basis for hemoglobin changes is discussed in Chapter 13. Population aspects of the hemoglobin diseases are found in Chapter 20.

PHENYLKETONURIA. With a frequency of about 1 in 11,000 in Northern European populations, there appear homozygotes for a simple recessive who lack an enzyme responsible for converting the amino acid phenylalanine into another amino acid tyrosine. As a consequence, phenylalanine and its by-products accumulate in the body and retard normal mental development. The retardation becomes evident six months after birth and the majority of patients who reach adulthood are clearly subnormal in intelligence, with an occasional rare case reaching borderline intellectual development. In addition, because tyrosine is necessary for the production of the pigment melanin and its normal intracellular synthesis is interfered with, affected individuals tend to have light hair and blue eyes.

In this instance, therapy is possible. If a newborn homozygote is placed on a diet with a low phenylalanine content, sufficient to provide the amino acid in the quantity necessary for normal development but not so great that it and its by-products reach high levels in the blood, along with additional tyrosine to make up for the amount missing because of the metabolic block, development may be more nearly normal. Such treated individuals may have normal intelligence, the degree of improvement depending on the speed with which the diet, and monitoring of the blood phenylalanine levels, is started after birth. Because the disease can be readily diagnosed shortly after birth by relatively inexpensive urine or blood tests, with only a small fraction of misdiagnoses, many states have mandatory test programs to determine the possible presence of phenylketonuria (PKU) in each newborn.

SIMPLE GENETIC CONDITIONS OF ANECDOTAL INTEREST. From time to time attention is called to the possibility that people of historical interest have suffered from some common genetic disease. The appearance of hemophilia, the disease responsible for excessive bleeding, in the royal families of Europe is well known and is discussed in detail in Chapter 7.

In other cases, the identification of the disease is highly speculative and more of conversational than medical interest. Nevertheless, the following have appeared in reputable medical journals. Abraham Lincoln, with his tall gaunt bearing, is suspected of having had the dominant Marfan syndrome, which is responsible for longer than normal bones, particularly of the digits. Nicolo Paganini, who was probably the greatest violin virtuoso of all time, was a tall, thin, even cadaverous man, with extraordinarily long fingers and extensible joints; it has been suggested that all these characteristics describe Marfan syndrome. The Greek Ypsilanti family, for whom a town in Michigan is named, was known to have the dominant debilitating muscular disease myotonic dystrophy, causing premature death; Toulouse-Lautrec, the French artist, may have inherited from both his first-cousin parents a recessive gene for pyknodysostosis, responsible for his short stature and weak bones.

Additional Reading

ALLISON, A.C. 1956. Sickle cells and evolution. *Scientific American,* **195**:87–94.

Benzyl esters as a desickling drug. 1980. *Science News,* **117**:102–103.

HARRIS, ANN 1981. On the track of cystic fibrosis. *New Scientist,* **89**:167–169.

MAUGH, T. H., II. 1981. A new understanding of sickle cell emerges. *Science,* **211**:265–267; Sickle cell (II): Many agents near trials. *Science,* **211**:468–470.

McKUSICK, V. A. 1978. *Mendelian Inheritance in Man: Catalogs of Autosomal Dominant, Autosomal Recessive, and X-linked Phenotypes,* 5th ed. Baltimore: Johns Hopkins University.

STANBURY, J. B., J. B. WYNGAARDEN, and D. S. FREDRICKSON, eds. 1978. *The Metabolic Basis of Inherited Disease,* 4th ed. New York: McGraw-Hill.

The row over sickle-cell. 1973. *Newsweek,* February 12, 63–65.

WIESENFELD, S. L. 1967. Sickle-cell trait in human biological and cultural evolution. *Science,* **157**:1134–1139.

Questions

1. Draw a pedigree, as extensive a one as you can recall, for your own family. Add a few complications to make it more interesting. Fill in some of the symbols so that the pattern is that of typical dominant inheritance. Repeat the same exercise, assuming segregation for some simple recessive characteristic.

2. In the pedigrees drawn in Question 1, indicate the genotypes and phenotypes of each person. Which are heterozygous and which homozygous? (Specify the persons by number as in Figure 4-2 in the text.)

3. Does your pedigree include any fraternal or identical twins? A propositus? Any consanguinity? Individuals of unspecified sex? If not, add the above features.

4. If an allele is dominant, does this mean that the homozygote (*DD*) has the same phenotype as the heterozygote *Dd*?

5. The chance that a father who is heterozygous *Dd* will produce a progeny with his *D* allele is 1/2, and there is a similar chance that the child will get the *d* allele. If this first child gets the *D* allele, does this mean that the second child must get the *d* allele, so that the 50 percent chance of each is satisfied? Present your answer with reference to chromosome behavior at meiosis.

6. If two parents are heterozygous *Dd*, the chance that their first two children will be *D*– is 9/16 (= 3/4 × 3/4). The first two might also be both homozygous *dd*, with a probability of 1/16 (= 1/4 × 1/4). The chance that there will be one *D*– child and one *dd* child in the pair is 3/8. Can you arrive at this last figure using simple probability calculations?

7. What do the polytene chromosomes of the fruit fly *Drosophia melanogaster* tell us about the structure of the chromosome?

8. How many alleles are found at a single locus on any one chromosome?
9. What is the scientific name of the garden pea? Why was Gregor Mendel's work with it so successful? How was Mendel able to make so many observations in such a short period of time on such a small plot of land?
10. Describe Mendel's two laws, using actual characteristics that he used in his crosses.
11. Describe the manifestation and manner of inheritance of the following diseases: (a) cystic fibrosis; (b) Tay-Sachs disease; (c) sickle-cell anemia; (d) thalassemia; (e) phenlyketonuria.
12. Mendel showed that the ratio of phenotypes expected from two pea plants both heterozygous for a recessive allele is $3:1$. Can you think of any reasons that this ratio might be more difficult to observe in humans? (Consider the problem of family size, for instance.)
13. The $9:3:3:1$ phenotypic ratio that Mendel found in *Pisum sativum* for two pairs of alleles (Y,y and G,g, for instance) is important in establishing the principle of independent assortment. Do you think that this ratio might also be important in humans?
14. What is the most common inherited disease in the United States? What genetic diseases are found predominantly in certain geographical regions or ethnic groups? Are there any good medical treatments for any of these diseases?
15. Can you argue that, from one point of view, the allele for hemoglobin S is recessive but, from another, dominant?
16. Why does the allele for hemoglobin S have a high frequency in certain parts of the world? What would you expect its frequency to be in an Eskimo population?
17. Many persons who are opposed to abortion in general would allow it in cases of incest. Is there any biological justification?
18. A laboratory testing for Tay-Sachs heterozygotes failed to determine that either member of a married couple was such a heterozygote, although, in fact, both later proved to be. They gave birth to an affected child. A lawsuit was brought against the laboratory on behalf of the child for emotional distress and loss of seventy-three years of life. Would you agree that this is a legitimate claim? Would you agree that any human life, however brief, is better than no life at all? Or would you reach another conclusion?
19. Two pediatricians in Detroit failed to diagnose PKU in an infant at birth. A jury awarded the parents $80,000 for the mental retardation that might have been prevented if the correct diagnosis had been made. If you were a lawyer, how would you try to defend the pediatricians?
20. In 1980 a student who was a sickle-cell heterozygote was expelled from the U.S. Air Force Academy on the grounds that his health might be endangered by the rigorous training at the high altitude of the school (7,000 feet). He sued and was reinstated. What is the difference between a person with the "sickle-cell trait," the heterozygote, and one with "sickle-cell anemia," the homozygote?

5

Gene Action

In 1902 the English physician Archibald Garrod published a paper describing the biochemistry of a disease known as *alkaptonuria*. Those affected with this disease have the curious phenotype of producing urine that turns black upon exposure to air. In this paper and in a later book, *The Inborn Errors of Metabolism*, Garrod attributed this defect to a breakdown in metabolism, noting that an increase in the intake of tyrosine led to an increase excretion of a substance called *alkapton* (hence the name), now known as *homogentisic acid*. He also cited the increased incidence within certain pedigrees, and particularly in cousin marriages, as evidence that the disease might be caused by a rare recessive gene. He did not, however, reach the ultimate conclusion, which we now know to be true, that specific genetic factors manufacture specific enzymes. His point of view was that of a physician, and it was not until a half century later that the contributions of many other workers in the field led to our present understanding of the way that genes control the synthesis of specific products.

INTERACTIONS WITHIN THE CELL. There are probably about 10,000 different proteins within a single cell, as well as many more compounds of simpler structure. Because the structure and function of a cell are transmitted from one generation to the next by way of the distribution of identical copies of DNA to daughter cells, it follows that most, if not all, of the complex reactions going on in the cell must be controlled by the DNA within it. We do

92

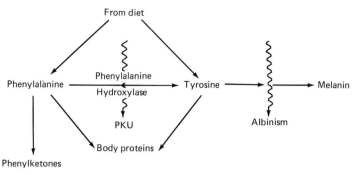

Figure 5-1 An abbreviated scheme of the biochemical pathways showing the utilization of phenylalanine. It is converted into tyrosine by the enzyme phenylalanine hydroxylase, which is missing, or low in activity, in persons with phenylketonuria (PKU). Tyrosine then goes through a number of steps to produce melanin. A single block in any one of several different steps will result in albinism.

not yet understand all the ways by which this is managed, but we do have a reasonably good insight into some of the major features of DNA activity.

If we look at the series of reactions (Figure 5-1) that involve the metabolism of the amino acid phenylalanine, we can see how it is transformed in one step into tyrosine, which, in its turn, is transformed in two steps to homogentisic acid. A block in the first reaction, found in persons with PKU, allows phenylalanine to build up in the body; the resulting production of an excess of phenylpyruvic acid, along with its by-products, is responsible for the ill effects on neurone development in people with PKU. The failure of homogentisic acid to be degraded leads to a similar buildup in alkaptonuriacs, with its immediate phenotypic consequence, as well as the appearance of arthritis in later years. Our question is how the genes that are found in the DNA can be responsible for these problems and, more generally, how the genes control all of the biochemical syntheses in the cell.

At the outset it should be noted that the DNA does not directly make (or fail to make) the chemical compounds found in the cell. The responsibility for the swift conversion of one compound into another by altering a chemical bond depends on enzymes. Enzymes are indicated at the arrows between the two compounds involved in the reaction (Figure 5-1) and can be quickly spotted because they usually end with the suffix *ase*. If the enzyme is not present, or is insufficient or inadequate, the task of converting one compound into the next, necessary for normal cell function, is impaired and the cell (and therefore the organism) may be affected. The DNA is ultimately responsible for the synthesis of such proteins as enzymes. We will see how the structure of DNA is able to direct the manufacture of so many different types. The simple model of DNA action that we consider first comes from studies of bacteria and viruses. Certain differences in gene structure and action between the primitive organisms and higher forms of life are discussed later.

Function of DNA

DNA REPLICATION. We have already discussed the general structure of the DNA helix; the role of the compounds A, T, G, and C in holding the two strands together, A with T as one *base pair* and G with C as another; and the

Figure 5-2 The steps by which a segment of DNA produces its final biological effect. The DNA makes a copy of itself with a slightly altered backbone, which is called *messenger RNA (mRNA)*. The mRNA is acted on by another type of RNA, *transfer or tRNA*, to produce a polypeptide. This polypeptide may by itself, or with other polypeptides, form proteins. The protein may have a function as an enzyme, as in Figure 5-1, or may be an end product in itself, as in the case of hemoglobin.

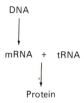

way in which the two strands, upon separating, are able to serve as a template for the synthesis of new strands forming pairs precisely like the original ones. This knowledge of the manner of replication of DNA answers one of the perplexing questions about the genetic material: how it reproduces itself precisely, practically without error, every time a cell goes through division.

STEPS IN PROTEIN SYNTHESIS. First, let us take an overall view of the way in which DNA operates (Figure 5-2). One strand of the double-stranded DNA of the chromosome, which we shall equate with the gene, manufactures complementary copies of itself, not as more DNA, but with a somewhat different structure using a ribose sugar in its backbone, instead of the deoxyribose sugar that is found in DNA itself. This is ribonucleic acid, or RNA. It serves as a messenger from the DNA for the synthesis of protein and is therefore called *messenger RNA*, or *mRNA*. The mRNA is then translated into protein by a series of smaller RNA molecules, which interpret the series of RNA subunits as specific building blocks to be used in putting together a highly specific protein. These smaller RNA molecules are called *transfer RNA*, or *tRNA*.

RNA SYNTHESIS. Because there is a correspondence between the structure of DNA and mRNA, the latter must also be composed of four subunits. They are given the symbols A, U, G, and C and have chemical structures similar to, but not identical with, the subunits of DNA.* The correspondences are given in Table 5-1. RNA is normally single-stranded and is synthesized in the nucleus enzymatically from one of the two DNA strands, according to the base-pairing rules given in the table. The production of an RNA molecule complementary to a length of DNA is known as *transcription* (Figure 5-3).

* The subunits of RNA are adenylic acid (A), uridylic acid (U), guanylic acid (G), and cytidylic acid (C), not to be confused with the DNA subunits. They are covalently linked through phosphodiester bonds between the 3'-carbon of one subunit and the 5'-carbon of the next.

DNA Subunit	Complementary RNA Subunit
A	U
C	G
G	C
T	A

Table 5-1 Base-pairing rules for DNA–RNA complementarity.

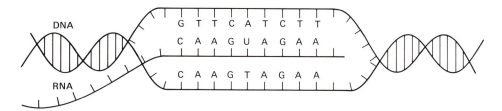

Figure 5-3 The transcription of a segment of DNA to produce the complementary mRNA. Notice that U is found in RNA as the complementary base to A, instead of the T found in DNA.

TRANSLATION OF RNA INTO PROTEIN. It is the transfer RNA (tRNA) that converts the information contained in the sequence of bases on the mRNA into proteins. This conversion performs the unusual task of determining which amino acids must be joined together in accordance with the sequence information of subunits on the mRNA. This conversion of the information of the sequence of the mRNA subunits into a specific chain of amino acid subunits is known as *translation*.

PROTEIN SYNTHESIS. There are twenty different amino acids in the long chains that make up proteins (Figure 5-4). Amino acids are characterized by having a terminal carbon atom which has both a carboxyl (COO⁻) group and an amine (NH_3^+) group attached (Figure 5-5). In addition, there is a portion (R) of variable constitution attached to the second carbon atom. Amino acids can be combined to form *polypeptides* of any length by a reaction involving the amine group of one and the carboxyl group of another, a reaction which releases a molecule of water at each juncture (Figure 5-5).

The total number of amino acids, the proportion of each type and their sequential order determine the properties of the polypeptide chain. These are determined by the order of the subunits A, U, G and C on the mRNA molecule which was synthesized on the DNA strand. However, because there are many more amino acids (twenty) than RNA subunits (four), some sort of system for specifying relationships of amino acids to the subunits, or *code*, is needed.

Let us think about how many RNA subunits would be needed in the code to specify each amino acid. A one-to-one relationship of one RNA subunit to one amino acid clearly would not be adequate, as then only four different amino acids could be specified, one for each RNA subunit. Pairs of RNA subunits would also not be enough for a unique determination of each amino acid because there are only sixteen (4 × 4) possible pairs to code for twenty different amino acids. On the other hand, three RNA subunits would allow sixty-four (4 × 4× 4) combinations, more than enough. In fact, it is the set of three or triplet of RNA subunits (called a *codon*) that codes for an amino acid. The correspondence is called the *genetic code* (Table 5-2).

The tRNA molecules are extraordinary in that they can translate a set of three subunits on the mRNA into specific amino acids. A codon comple-

Figure 5-4 The amino acids commonly found in polypeptides. Note that each has a specific constant structure (modified slightly in the case of proline) attached to a variable (R) group.

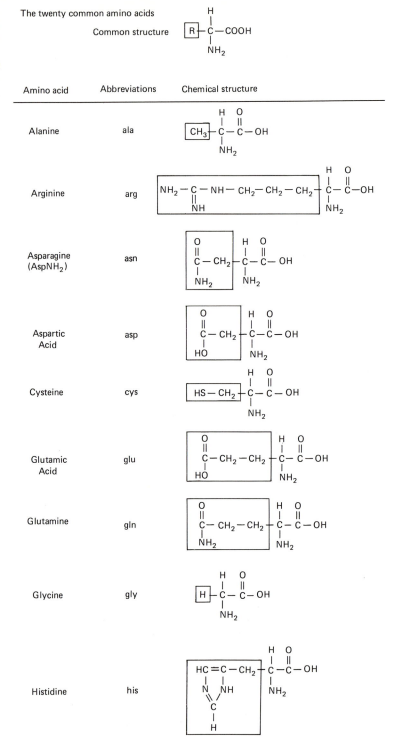

The twenty common amino acids (continued)

Figure 5-4 *(continued).*

Amino acid	Abbreviations	Chemical structure
Isoeucine	ile	$\boxed{CH_3 - CH_2 - CH}$ $\underset{CH_3}{\overset{H}{|}} \underset{NH_2}{\overset{O}{\overset{\|}{C}}} - C - OH$
Leucine	leu	$\boxed{\underset{CH_3}{\overset{CH_3}{|}} CH - CH_2}$ $\underset{NH_2}{\overset{H}{\overset{|}{C}}} \overset{O}{\overset{\|}{C}} - OH$
Lysine	lys	$\boxed{NH_2 - CH_2 - CH_2 - CH_2 - CH_2}$ $\underset{NH_2}{\overset{O}{\overset{\|}{C}}} - C - OH$
Methionine	met	$\boxed{CH_3 - S - CH_2 - CH_2}$ $\underset{NH_2}{\overset{H}{\overset{|}{C}}} \overset{O}{\overset{\|}{C}} - OH$
Phenylalanine	phe	$\boxed{\langle \bigcirc \rangle - CH_2}$ $\underset{NH_2}{\overset{H}{\overset{|}{C}}} \overset{O}{\overset{\|}{C}} - OH$
Proline	pro	$CH_2 - CH_2$; CH_2 $CH - \overset{O}{\overset{\|}{C}} - OH$; NH
Serine	ser	$\boxed{HO - CH_2}$ $\underset{NH_2}{\overset{H}{\overset{|}{C}}} \overset{O}{\overset{\|}{C}} - OH$
Threonine	thr	$\boxed{CH_3 - \underset{OH}{\overset{H}{\overset{|}{C}}}}$ $\underset{NH_2}{\overset{H}{\overset{|}{C}}} \overset{O}{\overset{\|}{C}} - OH$
Tryptophan	try	$\boxed{\text{indole} - C - CH_2}$ $\underset{NH_2}{\overset{H}{\overset{|}{C}}} \overset{O}{\overset{\|}{C}} - OH$
Tyrosine	tyr	$\boxed{HO - \langle \bigcirc \rangle - CH_2}$ $\underset{NH_2}{\overset{H}{\overset{|}{C}}} \overset{O}{\overset{\|}{C}} - OH$
Valine	val	$\boxed{\underset{CH_3}{\overset{CH_3}{|}} CH}$ $\underset{NH_2}{\overset{H}{\overset{|}{C}}} \overset{O}{\overset{\|}{C}} - OH$

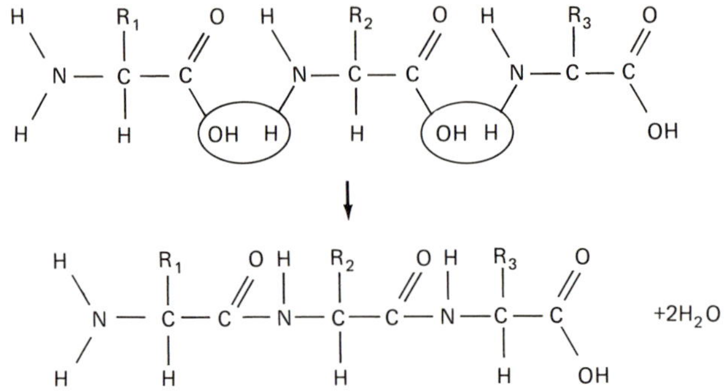

Figure 5-5 A group of three amino acids with different R groups combining to produce a polypeptide (in this case, tripeptide) chain and a molecule of water for each two amino acids joined.

mentary to the mRNA codon is part of the structure of the tRNA molecules. Thus one group of tRNAs has the subunits GUU near its center and has the amino acid *glutamine* (gln)—and none other—attached at one end (Figure 5-6). The GUU triplet bases on one of the tRNA molecules come into register momentarily with the complementary CAA of the mRNA, and glu-

Table 5-2 The genetic code. The bases are those found on the messenger RNA—the DNA bases would be the complements of these. The amino acid specified by each group of three bases, or codons, is given in small letters to the right.

First Base	Second Base U		C		A		G		Third Base
	UUU	phe	UCU	ser	UAU	tyr	UGU	cys	U
	UUC	phe	UCC	ser	UAC	tyr	UGC	cys	C
U									
	UUA	leu	UCA	ser	UAA	stop	UGA	stop	A
	UUG	leu	UCG	ser	UAG	stop	UGG	try	G
	CUU	leu	CCU	pro	CAU	his	CGU	arg	U
	CUC	leu	CCC	pro	CAC	his	CGC	arg	C
C									
	CUA	leu	CCA	pro	CAA	gln	CGA	arg	A
	CUG	leu	CCG	pro	CAG	gln	CGG	arg	G
	AUU	ile	ACU	thr	AAU	asn	AGU	ser	U
	AUC	ile	ACC	thr	AAC	asn	AGC	ser	C
A									
	AUA	ile	ACA	thr	AAA	lys	AGA	arg	A
	AUG	met	ACG	thr	AAG	lys	AGG	arg	G
	GUU	val	GCU	ala	GAU	asp	GGU	gly	U
	GUC	val	GCC	ala	GAC	asp	GGC	gly	C
G									
	GUA	val	GCA	ala	GAA	glu	GGA	gly	A
	GUG	val	GCG	ala	GAG	glu	GGG	gly	G

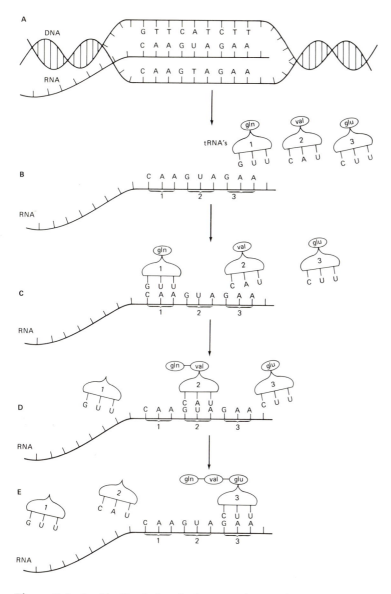

Figure 5-6 An idealized sketch showing the synthesis of a protein with a specific amino acid sequence determined by the coding on the DNA molecule. *A.* One DNA strand, at a region of separation from the other, determines the synthesis of a complementary messenger RNA (mRNA) strand (with the T of DNA replaced by U in RNA). *B.* The mRNA leaves the nucleus. Transfer RNAs (tRNAs) (three possible types are shown) have a set of three bases at one end, whereas the corresponding amino acid is at the other. *C.* The first three bases of the mRNA, CAA, code for *glutamine.* The specific tRNA, which has the complementary codon, or anti-codon GUU, pairs with it and thereby brings in a molecule of glutamine for the protein. The next codon on the mRNA, GUA, codes for the amino acid *valine.* The tRNA with the anti-codon CAU carries the amino acid valine into position to join with glutamine. The process is repeated for the third codon GAA which specifies the amino acid glutamic acid. The tRNA with the anti-codon CUU brings a molecule of glutamic acid into position to be added to the growing polypeptide chain. Not shown is the ribosome, a much larger body in which these events take place.

tamine is released from that tRNA to become part of a chain of amino acids. Another group of tRNAs have the triplet CAU at the center and the amino acid *valine* (val) at the free end. When CAU on one of these molecules comes into contact with GUA, the next triplet on the mRNA, the valine is released to add onto glutamine in the growing chain.

The chain of amino acids, the *polypeptide,* is constructed with the aid of a structure called a *ribosome.* A ribosome is composed of many proteins and several RNAs called *ribosomal RNA (rRNA).* An mRNA molecule becomes attached to a ribosome and moves across it, its codons being "read" in a manner analogous to a tape being read by a tape head. In this way, successive triplets are recognized by the recognition end of the appropriate tRNA, and the amino acid at the other end is attached to the growing amino acid chain (Figure 5-6). This process continues until the synthesis of the chain of amino acids is terminated. Thus the sequence of amino acids in a protein is uniquely determined by the sequence of the subunits in DNA.

PROTEIN STRUCTURE.   The specificity of proteins is a function of their three-dimensional structure, and this can be broken down into several orders of complexity. In the first place, the primary structure of the protein depends on the specific sequence of amino acids that make it up. On top of this is the secondary structure that comes about from a helical coiling (not to be confused with the helical structure of DNA) or bending of the polypeptide chain. This chain, then, is folded into a three-dimensional complex, the tertiary structure that exposes certain chemically active sites and makes the protein reactive, as, for instance, when a protein catalyzes a chemical reaction and is therefore an enzyme. Then, several different polypeptide chains may come together to form a more complex protein. This association is called the *quaternary structure.* This three-dimensional structure of the protein is further determined by electrical charges, hydrogen bonds, and bonds between the sulfur atoms of different amino acids. Finally, in forming its complex, the protein molecule may pick up an external compound or atom that may then be responsible for the unique property of that particular protein. Thus the protein molecule that makes up hemoglobin picks up *heme,* which contains iron. Heme becomes an integral part of that molecule and is responsible for its oxygen-carrying capacity.

DIFFERENCES IN DNA IN VIRUSES AND HUMANS.   Before considering some of the basic differences between the DNA of the higher organisms and primitive ones like viruses, a fundamental similarity should be emphasized. The DNA code that is used by the human cell to convert the DNA message into specific proteins is much the same as that used by the simplest forms of life. This common denominator that characterizes all living things on earth serves as a unifying principle in our concept of the nature of life itself. If evidences of life are ever found on some extraterrestrial body, regardless of the shape or form of that "living" creature, whether a microscopic blob or a bizarre monster with two heads and green cheeks, the first questions to be asked will be: does it contain DNA and, if so, does it translate it by means of the same genetic code?

By making a comparison of the amount of DNA in human cells and in viruses, where we know the total number of base pairs, we can calculate that the human haploid set has about 3×10^9 base pairs. On the other hand, our knowledge of the genetics of humans leads to the estimate of between 10,000 and 100,000 genes. We also know from measurements of the size of polypeptides that they may average 1,000 amino acids at most, or, because each amino acid is determined by a codon of three base pairs, about 3,000 (3×10^3) base pairs. Multiplying the maximum average size by the maximum number gives a value of 3×10^8 base pairs. It appears, however, that there may be some 10 times $(3 \times 10^9)/(3 \times 10^8)$ more DNA in the cell than would be necessary to take care of all the genetic activity of the cell. Because all of our figures were deliberately overestimated, the actual excess of DNA is probably considerably greater than tenfold.

Where is this excess to be found and what function does it perform? We can answer the first question in part; the second question has an answer that we can, at present, only guess at. Part of the excess DNA is found as DNA with a constantly repeating series of nucleotides; this has been named *repetitive,* or *satellite, DNA*. This type of DNA is found as an essential constituent of the heterochromatin found adjacent to the centromere (described in Chapter 2). These "constitutive heterochromatic" regions are responsible for the C bands, obvious when stained by special procedures, particularly in chromosomes 1, 9, and 16 and the Y chromosome. This type of DNA is not involved in the synthesis of specific proteins.

Excess DNA can be found in a completely unexpected place on the chromosome—within the gene itself. The RNA inside the nucleus is, for the most part, the direct transcript of the DNA nucleotide sequence. When it is measured, however, it appears to be far too long in relation to the size of the mRNA later found released into the cytoplasm of the cell. These large pieces make up the *heterogeneous nuclear RNA,* or *hnRNA*. The hnRNA, the immediate transcript of DNA, has an excess of RNA, and this is trimmed away enzymatically to produce the smaller mRNA molecules. Thus our simple diagram of mRNA synthesis must be modified for the higher organisms.

It would not be surprising to find that the excess DNA associated with the hnRNA was simply an extra piece attached to one end or the other. The surprise is that much of this excess appears to consist of a set of extraneous nucleotides that interrupt the sequence of essential genic nucleotides at intervals within the gene itself (Figure 5-7). These *intervening sequences,* found in the DNA and its immediate transcript (the hnRNA) are chopped out enzymatically, and the remaining pieces are spliced together so that the resulting mRNA has the nucleotides of the gene in uninterrupted sequence. It is this modified RNA, the messenger RNA, that is then translated outside the nucleus of the cell (Figure 5-8). These segments of the hnRNA that are chopped out, sometimes as much as 80 percent of the total nucleotides, are called the *introns* (from "*in*tervening"), and the fragments that are ultimately *ex*pressed as genes are called *exons*.

This peculiar structure of some genes of higher organisms, with several segments (one chicken egg protein has seventeen) of seeming nonsense material interspersed within the gene has raised a lot of new questions. How are these intervening sequences recognized and removed? What function do

Figure 5-7 Representation of a gene within the nucleus. The arrows mark the beginning and the end of the transcribed region. The immediate transcript RNA, the hnRNA, still contains the intervening sequences.

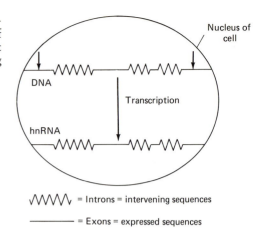

∿∿∿ = Introns = intervening sequences

——— = Exons = expressed sequences

they serve? Are they an indication of some important fundamental difference between the gene of a bacterium and that of a higher animal? Are they of some special significance in controlling gene action? Or do they have some important role in evolution? Or are both of the latter possibilities true?

TIMING OF GENE ACTION. Those genes that code for specific proteins, like enzymes, hormones, hair, and so on, are known as *structural genes.* It was first established in bacteria that there exists another category of genes: those responsible for turning the other genes on and off at the proper times. These are located adjacent to the structural genes and are known as *regulatory genes.* It seems self-evident that regulatory genes must also exist in higher organisms; in fact, it must be far more important to a multicellular organism like the human that certain genes function at one time in life and not at another, and that they function only in certain cells. For example, hemoglobin comes in several natural forms: the genes for one type are active during early embryonic life; another gene takes over during fetal life; and still another becomes predominant in postnatal life. We attribute this turn-

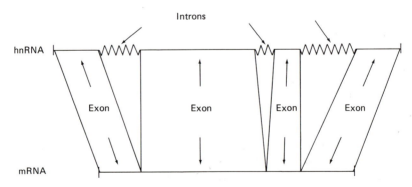

Figure 5-8 The elimination of the intervening sequences from the hnRNA to produce mRNA. This process is enzymatically mediated and takes place outside the nucleus.

ing on and off to regulatory genes. Thus, in our hemoglobin example, numerous cases have been recorded in which fetal hemoglobin has been detected in unusual quantities in adults—detected because it is not as efficient in transporting oxygen in an extrauterine environment. This is an inherited condition, and we can postulate that the defect is in the regulatory gene, which should have turned off the loci for fetal hemoglobin shortly after birth.

Normal Embryonic Development

During the nine-month period between the time of fertilization of the egg by a sperm and the birth of the infant, a series of most remarkable developmental changes takes place. The total number of cells increases phenomenally from one to countless billions, which then become transformed into hundreds of different types: one to become a heart muscle cell, another a nerve cell, and so on. In addition, these cells take up special positions, in some cases by migrating from one spot in the developing embryo to another, and in other cases by undergoing extra mitoses at those sites. In its final form, the newborn child is a creation of extraordinary complexity with a set of highly integrated systems that normally enable it to survive and cope with an almost infinite variety of situations for its long life span.

To accomplish this complex development, two basic forces are at work. The first is mitosis, which allows for the precise reproduction of all the essential contents of a single cell into two daughter cells. Also, as development proceeds, different cell lines take on new and characteristic structures and functions; this is called *differentiation.* Although differentiation has been studied in great detail from the fertilized egg to the adult in a wide variety of animals, our knowledge of the process is, by comparison, quite primitive. We are only now beginning to understand why one of the cell lines descending from the original fertilized egg becomes white blood cells and another cell line from that same egg becomes color-detecting cells in the retina of the eye. This area of research is one of the most active and promising in all of biology.

As the fertilized egg, the *zygote,* multiplies by mitosis, the cells differentiate to form an *embryo.* This, after two months of development, has acquired the basic structure of the human and is then referred to as a *fetus* until the time of birth. During the entire period of embryonic and fetal development, different proteins are being produced in the right places, following the directions specified in the genetic code of the DNA of the structural genes. At the same time other genes, the regulatory genes, provide the instructions by which the structural genes act at one time and not at another and by which they act in one cell line and not in another. Thus the locus for the fetal chain of hemoglobin is active primarily during the fetal period, and even then only in the cells that give rise to the erythrocytes and not in the hundreds of other cell types present in the fetus.

At the same time, all the active genes are interdependent on each other's products; they must act in harmony. If one fulfills its task inadequately, it may trigger a set of developmental errors that can affect the newborn in many different ways. The appearance of several apparently unrelated characteristics because of a single genetic change is called *pleiotropy*. Virtually all gene activity is pleiotropic. One case that is readily understood is that of PKU, where the homozygote shows defective mental development from the buildup of phenylalanine and its derivatives. At the same time, the affected person tends to have light hair and eyes because of the reduced tyrosine used in the production of melanin. Most pleiotropic effects are not that simple to understand. The reduction of hair and skin pigment in albinism is obvious; in addition, albinos in mammals tend to have abnormal development of the nerves leading to the retina of the eye. This latter peculiarity is thought to be responsible for cross-eyed Siamese cats (which are homozygous for an albino allele).

It is customary to refer to genes, alleles, or loci with respect to their obvious effect on the phenotype, as "the allele for albinism." This kind of labeling carries with it the unfortunate implication that there must therefore exist an alternative allele for normal melanin development that is entirely responsible for the production of melanin. Yet we know that melanin is produced only after a series of biochemical reactions, a few of which are shown in Figure 5-9.

DIFFERENT LOCI PRODUCING SIMILAR EFFECTS. Because of the complexity of the interacting biochemical reactions within the cell, it should come as no surprise that alleles at several different loci may produce somewhat similar phenotypic effects. Albinism is one such case. It is usually considered the result of homozygosity for a recessive at a single locus, but, in fact, the phenotype can be produced by changes in at least two different loci. We know this to be the case because occasionally two albinos marry and produce normally pigmented children, expected only if the cross is of the composition $aaBB \times AAbb \rightarrow AaBb$. A series of biochemical steps is necessary for the conversion of the simple molecules to the more complex pigment melanin. Some of these steps are controlled by different genetic loci, and mutations at these loci, then, may interfere with the production of the final product.

Although the end result is albinism in both cases, the two types can be differentiated biochemically by enzyme tests in the hair bulbs. In these cells, normal people have an enzyme, tryosinase, which speeds the changes of tyrosine into the pigment precursor, DOPA. The hair bulbs of some albinos have this enzyme; those of others do not (Figure 5-10). This clearly indicates that albinism may result from the interruption of melanin formation at a minimum of two different stages, and this implies that at least two different loci are involved. Albinos may be of either of these two types, which occur with about the same frequency (Figure 5-11).

Figure 5-9 The production of melanin from tyrosine by a set of several reactions. A reduction in the amount of tyrosine initially present also reduces melanin formation. For this reason, persons with PKU tend to have reduced pigment.

Tyrosine ⟶ Dopa ⟶ Dopa-Quinone ⟶ Melanin

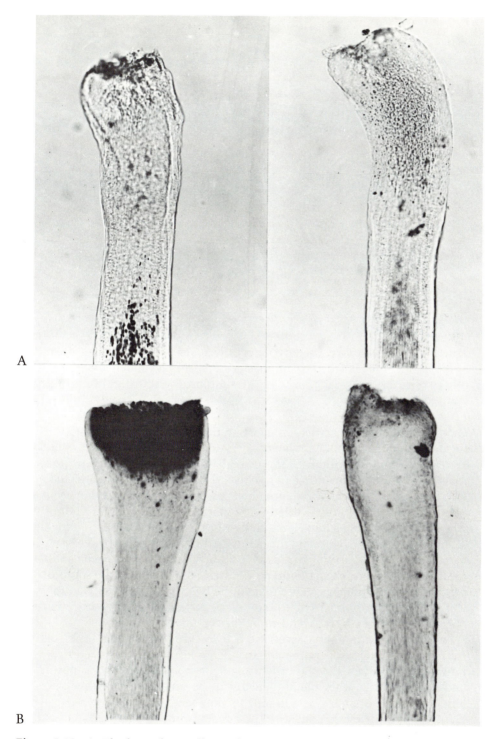

Figure 5-10 *A.* The hair of two albinos, both lacking pigment. *B.* After addition of the pigment precursors, one of the hair bulbs manufactures pigment readily (left) and the other does not. (Courtesy of C. J. Witkop, from Albinism, from *Advances in Human Genetics,* Eds. Harris and Hirschorn, Plenum Press, New York, London, 1971.)

Figure 5-11 The two kinds of albinos (unrelated) produced by recessives at the two different loci differentiated by the test shown in Figure 5-10. (Courtesy of C. J. Witkop.)

In addition, there are more than half a dozen other genetic defects, with other pronounced morphological defects, that also show a decreased amount of melanin. This fact further emphasizes the interdependence of the multiplicity of reactions taking place at the cellular level during the early developmental stages.

GENE EXPRESSION. Because of the large number of variables interacting during development, the phenotype resulting from a specific genetic condition varies from one individual to the next. It may even vary within the same individual. In the simple case of polydactyly, for instance, the affected individual has the dominant allele for an excess number of digits; however, the two hands may look quite different. Sometimes one appears quite normal, whereas the other has additional digits. In these cases, X-ray photography often reveals extra bones in the hand that appears normal. This type of variation is called *variable expressivity*. The variability may even extend to the extreme where the phenotype does not appear at all in some of the persons with the proper genetic makeup. When a genotype always expresses itself phenotypically, we say it is *completely penetrant;* when it sometimes fails to express itself, it is said to be *incompletely penetrant.*

Even such a simple characteristic as eye color may cause problems because of incomplete penetrance. In a small frequency of genotypically brown-eyed individuals, the brown pigment develops poorly or not at all, and the person appears phenotypically to be blue-eyed. However, with a blue-eyed spouse, such a person may produce one or more brown-eyed children. This result could lead to some concern on the part of the parents if they have been told categorically that the allele for blue eyes was a simple recessive and that, therefore, two blue-eyed parents can never have brown-eyed progeny (unless, of course, by the process of mutation, an allele changes from blue-eye-determining to brown!).

DEVELOPMENTAL MIMICS. At times, offspring appear with a developmental oddity. The first question that arises is whether that change is, in fact, a genetic one or whether it is simply a developmental change, perhaps induced by disease, drugs, or some unknown condition in the mother during pregnancy. Phocomelia, for instance, the condition in which the arms or legs are so short that in extreme cases the hands or feet appear to be attached directly to the trunk, may be induced prenatally by thalidomide but is also found as part of the phenotype of a genetic defect. Making a distinction between genetically and environmentally caused or chance changes can be difficult, sometimes impossible. Such changes of a nongenetic nature that mimic known genetic types are known as *phenocopies* (Figure 5-12).

GENE ACTION IN DIPLOIDS. In our description of DNA transcription and RNA translation, and the resulting gene product, it was not necessary to take into consideration the obvious fact that in diploids there are two similar loci present, one on each of the two homologous chromosomes. The relative activity of two alleles in heterozygotes led Mendel to call them dominant versus recessive. Whether a specific allele is dominant or recessive may be a matter of the gravest importance to the individual who carries it, to say nothing of his parents, sibs, and offspring. This is clearly true for an allele with devastating developmental defects, in which the fortuitous presence of that single allele may mean the difference between survival to a ripe old age, if recessive and death or incapacitation at an early age, if dominant.

Let us consider a few of the simplest ways in which an allele may appear to be either dominant or recessive. There are many genetic defects caused by a biochemical deficiency in which an enzyme is either absent or relatively inefficient, so that it does not produce enough (or any) of its usual product or produces a slightly defective product. In the case of albinism and many doz-

Figure 5-12 A phenocopy in an Ecuadorian Indian. The child, who at first sight appears to be an albino, is in fact the victim of kwashiorkor, a disease of protein deficiency. (Courtesy of C. J. Witkop; original photograph from the International Committee for Nutrition for National Development, National Institutes of Health, Bethesda, Maryland.)

ens of well-known human defects, when there is a normal allele present on one of the two homologues and a defective allele on the other, the normal one produces sufficient enzyme for the individual to be essentially normal. In these cases, the defective allele appears to be recessive. It can be stated as a general rule that mutant genes that are responsible for enzymatic deficiencies will behave as recessives in their inheritance. This does not mean that the heterozygote *Aa* is absolutely identical to the homozygote *AA;* in fact, for many traits classified as simple recessives, the heterozygote *Aa* may be distinguished from the homozygote *AA* after close study or biochemical tests. Furthermore some aspects of the phenotypes of some genetic conditions may be detectable in the heterozygote (i.e., dominant) and others not (i.e., recessive). Thus the severe anemia of sickle cell disease is found in homozygotes and therefore is a recessive effect. However the sickling of the red cells under conditions of oxygen deprivation is detectable in the heterozygote and might therefore be referred to as a dominant effect.

On the other hand, consider the consequences if a change takes place in a regulator gene that causes it to send an incorrectly timed message to another gene at another locus, so that the latter functions at a time when it should not or does not function when it should. A mutation of such a gene may very well behave as dominant because only one member of the two alleles present in the cell need give the incorrect signal. This may explain why many of the defects of embryological development (polydactyly, syndactyly, and so on) are dominant in inheritance.

CLASSIFICATION OF ALLELES AS DOMINANT OR RECESSIVE.   The dominant or recessive labels we attach to alleles are a matter of convenience, generally dictated by the appearance of some clearly obvious or important aspect of the phenotype in the heterozygote, and not by the detection of any difference between the homozygote normal *AA* and the heterozygote *Aa.* Thus, although we may be able to detect heterozygotes for the sickle-cell anemia by simple biochemical tests, the allele is properly referred to as recessive because it is the homozygote who is at serious risk of early death. Perhaps the application of these descriptive words to any genetic condition should be made with more consideration of the impact on affected persons and their relatives than of the medical or biochemical details of the action of the allele. Certainly, in one case (that of sickle-cell anemia) where this rule was not followed and medical terminology was applied instead, the result has been endless confusion, needless apprehension, and in some cases outright injustice (Chapter 9).

Generally speaking, those anomalies recognized as recessives are more severe than those classified as dominant. A severe recessive may exist hidden in heterozygotes for many generations until a couple who happen to carry similar defective alleles become parents and produce seriously affected children. A dominant allele with similar severity would eliminate itself quickly because of the low viability of the individuals carrying it. If it were not so severe in its effect, it might take more time, but eventually it would disappear from the population. Therefore defective alleles transmitted from one generation to the next in the human population are not randomly selected from all possibilities but are restricted to recessives and to the more viable dominants.

Additional Reading

Brown, D. D. 1981. Gene expression in eukaryotes. *Science,* **211:**667–674.

Crick, F. H. C. 1954. The structure of the hereditary material. *Scientific American,* **191:**54–61.

Crick, F. H. C. 1962. The genetic code. *Scientific American,* **207:**66–74.

Crick, F. H. C. 1966. The genetic code III. *Scientific American,* **215:**55–62.

Crick, F. H. C. 1979. Split genes and RNA splicing. *Science,* **204:**264–271.

Geography of the gene. 1979. *New Scientist,* **23:**591.

Lagerkvist, U. 1980. Codon misreading: A restriction operative in the evolution of the genetic code. *American Scientist,* **68:**192–198.

Lewin, R. 1979. Why split genes? *New Scientist,* **82:**452–453.

Lewin, R. 1981. How conversational are genes? *Science,* **212:**313–315.

Penderson, T. 1981. Messenger RNA biosynthesis and nuclear structure. *American Scientist,* **69:**76–84.

Rogers, J. 1978. Genes in pieces. *New Scientist,* **77:**18–20.

Questions

1. What is an amino acid? How many common types are there? How do several amino acids join to form a polypeptide? What difference is there, or may there be, between a polypeptide, a protein, and an enzyme?
2. Is the immediate product of a gene an amino acid? How does a person obtain the amino acids that are necessary for protein synthesis?
3. What is the immediate product of a gene? How does it differ chemically from the DNA that models it? What is the process called?
4. How do mRNA and tRNA differ? How do they collaborate to produce a polypeptide? What is the name of this process?
5. Describe the genetic code. What is a codon? Is this code the same for humans and bacteria? How many bases determine the amino acid? Could coding be managed with two or with four bases? Explain.
6. Proteins seem to come in an almost infinite variety of sizes and shapes. Can you explain why this is so?
7. What is meant by *heterogeneous nuclear RNA?* How does it differ from mRNA? What are intervening *sequences? Introns? Exons?*
8. What is the basic difference in function between *structural* and *regulatory* genes? About which do we have the most information in humans?
9. Would differentiation during embryonic development usually be the primary result of structural or regulatory genes?
10. What is the difference between a zygote, an embryo, and a fetus?
11. What is meant by the observation that "virtually all gene activity is pleiotropic"?
12. What is the difference between variable expressivity and incomplete penetrance? Can you illustrate with examples?
13. What is meant by the term *phenocopy?* Give an example.
14. Can you explain why enzyme defects in humans are generally recessive in inheritance, whereas developmental defects are more often dominant? Does the fact that humans (like most species) are diploid enter into the explanation?

15. Imagine that you notice that a new and interesting inherited characteristic is present in some members of your family. On investigation it appears that the most pronounced manifestation is found in apparent homozygotes, but that on close inspection heterozygotes can also be detected. You appear to have some latitude in describing this characteristic as either dominant or recessive. What kinds of nonscientific considerations might lead you to make a decision one way or the other?

6

Sex Determination

Definition of Sex

Sex may be broadly defined as the combination of the genetic material of two different individuals to form a new individual. From this scientific definition, it is clear that there is no restriction on the physical form of the sexes—they may be identical to each other (and often are in primitive plants and animals)—and there is no limitation on the number of sexes that may occur in one species, although our preoccupation with the condition in humans and other higher animals leads us to think of two morphologically different sexes, male and female. In fact, in some species of lower animals and higher plants, there may exist a large number of different "sexes." If diploidy is an almost invariable feature of life, sex is even more so, being found in all plants and animals, with the possible exception of a group of primitive organisms, the blue-green algae.

Evolutionary Advantages of Sex

Except for the rare cases of identical twins, it is likely that no two humans who have ever existed on the face of this planet since the dawn of time have been genetically identical. The allelic variation in a large fraction of the tens of thousands of loci that human chromosomes carry, along with the process of recombination that occurs during meiosis in both sexes and the chance union of two dissimilar gametes, gives every zygote its own genetic constitution. From this point of view every individual represents a natural and unique experiment that tests the potentialities of the virtually unlimited number of different gene combinations possible in the species. There is an opportunity for beneficial gene or chromosomal changes that occur independently in the germ line of different individuals to meet in an even more advantageous combination in their progeny. In addition, allelic changes with apparently neutral effects may combine with similar alleles in other lines to produce some new, unusual, and possibly advantageous phenotypes.

On the other hand, consider the consequences if some women were to acquire the capability of reproduction without benefit of fertilization of the eggs by sperm, that is, of *parthenogenesis*. They would be more efficient in reproducing their kind than ordinary females, and they should rapidly increase in numbers in the population, eventually making up the entire population. The fact that we see no indication that this is happening in any mammalian species—and, indeed, in only a few vertebrates—is strong evidence that there exists a strong force in favor of sexual reproduction.

Furthermore each offspring of a parthenogenetic mother could contain nothing but the genetic material of the mother and would in most cases be a carbon copy of her. This would happen for generation after generation, with an occasional spontaneous genetic change altering each line. Such a change, once established, would become a permanent feature of the line. Other maternal lines would likewise slowly diverge in different ways, and after a period of time each maternal line would have its own special characteristics, gradually deviating from what we now consider humanness.

The phenomenon of sexuality, then, has the great advantage of reshuffling genes to try out new allelic combinations in every individual produced. Furthermore, as a thought more philosophic than scientific, it can be argued that sexual reproduction has the effect as well of reaching into the common "gene pool" that characterizes humankind and of constantly reestablishing the common basis for the unique properties of the human species. This maintenance of a common gene pool is, of course, a consequence of sexuality not just for humans but for all living species.

Forms of Sexuality

PROTOZOANS.   In one-celled organisms it is common for the two sexes to look alike. Indeed, the knowledge that sexes exist may be obtained only by experimental testing. If two different populations of one-celled animals, protozoans, are each derived from a single ancestor by repeated mitotic divisions, an individual from culture A and an individual from culture B may fuse in a process of *conjugation* and exchange genetic material. This event does not usually occur between two individuals of the same culture. Such a mating may not produce a new individual immediately, but those progeny subsequently produced by mitosis, or *fission,* of the mated pairs now have a new genetic constitution. By testing large numbers of independently derived strains, it is possible to find out how many independent, self-sterile groups or mating types (i.e., sexes) exist in that species; it has been found that for one species of paramecium, for instance, there are at least four different mating types, and for another, six.

PLANTS.   In molds, the sexual process takes place between two groups that are morphologically indistinguishable. The only difference between them appears to reside at one locus with two different alleles; these are referred to simply as + and −. Many plants have male and female gametes produced on different individuals—even the ancient Babylonians recognized the necessity for having a male date palm growing near a female date palm to produce dates. In fact, they practiced artificial pollination to ensure a good crop. Other plants—for instance, tobacco—have both male and female sex organs on the same plant but are self-sterile; that is, the pollen does not function on a plant of the same constitution or on one of identical genetic constitution with respect to a particular locus. This is a fact of obvious importance to agriculturalists. An orchard of cherry trees consisting of all the same variety of tree would produce no fruit because cherries are self-sterile, and all trees of a specific commercial variety are sterile with each other. The knowledgeable horticulturist will plant several varieties of tree within the same area so that, in the spring, the bees will readily transport pollen back and forth from one strain to another. Because the desirable characteristics of the cherry (flavor, size, and appearance) are purely maternal in origin, the exact variety of cherry contributing the pollen is of no importance. However, to perpetuate the good qualities of a strain, planting seeds would be of no use because the seeds would always be hybrids; instead, cuttings from the desired variety are induced to form roots and develop into young trees with precisely the same genetic constitution as the original variety.

ANIMALS.   A few lower animals are *hermaphroditic,* with the sex organs of both male and female in each individual, as in the earthworm and the snail. Clearly such individuals do not need any specific genetic mechanism of sex determination, because the male and female cells have the same genetic composition. A basically hermaphroditic animal may preferentially differentiate into one sex or the other depending on external conditions, a circumstance that makes it appear as though the sex is, in fact, being environmentally determined. A classic example of this sort of determination is found in the marine worm *Bonellia,* in which fertilized eggs develop into females when there are no other females in the vicinity during development; if there are females present, the eggs develop into males, which live parasitically within the bodies of the females.

In the hymenopterans, bees and wasps, sex is generally determined by the number of chromosome sets. Queens (fertile females) and workers (sterile females) come from fertilized eggs that have the diploid number of chromosomes, thirty-two. The difference between the queens and the workers is not genetic but developmental, depending on the kind of food the immature individuals are given during early development. On the other hand, drones (males) are produced parthenogenically, that is, without benefit of fertilization of the egg by a sperm, and have the haploid number of sixteen. The meiotic divisions of the drones are modified so that the sperm have the adult number, sixteen. Further work has shown that the distinction between female and male has a more subtle basis, that the female has two different alleles (i.e., is *heterozygous*) at a certain locus that makes her female. The male, being haploid, has only one allele, which causes the developing egg to become a male. If bees are closely inbred experimentally so that a fertilized egg gets two identical alleles, that individual, although diploid, will develop into a male because of the homozygosity for the sex alleles.

Sex Determination in *Drosophila*

By far the most detailed work on sex determination has been carried out with the fruit fly *Drosophila*. It was established very early that its females have two X chromosomes and males have an X and a Y, just as humans have, and also that the Y chromosome is necessary for fertility in the male but has no other effect. When nondisjunction occurs and an XXY egg is produced, it develops into a fertile female indistinguishable from normal, and a single-X egg similarly produced, without a Y, develops into an infertile male. The first rule for sex determination in *Drosophila* seemed simple: a zygote with two X chromosomes developed into a female, one with only one X into a male. Thus it was clear that the X chromosome was sex-determining and that because the properties of a chromosome must come from the genes carried on it, there must exist sex-determining loci somewhere on the X. However, a search for specific sex-determining loci on the X chromosome, made by fractioning it into bits and pieces with X rays, proved fruitless, although it occupied the attention of a number of investigators for several years. Instead, it appeared that the action of the additional X in producing femaleness was not related to few specific sex-determining loci but was a generalized, nonspecific effect of sections of chromosomes.

TRIPLOIDY IN *Drosophila*.   It has been shown, however, that sex determination does depend on a balance between the autosomal complement and the sex chromosomes. When the number of autosomal sets equals the number of sex chromosomes, the resulting individual is a typical fertile female. Thus a *triploid Drosophila* female with three haploid sets of autosomes and three X's is virtually indistinguishable from a normal diploid female with two sets of autosomes and two X's. A much rarer class, consisting of tetraploid females, with four sets of autosomes and four X chromosomes, also proves to be relatively normal and fertile. We have already discussed the case of polyploidy in humans, where triploids and tetraploids are inviable and represent a major category in aborted fetuses with chromosome abnormalities.

Triploid females of *Drosophila* produce gametes with a wide variety of different chromosome constitutions, because there are three chromosomes of each type instead of the normal two. As a rule, each set of three solves the problem of segregation by sending two of the three to one of the two poles at meiosis and the other to the second pole. Because this phenomenon occurs independently for all groups of three homologues, a wide variety of chromosome numbers is found in the gametes. All zygotes that have unbalanced

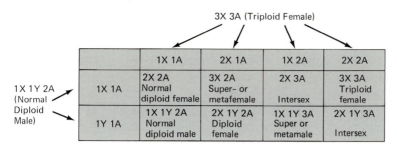

Figure 6-1 The different kinds of progeny produced when a triploid *Drosophila* female is mated to a normal male. The arrows indicate the type of gametes that will produce viable offspring; these carry either one complete set (1A) or two complete sets (2A) of autosomes, each with either one or two X chromosomes.

sets of autosomes die early in development. For a zygote to survive and develop it must have the same number of each of the large autosomes; the X's, however, can vary in number. In this way progeny can be produced with different ratios of X chromosomes to autosomal sets (Figure 6-1). When the ratio is one X to two autosomal sets, maleness results; a ratio of two X's to two autosomal sets results in femaleness; and any intermediate ratio, such as two X's to three autosomal sets, gives an intermediate phenotype called an *intersex*; which has male and female characteristics together in variable degrees and is sterile. Individuals with three X chromosomes but only two sets of autosomes are highly abnormal females of very low viability, are extremely infertile, and have gross abnormalities. Such females were originally named *superfemales* because of the high ratio of sex chromosomes to autosomes; however, they are not the kind of individual such a term would imply. Another name, *metafemale*, has been suggested as an alternative. At the other end of the scale are those *Drosophila* with only one X chromosome but three sets of autosomes. This ratio is less than that of ordinary maleness; such individuals are malelike but morphologically abnormal and completely sterile. Because of the low ratio, they are called *supermales* (or *metamales*). These ratios are summarized in Table 6-1.

The concept that the determination of sex is primarily a matter of balance between the autosomes and either two or one X chromosome, with the Y imparting fertility to the male but with no other essential function, was thought for a long time to represent the model for all animals with an X–Y sex-determining mechanism. In 1959 it was shown that most human beings

	Constitution	
	Sex chromosome/autosome	*Ratio*
Superfemale	3X:2A	1.5:1
Tetraploid female	4X:4A	1:1
Triploid female	3X:3A	1:1
Diploid female	2X:2A	1:1
Intersex	2X:3A	.66:1
Diploid male	1X:2A	.5:1
Supermale	1X:3A	.33:1

Table 6-1 The relationship of the sex chromosome/autosome ratio to sex in *Drosophila melanogaster*. Note that as the number of sex chromosomes relative to the sets of autosomes decreases, the sexuality shifts from femaleness toward maleness.

characterized by Klinefelter syndrome were in fact XXY; as we have already seen, these persons are essentially male. The contrary class, the XO condition, that of the female with Turner syndrome, was also described cytologically in 1959. These results are quite contrary to the situation in *Drosophila:* In humans the Y chromosome and not the number of X chromosomes is the primary sex determinant. This point is so important that it merits reemphasis: a human without a Y chromosome is usually a female and one with a Y chromosome, regardless of how many X chromosomes may be present, is a male.

The Puzzle of the Missing X

The X chromosome in humans is unusual in still another way. From the previous discussions of chromosome imbalance involving the autosomes, either trisomies or monosomies, it is obvious that the addition or deletion of even a small chromosome from the normal diploid complement has severe consequences in the development of the embryo and the fetus. On the other hand, this rule does not seem to apply to the sex chromosomes, for the subtraction of an X chromosome from a female (as in the case of Turner syndrome) has effects on the phenotype that are perceptible only to one experienced in such diagnoses. It seems reasonable to conclude that the X chromosome does not follow the same rules as the autosomes with respect to the effects of aneuploidy, particularly as it is a large chromosome, distinguishable from other members of the C group in a karyotype only by special staining techniques. There must be some special explanation for this unusual developmental behavior of the human X chromosome.

DOSAGE COMPENSATION.   The bulk of the synthetic machinery of the cell is derived from genes carried on the autosomes, which, always occurring in pairs in normal cells, operate normally in the diploid state. The genes on the X chromosome, however, are in a unique situation because the male cell with only one X chromosome must perform at essentially the same level as the female cell with two.

Clearly the logic of this problem can be approached from either of two directions to make male and female cells synthetically equivalent. Either the single X chromosome of the male can work twice as hard so that it performs at the same level as the two of the female, or the two Xs of the female can work half as hard to perform, jointly, at the same level as the single X of the male. This adjustment of the cell to take into account the inequalities that result from having two X chromosomes present in one sex and only one in the other is referred to as *dosage compensation.* Of the two solutions suggested, *Drosophila* follows the first course: a single X in the male works much harder and approaches the two-X condition of the female, so that the autosomes in both sexes sense the equivalent of two X chromosomes. In humans one of the two X chromosomes in the female "lies down on the job," that is, most of its loci are turned off synthetically. Male and female human cells are similar in having only one completely functional X chromosome.

The Significance of the Sex Chromatin

As early as 1891 it was observed that nuclei of cells in the resting stages have an undifferentiated mass of *chromatin,* which was referred to as *X.* Over the next twenty years it became apparent that this body was a chromosome that played a role in sex determination. The interest in the X chromosome, as it was henceforth called, then focused on its meiotic behavior—particularly in the male, where the homologue was not another X but a Y chromosome—and on its developmental effect in sex determination.

X-CHROMOSOME INACTIVATION. The sex chromatin body (as distinct from the sex chromosome) was described and illustrated in 1909 by Santiago Ramon y Cajal, a Spanish nerve anatomist, in his descriptions of nerve cells of dogs, cats, and humans. However, he was not aware of the relationship of the presence or absence of this body to the sex of the individual from which the cells were taken. In 1937 a German cytologist, L. Geitler, pointed out that in some insects the two sexes differ in their manifestation of the sex chromatin bodies. These observations, however, did not appear meaningful until 1949, when two Canadian workers, L.F. Bertram and Murray L. Barr, reported that in the nuclei of nerve cells of cats the two sexes are characteristically different with respect to the sex chromatin bodies. When a sex chromatin body appears, the cell is from a female; the male never shows any (Figure 6-2). In 1959 L.B. Russell and S. Ohno independently proposed that the sex chromatin resulted from the inactivation of a single X chromosome, and in 1961 Lyon suggested further that the X that is inactivated may be either maternal or paternal in origin, and that the inactivation occurs during early embryonic development, resulting in groups of cells with the same X chromosome active. These conclusions were based on observations in rodents; at about the same time E. Beutler, a human geneticist, working with

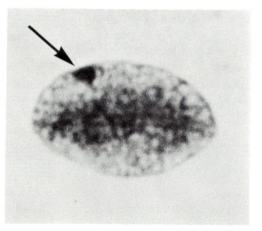 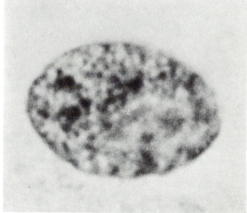

Figure 6-2 The X-chromatin body in a human female (left) and the absence of such a body in the male (right). (Courtesy of B. Kaiser-McCaw, The Genetics Center of the Southwest Biomedical Research Institute.)

the levels of enzymes produced by genes on the X chromosome, concluded that one X chromosome in human females must be inactive.

This idea of X inactivation is sometimes referred to as the *Lyon hypothesis,* and the inactivation process as *lyonization.* We shall refer to this phenomenon also as *X-chromosome inactivation* and to the inactive X visible at interphase, sometimes called the *Barr body,* as the *X chromatin.* When a cell exhibits X chromatin, it is said to be chromatin positive.

How sex chromosome inactivation occurs is not at all clear. After fertilization of the egg, the cells proceed through the cleavage divisions, and at about the eighth or tenth mitotic division, about twelve to sixteen days after fertilization, a "decision" is made by each cell that if there is more than one X chromosome present, only one will be functional. Thus, if there are two Xs within a cell, one of them, selected in a lottery not yet understood, will become the functional X and the other nonfunctional. This information (which X has been chosen for functionality) is transmitted to the daughter cells, and as the cells go through subsequent mitoses, each cell present at that early stage gives rise to a *clone* of descendant cells, all of which have the same functional X (Figure 6-3). A female is thus a mosaic, part of her cells having one X functional and part having the other homologue functional.

The Mosaic Female

If there existed allelic differences for skin pigmentation or some similar obvious phenotype on the X chromosome, X-chromosome inactivation would have been immediately obvious because the heterozygous female would have a blotchy appearance, resulting from the natural grouping of the cells from each clone derived from single cells with a specific X inactivated. In this case, appreciation of the unusual developmental role of the X chromatin would not have been delayed until about 1960. Unfortunately such clear morphological characters do not exist in humans. We can, however, see sex chromosome inactivation at work in the cat, where such factors do exist.

The type of cat that exhibits this characteristic is the well-known *calico* or *tortoiseshell.* There is a locus on the X chromosome that affects coat color. One allele, *o,* at this locus promotes the normal development of mature pigment, whereas another allele, *O,* inhibits melanin pigment production. This inhibition results in a yellow color. A calico cat is heterozygous for these two alleles and therefore has the genotype *Oo.* (Figure 6-4) The sequence of events that gives rise to a calico cat is shown in Figure 6-3. In different patches of tissue, one or the other X chromosome is inactivated. Those patches of cells in which the chromosome carrying the *O* allele is inactivated give rise to fur with melanin pigment that appears either black or brown. Those patches with the chromosome carrying the *o* allele inactivated have the *O* allele active; because the *O* allele inhibits melanin formation, the patches are yellow. (The white fur, in which color pigment is inhibited, is caused by an independent autosomal gene and is irrelevent to our present discussion.) In the tortoiseshell mosaics the black and yellow fur

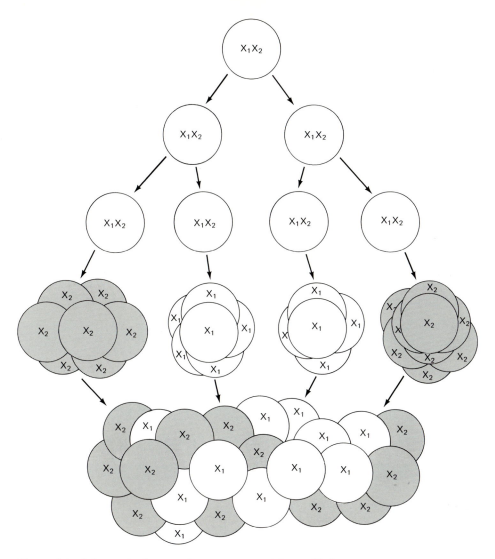

Figure 6-3 Schematic showing how a zygote with two X chromosomes gives rise to a mosaic individual, with approximately half of its cells expressing one of the two X's and half the other. After several cleavage divisions (only two are shown here for simplicity), one of the two X's, randomly selected, becomes inactive. Further mitoses produce clones of cells with the same X inactive; because these divisions occur in a mass of cells with some mixing of individual cells, a mosaic is produced with grouping of similar cell types, but also with some intermixing.

patches are more mixed, without the large patches of each, and usually with less of the extraneous white coloration.

Thus, if one were to see a calico cat from a distance, it could be reasonably predicted that the cat is a female. Male calicos occur with a very low frequency and are sterile, because the calico characteristic requires two X chromosomes and maleness requires a Y, making the male calico the equivalent of a human with Klinefelter syndrome.

A

B

Figure 6-4 The classic visual example of X-chromosome inactivation: the *calico* or *tortoiseshell* cat. In addition to the black and orange (reproduced here as gray), these two cats also have an independent autosomal gene that removes the pigment entirely from large sections, making them partially white. It is entirely separate from the mosaic phenomenon. When the black and orange colors appear in patches, the pattern is called *calico*; when the colors are more mixed, it is called *tortoiseshell*. The cat in B has, in addition, a gene for tabby, which superimposes a banded pattern. (Courtesy of P. G. N. Kramers, University of Leiden.)

It is possible to obtain evidence for the mosaicism of the human female heterozygous for a few characters determined by alleles carried on the X chromosome. A condition known as *anhidrotic dysplasia* is responsible for an absence or a low frequency of sweat glands, and this condition can be detected in areas of the skin by an unusually high resistance to electricity in affected areas. In Figure 6-5 three generations of heterozygous women have been "mapped," showing the areas where the chromosome carrying the normal allele is active and where the homologue carrying the allele for the absence of sweat glands is active.

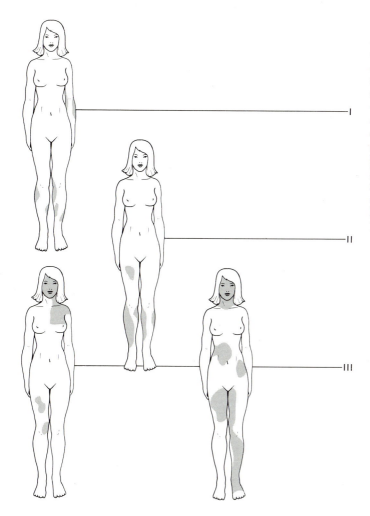

Figure 6-5 A mother, daughter, and two identical twin grand-daughters all heterozygous for an allele reducing the number of sweat glands in the skin. The stippling indicates areas where one X chromosome (that with the allele for anhidrotic dysplasia) is active and the clear areas represent skin with the other X active. (From A. H. Kline, J. B. Sudbury, Jr., and C. P. Richter; The Occurrence of Ectodermal Dysplasia and Corneal Dysplasia in One Family; *J. Pediatr.* **55**:355–66, 1959.)

In a second demonstration of this mosaicism, a woman is heterozygous for two detectably different alleles producing an enzyme, G-6-PD, that modify a sugar molecule, glucose, and appears to carry both enzymes in her cells (Figure 6-6). However, when the individual cells are isolated, and their descendants tested, all the clones carry either one of the enzyme types or the other, but not both, indicating that in some cells one X is active and in the rest the other X functions.

The Presence of Multiple X Chromosomes

In cells with more than two X chromosomes one X remains active and all others become inactivated. However, because the inactive Xs do not always appear as unambiguous X-chromatin bodies, the number of such bodies actually visible may be highly variable (Table 6-2).

Because of X inactivation, the human can tolerate more variation in X-chromosome number than in autosomal number. An XXX (triplo-X) female is

Figure 6-6 Cell cultures from a woman heterozygous for two sex-linked alleles that produce detectably different enzymes. When a mixture of the two is placed in an electrical field, one *(A)* migrates at a slightly faster rate than the other *(B)*. The original heterogeneous cell culture from the woman indicated that both enzymes were present, but when nine single cells were used to start nine clones, six of the clones produced only variant *B* and the other three variant *A*, showing that any one cell, and its progeny, will carry only one active X chromosome. (R. G. Davidson, H. M. Nitowsky, and B. Childs, *Proc. Natl. Acad. Sci. U.S.A.* **50:**481–85, 1963.)

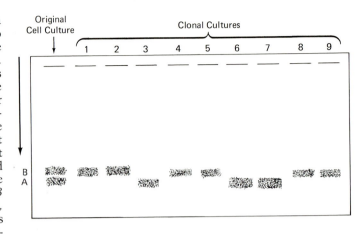

not ordinarily distinguishable from a two-X female; such persons are fertile and otherwise normal, except that their psyche may be disturbed as indicated by an IQ slightly lower than normal and a predisposition to mental illness.

In fact, the degree of deviation from the normal sex chromosome complement can be rather extreme, as is seen in Table 6-2. In this table is a list of some of the chromosome numbers that have been observed (sometimes as mosaics) in individuals with more than one X chromosome (in at least some cells), both with and without a Y chromosome. It can be seen that in some individuals the number of X chromosomes can reach a total of four and in mosaics even as many as five.

Table 6-2 The number of X-chromatin bodies in cells of persons with various numbers of X chromosomes. (Adapted from W. M. Davidson, Sexual Dimorphism in Nuclei of Polymorphonuclear Leukocytes in Various Animals. In K. L. Moore, *The Sex Chromatin*, W. B. Saunders Company, Philadelphia, 1966.)

Sex Chromosomes	Number of Cases	0	1	2	3	4
		X-Chromatin Masses (percentage)				
XX	12	55	45			
XXX	12	24	48	28		
XXXX	3	7	21	49	23	
XXXXX	1	10	20	32	30	8
XY	12	100				
XXY	12	45	55			
XXXY	5	24	50	26		
XXXXY	8	13	35	31	21	
X0	12	97	3			
XYY	1	100				
XXYY	1	55	45			
XYYY	1	100				

On the other hand, it is obvious that the additional X chromosomes cannot be completely inactivated under all circumstances; otherwise we would expect an XO individual to be exactly like an XX. Also the XXY Klinefelter should be a male normal in all respects if all Xs beyond one were completely inactivated. The hypothesis of X-chromosome inactivation must therefore be qualified to account for the observation that there is, in fact, some slight influence of the additional X chromosomes beyond the first, particularly with regard to intellectual and sexual development. Inactivation may be the rule for most somatic cells, but there are some special cell lines that do not undergo inactivation. Because single germ cells of the female can be shown to possess both of the enzymes produced by different alleles on the two X chromosomes, we can argue that X chromosomes in such cells either do not undergo inactivation or become reactivated just before meiosis, after having been inactivated. It has been suggested that inactivation may occur late enough in embryonic development so that the early action of the two X chromosomes together influences subsequent development. We do know that the X chromosome is not entirely inactivated but remains active in a small segment, and that it is the genetic imbalance of this small segment that, when either absent (in the XO female) or present in excess (in the XXX female and other multiple-X types), is responsible for some of the differences of these females from the normal XX.

From the preceding discussion, it may appear that the X-chromatin bodies are invariable features of cells with more than one X and that a definite determination of a person's X chromosome constitution could be made by looking at a single cell. Unfortunately this is far from the case. In one investigation, counts of cells of triplo-X females gave 24 percent with no sex chromosome bodies, 48 percent with one, and 28 percent with two. Similar counts of tetra-X (XXXX) cells gave 7 percent with none, 21 percent with one, 49 percent with two, and 23 percent with three. Other cell types with percentages are given in Table 6-2. Clearly, the presence or absence of X chromatin in a single cell or in a few cells would not be sufficient to make a definite determination of the X-chromosome constitution of an individual.

The Y Fluorescing Bodies

Human chromosomes stained with the fluorescent dye *quinacrine* generally show a very prominent Y chromosome at metaphase (Figure 6-7). The brightness is detectable not only in ordinary metaphase chromosomes but also in cells at interphase and even in mature sperm cells; in the latter two types of cells, the fluorescent spot is referred to as the *Y-chromatin body*, which makes it possible to make a specific analysis for sex chromosomes in addition to the X-chromatin tests. The X-chromatin tests can distinguish among X, XX, and XXX individuals, but not between XY and XO or between XX and XXY. The detection of one or more Y chromosomes by means of fluorescent dyes makes it possible to describe the sex chromosome complement of a cell completely. These techniques provide a powerful tool for ascertaining the sex of an individual without actual karyotyping in those cases where there is some ambiguity, as sometimes happens at the time of

A   B

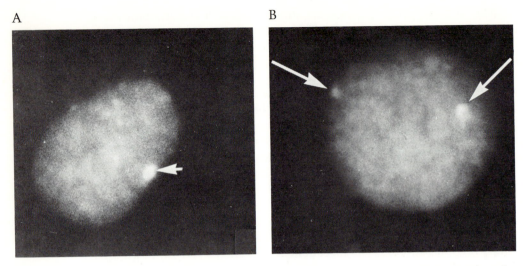

Figure 6-7 The brightly fluorescing Y chromosome. *(A)* The single spot found in the cells of a normal male. *(B)* The two Y chromosomes of an XYY male. (Courtesy of B. Kaiser-McCaw, The Genetics Center of the Southwest Biomedical Research Institute.)

birth. Cells scraped from the lining of the mouth are quite useful; the determination is fast and simple, requiring only an immediate staining procedure and microscopic examination. It should be pointed out, however, that there do exist some Y chromosomes that characteristically do not fluoresce, as well as some autosomes that do. That these variants are innately nonfluorescent is shown by their similar appearance in parents and offspring. Such irregularities lead to the possibility of a small degree of error in such determinations. When conditions warrant, a complete karyotype provides a more reliable answer.

Human Sexuality

When the word *sex* is used with reference to humans, it has implications beyond those covered by our original biological definition. These can be categorized as cellular, organic, or behavioral.

CELLULAR SEX. At the cellular level there are two aspects of sexuality; the chromosome and the genetic constitution of the individual. Basically it is the Y chromosome that is responsible for the initiation of male sexual differentiation. Under ordinary circumstances any individual without a Y chromosome will develop into a female and any with a Y can be assumed to be male. However, in addition to the chromosomal basis, the genetic constitution of the individual may play a role. There are rare individuals who clearly have two X chromosomes, without a Y chromosome, who are males in appearance; in such cases the male sex-determining loci on the Y chromosome may have been translocated to an autosome.

ORGANIC SEX. The organic manifestation of sex has three aspects: internal sex, external sex, and hormonal sex. During the course of early prenatal development the presence of the Y chromosome in the male ordinarily suppresses the development of the ovaries and female ducts and promotes the development of the testes and male ducts. In the absence of the Y chromosome, the reverse happens.

From the time of the fertilization of the egg until the seventh week of embryonic development sexual development is neutral. The primitive gonad consists of an internal mass of cells, the *medulla* and an outer layer of tissue, the *cortex*. In addition there are two pairs of ducts present, the *Mullerian* and the *Wolffian* ducts. If a Y chromosome is present, cell division is accelerated and during the seventh week the medulla differentiates into a testis. If the medulla does not undergo this differentiation, the cortex then develops into an ovary. When an embryonic testis is present, the Wolffian ducts differentiate into the sperm ducts of the male and a hormone produced by the testes causes the Mullerian ducts to disappear.

In the absence of an embryonic testis, the Wolffian ducts degenerate and the Mullerian ducts become the oviducts and the uterus of the female. The male hormones produced by the testes are responsible for the further embryonic differentiation of the male sex. In the absence of these male hormones, sexual development is female.

Thus the initial neutrality of the early embryo, the presence of primitive structures for both sexes during the second month, the suppression of the set for one sex and the promotion of those for the other depending upon the presence of the Y chromosome and the hormones produced by the embryonic testes, make it possible to find a wide variety of abnormalities of sexual development. Individuals with sets of organs characteristic of both sexes, with an ovary on one side and a testis on the other, with one gonad which simultaneously appears to be both male and female, with external genitalia that are ambiguous, and so forth, can result from simple deviations from the path of normal sexual development.

Perhaps the most commonly used criterion of sex is the structure of the external sex organs. Not only is this criterion simple, direct, and usually determinable at birth, but also the behavioral patterns the individual is expected to show (instinctively or by training) depend on this sex assignment (Figure 6-8). In most cases this characteristic is obvious at birth, but in a few it is not and the newborn may be of ambiguous sexuality. As parents of a newborn are anxious to learn of its sex immediately after birth and are understandably concerned if no definite statement can be made, a check for sex chromatin bodies can be made. The presence or absence of Y fluorescing bodies can be determined also to make the diagnosis more conclusive. Nevertheless there are many cases where such efforts fail and where infants who appear to be XX develop into males or who appear to be XY develop into females. There exist also cases where there is an apparent conflict between the sex of the gonads and that indicated by the appearance of the external genitalia.

The third criterion at the organic level is the hormonal sex. The changes induced by the sex hormones become most obvious at puberty, when both male and female sexes develop the specific characteristics that make them

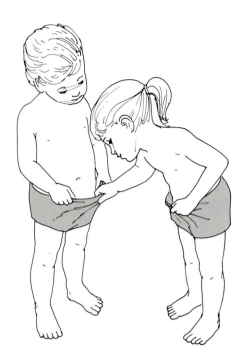

Figure 6-8 An illustration of the universal recognition of organic sex as the most important criterion for differentiating the sexes. (Courtesy of Leonie Pera and G. H. Valentine, *The Chromosome Disorders*, Heinemann Medical Books, London, 1966.)

THERE IS A DIFFERENCE!

most distinctive. In those cases where the development of the secondary sexual characteristics is impeded (as in Turner syndrome), the application of additional female hormones will promote their development, although the problem of sterility cannot usually be helped.

The adrenal gland is responsible for the production of male hormones, or androgens. An enzyme defect in this gland results in the production of great excesses of male hormones during fetal life. Studies of girls born with this defect show that they are extreme tomboys, are much like boys in their play and orientation, and spend relatively little time at activities like playing with dolls. In an opposite kind of defect some males have an enzyme deficiency that prevents the normal development of male sex organs during fetal life. Externally they appear to be female at birth and may be raised as such. However, at puberty the male sex hormone production is normal, male genitalia start to develop, and such persons assume male sexual identities.

The personalities of children whose mothers were given hormone treatments during pregnancy have been compared with those of their untreated sibs. Those girls whose mothers had received treatment with a male hormone while they were *in utero* were significantly more independent, individualistic, self-assured, and self-sufficient than their sisters. On the other hand, the boys whose mothers had received a female hormone were significantly less individualistic and less self-sufficient than their brothers. Finally, some work on the development of the nerves of the brain of very young rats has shown different patterns of nervous connections in males and females. If newborn males are deprived of male sex hormone by castration, their pattern of development is female, and contrarily, if females are given the male sex hormone testosterone, their pattern is male. Similarly, it has been shown in humans that brain lesions localized to one side produce very different effects on intelligence test scores of men compared to women. Observations such as these are considered significant by those who are inclined to emphasize the biological contribution to sex orientation over the environmental one.

BEHAVIORAL SEX. The behavioral traits that are commonly considered either masculine or feminine can be categorized as psychological sex. Every individual has characteristics of both sexes to varying degrees, and these may be expressed by actions or attitudes that may not be consistent with the other criteria for sex labeling. In extreme cases, an individual may take on, in toto, the behavior of the opposite sex. Another aspect of behavioral sex is social sex, referring to the influences that individuals may be exposed to early in life that cause them to adopt the attitudes and behavioral patterns expected of them.

OVERALL SEX. In the vast majority of cases, the various indicators of sex are in substantial agreement in one direction or the other. On the other hand, at each step in the development from the zygote to the mature individual, there is the possibility of a deflection from the typical pattern; it is not surprising that there should exist such a wide array of differences in sexuality. Our concern is with the biological basis for these differences, when such a basis exists. With respect to aberrant behavior, for instance, probably the most interesting and important aspect of sex, little can be said here. Some persons exhibit behavioral patterns of the opposite cellular sex; these are transsexuals. Transvestites, usually males, show a predilection for clothes of the other sex. Homosexuals show a sexual preference for members of the same sex. In none of these cases is there unambiguous evidence for a genetic basis.

INTERSEXUALITY. In some cases a human of one sex embodies some of the physical characteristics of the other sex and is therefore referred to as an *intersex*.

The hormonal influence may occur prior to birth. The classic example is the so-called freemartin in cattle. When twins of opposite sexes develop in the cow, the hormones of the male (and in many cases some cells as well) reach the developing female embryo and induce that female to become malelike. When newborn, the female may resemble a bull calf, but the external genitalia are those of a female and the internal sex organs are usually poorly developed. This does not occur in humans: a twin of one sex is unaffected by the presence of one of the other sex in the womb during development. However, male hormones may sometimes influence an otherwise normal female fetus to take on male characteristics, as has been observed repeatedly after women have been given such injections after repeated miscarriages. Such a child may be identified as a male at birth and may be raised as a male until puberty, when female secondary sexual characteristics start to develop.

HERMAPHRODITISM. Hermaphroditism is part of the larger category of intersexuality. Strictly speaking, a *hermaphrodite* is an individual with both male and female gonads, and as has been noted earlier, in other species both of the sexes may be functional. In humans, hermaphrodites usually have ambiguous genitalia or, if they have the genitalia of one sex, have a mixture of the two types of internal sex organs. A contribution to this class might come from mosaics in which the somatic cells are of two different compositions. One of the common types of mosaics, for instance, comes about when a Y chromosome is lost from a male cell during early cleavage, giving rise to two cell lines, one XY and the other XO. This combination can give rise to an individual who appears to be part male and part female. However, sharp differences are not usually found in such cases, because the direction of differentiation may not depend entirely on the chromosome constitution of the individual cells; rather it may be determined by the levels of hormone in the circulation.

In a group referred to as pseudohermaphrodites, the internal gonads are not mixed but are either male or female. The external genitalia, however, are either ambiguous or primarily of the sex opposite that of the gonads.

TESTS FOR SEX OF ATHLETES. With the exception of shooting and most equestrian sports, women are at a physical disadvantage in competition with men. Because men have a 10 percent or greater advantage in track, swimming, and field events, persons who may consider themselves women but who have some biological properties of males—like intersexes, hermaphrodites, and some mosaics—might have an unfair advantage over other women. In fact, it has been estimated by one authority that in 1968, as many as 50 percent of all holders of European and world records in women's field athletics may have been intersexes of one sort or another!

Such intersexes usually appear to be normal females, have female external genitalia, and, in the vast majority of cases, are not aware of their ambiguous sexuality until tests reveal it.

All female athletes in the Olympic games must now have a certificate of femininity from a medical group qualified by the International Olympic Committee. Those not having such a certificate must submit to an examination of the number of Barr bodies and Y flourescing bodies in cells from the lining of the cheek or the roots of the hair. If the results are ambiguous, these persons may then be karyotyped. The results of such tests unfortunately received a good deal of publicity in the early years of such sex checking. Such sensitive medical information should be kept confidential, and safeguards are now taken to see that the privacy of the individual concerned is respected.

Additional Reading

DALY, M., and M. WILSON. 1979. Sex and strategy. *New Scientist*, **81**:15–17.

GERMAN, J., et al. 1978. Genetically determined sex-reversal in 46, XY humans. *Science*, **202**:53–56.

GRIBBIN, M. 1979. Hetero, homo, trans: Sex is in the mind. *New Scientist*, **82**:436–437.

Hutt, Corrine. 1978. Biological bases of psychological sex differences. *American Journal of Diseases of Childhood,* **132:**170–177.

Inglis, James and J. S. Lawson. 1981. Sex differences in the effects of unilateral brain damage on intelligence. *Science,* **212:**693–695.

Kolata, G. 1979. Sex hormones and brain development. *Science,* **205:**985–987.

Mohandus, T., R. S. Sparkes, and L. J. Shapiro. 1980. Reactivation of an inactive human X chromosome: Evidence for X inactivation by DNA methylation. *Science,* **211:**1263–1323.

Naftalin, Frederick. 1981. Understanding the bases of sex differences. *Science,* **211:**1263–1264. See related articles in same issue.

Reinisch, J. 1977. Prenatal exposure of human foetuses to synthetic progestin and oestrogen affects personality. *Nature,* **266:**561–562.

Smith, J. M. 1978. *The Evolution of Sex.* Cambridge: Cambridge University Press.

Tonz, O., and E. Rossi. 1964. Morphological demonstration of two red cell populations in human females heterozygous for glucose-6-phosphate dehydrogenase deficiency. *Nature,* **202:**606–607.

Trotter, R. J. 1977. The transsexual riddle: A hypothesis. *Science News,* **111:**236–238.

Wachtel, S. S. 1977. H-Y antigen and the genetics of sex determination. *Science,* **198:**797–799.

Questions

1. How does the scientific definition of sex differ from the popular view?
2. About fifteen years ago in England a woman claimed that her newborn son had been produced parthenogenetically after she had gone swimming in the ocean and presumably exposed herself to salt water. Can you find anything wrong with her argument?
3. In what way does the manifestation of sex differ in the lower forms of the plant and animal kingdoms compared with the higher forms? Are there cases where the sexes are physically identical? Are there ever more than two sexes?
4. Would you say that sex exists in a hermaphroditic organism (the snail, for instance)? How about a parthenogenetic one?
5. How is it known that in *Drosophila* the autosomes play a part in sex determination, in addition to the influence of the sex chromosomes?
6. What does Turner syndrome tell us about the mechanism of sex determination? Klinefelter syndrome?
7. What is meant by dosage compensation? Why is it necessary?
8. How is the sex chromatin related to dosage compensation?
9. Why is sex chromosome inactivation not obvious in the appearance of humans? Is there any animal in which it is quite clear?
10. What is meant by *clone?* In what sense is a human female a mosaic consisting of a large number of clones of cells, whereas a male is not?
11. Have any human adults been seen with as many as three or four sex chromosomes? The X chromosome has more genetic material than chromosome 21; why, then, does an extra X chromosome (or, for that

matter, several extra X chromosomes) have less impact on the phenotype of a human than the presence of only one extra chromosome 21?

12. Many characteristics are commonly attributed to the male sex (e.g., aggressiveness and physical prowess) and others to the female sex (e.g., gentleness and maternal interest). List as many of these as you can think of, and indicate which of these, if any, are probably inherently characteristic of that sex and which, if any, are imposed on that sex by society.

13. Occasionally newborn children are found with ambiguous sex organs, so that it is difficult to determine the sex at birth. Explain how it might be possible to shed some light on this problem, taking advantage of the sex chromatin bodies, and tell why it might be desirable to supplement these tests with a test for Y-chromosome fluorescence.

14. What are the different criteria by which a person may be classified into one or the other of the two sexes? Which do you consider more important than the others?

15. Do parents have a right to know the sex of their unborn fetus? If they feel that there is a compelling reason (to them) for wanting a specific sex, should they be allowed the option of asking for the abortion of a perfectly normal fetus simply because it is the "wrong" sex?

7

Sex-Linked Inheritance

The sex chromosomes have an unusual pattern of inheritance from one generation to the next. The X chromosome of the father goes only to his daughters, his Y chromosome only to his sons. With this in mind, it can be expected that genetic traits determined by the genes carried on those chromosomes will also exhibit an unusual pattern of transmission. This pattern has been referred to over many years as *sex-linked inheritance,* although the phrase is misleading because the transmitted traits are not strictly linked to the sex of the individual. Rather, they follow the inheritance of the sex chromosomes and therefore affect both male and female progeny, but unequally. Traits that follow the transmission of the X chromosome from one generation to the next are called *X-linked;* those whose pattern of inheritance comes from the Y chromosome are *Y-linked* or *holandric.*

Y Linkage

The Y chromosome has a very important function in determining the differentiation of the embryo into maleness. There are several functions attributed to the Y chromosome: one that determines the differentiation of the primitive gonads into testes, another that governs the production of antibodies (the H-Y antigen), and a third that controls spermatogenesis. In addition, there is some evidence that genes controlling stature and other aspects of maturation in the male are carried by the Y chromosome. It is not yet known definitely how many loci are involved in these functions.

After a little reflection, however, we might not expect to find very many loci for other characteristics on the Y chromosome, for several reasons. For one thing, we have seen that individuals with two Y chromosomes are, in many cases, indistinguishable from XY males, suggesting that the number of genes duplicated in XYY individuals is relatively small. In addition, the female is clearly a normal individual in all respects, yet lacks a Y chromosome completely. This indicates that the Y-chromosome is not necessary for normal development.

It is not surprising, then, that except for a few factors concerned with maleness itself, we are hard-pressed to find an array of genetic characteristics transmitted by the Y chromosome. When they occur, there is no difficulty in identifying them because they appear in a pedigree as in Figure 7-1, where the father, brothers, and sons of every male with a particular trait are similarly affected, but the mother, sisters, and daughters are unaffected. This type of inheritance is known as *holandric*.

X Linkage

Characteristics determined by alleles known to be located on the X chromosome, on the other hand, not only are frequent but also represent some of the most interesting and striking of those known in the human population. Partial color-blindness involving red or green vision is the one most common genetic variant present in the human race, present in more than 5 million males in the United States. Hemophilia, another X-linked characteristic, although not nearly so common, has a fascinating history in the royal families of Europe.

Some of the characteristics of X-linked inheritance have been known for centuries. More than twenty-five centuries ago it was stated in the Jewish Talmud that male children should not be circumsized when their male sibs had difficulties because of excessive bleeding. This prohibition applied also if the newborn had affected cousins and if the cousin relationship stemmed from the fact that the mothers were sisters.

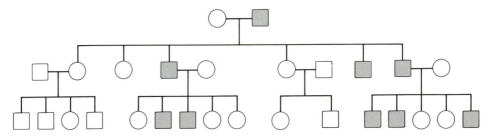

Figure 7-1 Pedigree showing the manner of transmission of a Y-linked, or *holandric*, character. The sons of all affected males are similarly affected; daughters are never affected. Note that this type of transmission parallels precisely the transmission of the male sex.

These empirical rules were not really understood until 1910. At that time, T. H. Morgan, an embryologist at Columbia University who had been an eloquent anti-Mendelian and harshly critical of the growing number of biologists who were interpreting all inherited traits as simple single-gene effects, observed an unusual event in some cultures of fruit flies he had been growing in his laboratory on rotting bananas. He found that some of the flies had eyes that were completely white instead of the normal red (Figure 7-2). In working out the details of the inheritance of this characteristic, he showed that they corresponded exactly to that of such human characteristics as red/green color blindness and hemophilia. Interesting as this observation was, its true importance rests in the relevance of scientific observations on lower organisms to human problems, for that simple experiment with fruit flies led to an explosion of research effort in genetics, involving many of Morgan's colleagues and students, who laid the foundations of our present highly sophisticated knowledge of genetics.

SPECIAL PROPERTIES OF X-LINKED LOCI.   It is clear that because of the gross difference in size of the two chromosomes, most of the loci on the X cannot be present on the Y, and therefore the male must be effectively haploid with

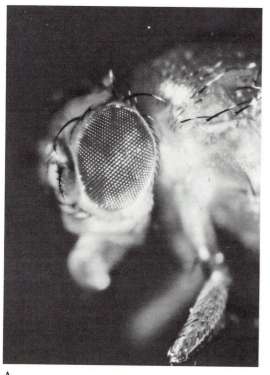

A   B

Figure 7-2   The eye of *Drosophila melanogaster*, at high magnification so that the individual elements of this compound eye are visible. *(A)* The normal or *wild-type* red color; *(B)* the mutant *white*, the first striking mutant to be discovered in this species.

respect to most (if not all) X-chromosome loci. For this reason the terms *dominant, recessive, homozygous,* and *heterozygous* do not apply in the usual way to these loci in the male, because those terms are meaningful only in describing the relationship of two alleles to each other. Because the male has only one allele of each X chromosome locus, he is said to be *hemizygous* for those alleles. The female, with two X chromosomes and thus two alleles for each X locus, might be described as homozygous or heterozygous; however, in most of the cells of her body, one or the other of her X chromosomes is inactivated, making her a mosaic for those characteristics for which her two Xs have different alleles. If, at a particular locus, a woman has different alleles on her two X chromosomes, the allele responsible for the essential phenotype of the mosaic female is referred to as the dominant allele.

Transmission of Dominant X-Linked Alleles

Some persons have a low level of phosphorous in their blood serum and are described medically as having hypophosphatemia. When this condition is determined by an allele on the X chromosome, it is called *X-linked hypophosphatemia.* Affected persons may develop a type of rickets (Figure 7-3) that cannot be cured by the administration of vitamin D and they are therefore sometimes described as having "familial vitamin D-resistant rickets." This characteristic can appear in both males and females, although it is generally more extreme when found in males. A dominant allele is responsible for X-linked hypophosphatemia; that is, heterozygous women are affected.

Because the distribution of the X and Y chromosomes in the progeny of affected individuals is important to an understanding of the mode of inheritance of these alleles, it is convenient to depart from the usual system of using single capital and lowercase letters to represent the dominant and recessive alleles, respectively, and instead to use these letters as superscripts after the symbol of the chromosome carrying them.

Thus, we can define the following genotypes with their corresponding phenotypes:

$X^R Y$ = a male with the dominant allele for vitamin D-resistant rickets.

$X^r Y$ = a normal male

$X^r X^r$ = a normal female

$X^R X^r$ = a heterozygous female, with vitamin D-resistant rickets

$X^R X^R$ = a female homozygous for vitamin D-resistant rickets

We can easily see the pattern of inheritance of this allele in abbreviated form (Figure 7-4). Men with this dominant allele have only affected daughters and normal sons, because their X chromosomes with the defective allele go only to their daughters. On the other hand, women who carry the allele in the heterozygous state should produce all four possibilities, affected and unaffected, males and females, with equal frequency. Homozygous af-

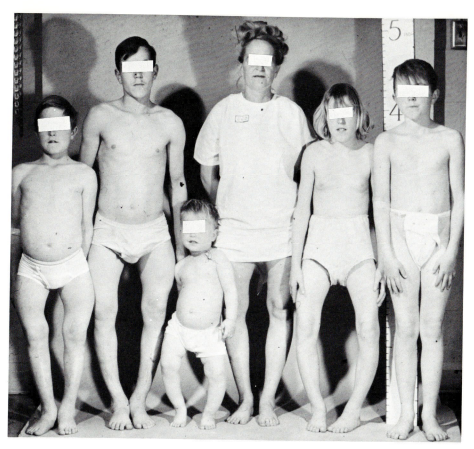

Figure 7-3 Vitamin D-resistant rickets (X-linked hypophosphatemia) in a heterozygous mother and some of her offspring. (Courtesy of E. W. Lovrien, University of Oregon Health Sciences Center.)

fected women will produce, of course, only affected offspring: hemizygous males and heterozygous females. A pedigree showing the inheritance of such a dominant X-linked allele is given in Figure 7-5.

TRANSMISSION OF RECESSIVE X-LINKED ALLELES.   A good example of a recessive X-linked allele is that for hemophilia A, the most frequent form of hemophilia. It is defined as recessive because the heterozygote appears to be normal. Once again, we can specify the genotypes and phentoypes of all

A. Normal mother x affected father.

B. Heterozygous affected mother x normal father.

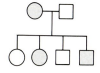

C. Homozygous affected mother x normal father.

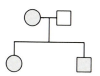

Figure 7-4 The pattern of inheritance of an X-linked dominant allele. The diagnostic feature is shown in *A:* an affected male and an unaffected female will produce all affected daughters and no affected sons.

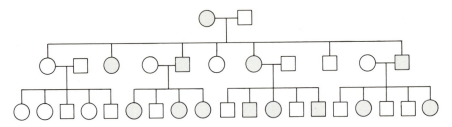

Figure 7-5 A pedigree showing the inheritance of an X-linked dominant like hypophosphatemia.

possible combinations, using a notation that indicates the chromosome that the allele is carried on:

$$X^H Y = \text{normal male}$$

$$X^h Y = \text{hemophilic male}$$

$$X^H X^H = \text{normal (homozygous) female}$$

$$X^H X^h = \text{normal (heterozygous) female}$$

$$X^h X^h = \text{hemophilic female}$$

Figure 7-6 shows what to expect from a series of matings of persons with sex-linked recessive alleles. If the male has the abnormal allele and the female is homozygous for the normal allele, then none of the progeny will be affected although all daughters will be heterozygotes (carriers). If the female is affected—that is, is homozygous for the abnormal allele—and the male is normal, all sons will be affected and all daughters will be normal but will be heterozygous. Finally, if the woman is heterozygous and the man normal, then all daughters will be normal, but half will be carriers, and the sons will be equally normal and affected.

Red/Green Color Blindness

The most common genetic defect in the human population is color blindness for red and green colors; the genes responsible are found on the X chromosome. A pedigree showing typical X-linked inheritance is shown in Figure 7-7. About 8 percent of all males are either red or green color-blind. There are two loci located close together on the X chromosome responsible for producing visual pigments in the cones of the retina; one produces pigment sensitive to red light and the other, pigment sensitive to green. Most persons who are color-blind have alleles at the locus for green pigment that are defective, and only about a quarter have defective alleles at the red locus (Table 7-1). The alleles responsible for red and for green color blindness are recessives.

Because red and green color blindness are determined by recessive alleles at different loci, a woman will appear normal if she has one chromosome with the allele for red blindness and the second with the allele for green blindness. The first chromosome, with the allele for red blindness, carries the normal allele for green vision, and the second, with the green color-blind allele, carries the normal allele for the red. The mosaicism that is expected

A. Homozygous normal mother X hemizygous affected father

AA X A'Y

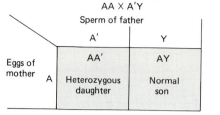

	Sperm of father	
Eggs of mother	A'	Y
A	AA' Heterozygous daughter	AY Normal son

=

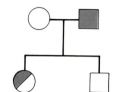

B. Homozygous affected mother X normal father

A'A' X AY

	Sperm of father	
Eggs of mother	A	Y
A'	AA' Heterozygous daughter	A'Y Affected son

=

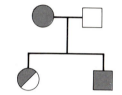

C. Heterozygous mother X normal father

AA' X AY

	Sperm of father	
	A	Y
Eggs of mother — A	AA Homozygous daughter	AY Normal son
A'	AA' Heterozygous daughter	A'Y Affected son

=

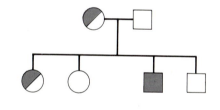

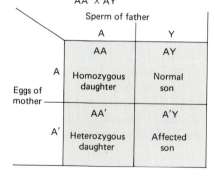
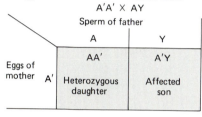

Figure 7-6 The transmission of characters determined by recessive alleles on the X chromosome when *(A)* the father is hemizygous, *(B)* the mother is homozygous, and *(C)* the mother is heterozygous. The Y chromosome of the male is indicated by a Y. The "sibships" are not meant to indicate actual families; instead they show the theoretically expected types of progeny in the proportions ideally expected: thus there are two different types of progeny expected equally frequently in *(A)* and *(B)* but four in *(C)*. Heterozygous females are indicated by circles only half filled in.

from random X-chromosome inactivation during early development apparently involves such small areas of the retina that the visual image appears normal to the heterozygote.

Hemophilia

Perhaps the most infamous inherited congenital disease is the "bleeding disease," hemophilia A, caused by an X-linked recessive allele that results in the failure of the blood to clot properly. It appeared in the royal families of

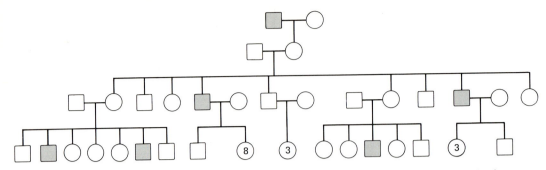

Figure 7-7 A pedigree of the X-linked recessive condition ichthyosis, which causes excessive dryness and scaliness of the skin. This example is taken from a much more extensive Dutch pedigree, showing the simple rules of sex-linked inheritance. (L. N. Went et al. X-Linked Ichthyosis: Linkage Relationship with the Xg Blood Groups and Other Studies in a Large Dutch Kindred, *Annals of Human Genetics*, **32**:333–45, 1969. Published by Cambridge University Press.)

Europe via mutation occurring either in Queen Victoria (1819–1901) or in one of her parents. We consider it likely that it is a new mutational event because there is no record of any occurrence of this disease in the royal family prior to this time.

Of her children, three daughters (Victoria, Alice, and Beatrice; see Figure 7-8) were heterozygous for the gene—hence were unaffected carriers—and only one son, Leopold, was hemophilic. All of Queen Victoria's other children were free of the defective allele. Because the present royal family of England is descended from one of the normal sons, Edward VII, it can be safely assumed to be free of the disease. However, Alice and Beatrice married into other royal families of Europe and transmitted the gene to their progeny, so that the next several generations of European royalty had an undue proportion of hemophiliacs (Figure 7-9). The last carrier, a grand-daughter, died in 1981.

Perhaps the best-known case involved the wife of Czar Nicholas II of Russia, Czarina Alix, who was a granddaughter of Victoria. She had a son, Czarevitch Alexis, who was hemophilic. It is of historical interest to note that the monk Rasputin claimed to have supernatural powers that could cure Alexis and thus succeeded in becoming a powerful figure in the Russian royal court. It was partly as a result of Rasputin's influence that the Russian monarchy became disorganized and eventually suffered a complete

Table 7-1 Frequencies of the four common types of red and green color blindness in the European population. (Adapted from R. Cruz-Coke, *Colorblindness, An Evolutionary Approach*, 1970. Courtesy of Charles C. Thomas, Publisher, Springfield, Ill.

		Type of Cone Affected	
		Green (percentage)	*Red (percentage)*
Severity of Defect	*Mild*	53	8
	Severe	19	20

Figure 7-8 Some of the royalty of Europe surrounding Queen Victoria, as they posed for a group photograph during a family wedding in 1894. Queen Victoria *(A)* was the initial heterozygote; her youngest daughter, Princess Beatrice *(B)*, and two grand-daughters, Princesses Alix *(C)* and Irene *(D)*, all received the defective allele from her. Beatrice transmitted the allele to the Spanish royal family and Alix to the Russian. (Courtesy of the Gernsheim Collection, Humanities Research Center, University of Texas.)

collapse. Although there were certainly additional sociological factors of greater weight, Rasputin (and thus hemophilia) can be said to have been partly responsible for the downfall of the Russian czardom and the institution of the present soviet government.

The term *hemophilia* is, strictly speaking, a misnomer, because its literal translation is "liking for blood," a disposition that might better describe a vampire than the victim of a well-known sex-linked defect. The disease was first definitively described by a U.S. physician, Dr. John Otto of Philadelphia, who in 1803 outlined its clinical manifestations as well as noting that only males were subject to the disease, and that the females of the same family were unaffected but could still transmit it to their male children.

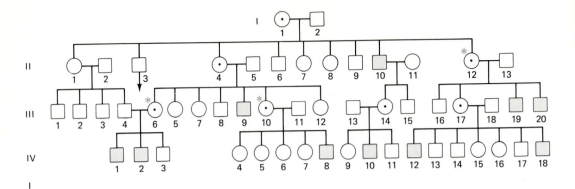

Figure 7-9 An abbreviated pedigree of Queen Victoria and her descendants, showing the distribution of hemophilia in royal families of Europe. Known heterozygotes have dots within the female symbols; the three heterozygous descendants (II-12, III-6, and III-10) depicted in Figure 7-8 are indicated here by asterisks.

       I1. Queen Victoria.
       I2. Prince Albert.
      II3. King Edward VII of England.
      II4. Princess Alice of Hesse Darmstadt.
    II10. Prince Leopold, Duke of Albany.
    II12. Princess Beatrice of Battenberg.
    III6. Princess Irene of Hessen.
  III10. Princess Alix, wife of Czar Nicholas of Russia.
  III14. Princess Alice, wife of Alexander, Prince of Teck.
  III17. Princess Victoria, wife of King Alfonso of Spain.
      IV8. Czarevitch Alexis of Russia.
IV10, 12, 18. Lord Trematon, Prince Alfonso, and Prince Gonzalo, respectively; all three died after automobile accidents.

Medical Aspects

The primary characteristic of hemophilia is the greatly increased clotting time of the blood. Normal clotting of the blood results after a series of reactions involving at least thirteen different factors. If these reactions are interrupted, or if one of the factors is present in a quantity less than normal, the clotting time may be extended from the usual few minutes to an hour or longer. In the common form of hemophilia (hemophilia A), the eighth factor (factor VIII) is missing or is present in much reduced amount. In another, less common, form of hemophilia (hemophilia B, or Christmas disease), factor IX is affected. Like color blindness, which has two different loci on the X chromosome, hemophilia A and B are also caused by defective alleles at different loci.

The hazards of this condition can be easily appreciated. A cut that might be completely innocuous to a normal person could bleed for days in a hemophiliac. Minor injuries that are not usually even noticed in unaffected people, such as bruises, can produce serious hemorrhaging in affected persons. Simple operations such as dental extractions can be fatal. Internal bleeding, especially in such vulnerable areas as the soft palate of the mouth, represents a special hazard because of the difficulty of stopping the flow of blood inside the body. Strains and sprains of the arm and leg joints can result in accumulations of blood inside these relatively delicate areas inhibiting movement and damaging the joint.

Hemophilia occurs with a frequency of 1 in 10,000 males. Today more than 95 percent of affected males survive beyond the age of sixteen. Treatment for hemophilia consists of reducing the coagulation time, which can be done by transfusion of fresh, normal clotting blood, or by injection of antihemophilic globulin derived from normal blood. The techniques for this latter treatment have become sufficiently standardized so that affected individuals can maintain their own supply of globulin in refrigerators and inject themselves at regular intervals. It is nevertheless a serious disease with many potential dangers, especially during youth when injuries are more frequent, and these can accumulate with additional side effects to decrease the affected person's general well-being.

MOSAICISM IN HETEROZYGOUS FEMALES. A male carrying the allele for hemophilia A on his X chromosome produces very little of the antihemophilic factor (AHF) necessary for the normal clotting of the blood, with serious, sometimes fatal, consequences. A homozygous female with two such alleles, one on each of her X chromosomes would have similar problems. However, in a heterozygous female, on the average, half of the cells with the chromosome carrying the normal allele are active, while in the other half of her cells the other chromosome with the allele for hemophilia is active. That half with the active normal allele produces enough AHF so that her clotting time is normal, or nearly so. Because she is a heterozygote with one allele normal and the other the allele for hemophilia, but with a normal phenotype, the normal allele is considered dominant and the hemophilic allele recessive.

HEMOPHILIA IN FEMALES. Because, at the time of menses and childbirth, women suffer unusual hazards from slow clotting, one might expect to find that hemophilic women die young. In fact, adult females occur only with a low frequency. Some of these are heterozygotes in which the amount of factor VIII or AHF is low. Others carry one of the autosomal loci that produce hemophilia, either as homozygotes for an autosomal recessive or, more commonly, as heterozygotes for an autosomal dominant (von Willebrand disease). However, a few cases have been described in which an apparently homozygous daughter has been born of a hemophilic father and a heterozygous mother, and that affected daughter has produced offspring.

Muscular Dystrophy

Muscular dystrophy is characterized by slow muscular deterioration; the afflicted person begins to show signs of the disease at ages three to six and is usually chair-ridden by the age of twelve; death occurs in the teens, usually by failure of the heart or respiratory muscles. There are ten or more types of muscular dystrophy. Five of these are X-linked; the remainder are autosomal.

One of the X-linked forms (Duchenne) is by far the most common, affecting about 3 males per 10,000 births. Occasionally women manifest the disease, usually mildly. Some of these women are heterozygotes. A few such affected females have the Turner syndrome. Because they have only one X chromosome, if that chromosome carries the defective allele, they will manifest the phenotype just as will a male (who also has only one X).

Heterozygous women can sometimes be detected because they have higher-than-normal levels of an enzyme (creatine kinase) and other blood and muscle components. On occasion, heterozygous women manifest this disease to some extent; perhaps because of the inactivation of the normal X chromosome in an unusually high proportion of the cells of the heterozygote.

Frequencies of Alleles in Populations

It seems fair to assume that the chromosomes that males carry are a random selection of all the X chromosomes present in the human population. If men carry a random sample, then we can come to some simple conclusions about the relative overall frequencies of various combinations by examining the relative frequencies with which they occur in the male sex. Thus, if in the population of the United States, 6 percent of all males are color-blind because of a defect at the locus involved in vision of green wavelengths, then it is reasonable to conclude that 6 percent of all X chromosomes, irrespective of whether they occur in males or females, carry the allele for this type of color blindness.

Now, we can futher make the simple statement that the chance that any particular chromosome, whether it be found in the male or the female, will carry the recessive allele (c) for color blindness involving this locus is 6 percent or .06. We can also state that the chance that an X chromosome will carry the normal allele (C) is $1 - .06$ or .94.

If we know that the frequency of the c allele is .06 and the frequency of its alternative C is .94, then we can ask what the relative frequencies of the different genotypes might be on the basis of simple algebra. This approach suggests that we can regard each person as being a random selection of X chromosomes (two for the female, one for the male) drawn from a much larger pool of chromosomes, in which the chance that any chromosome of a given type will be drawn is given by its overall frequency in that pool. Let us consider the probabilities of the different combinations when we take two chromosomes from this large supply of X chromosomes in order to constitute a female. The chance that any one chromosome will be normal, with the C allele, is .94. The chance that two in succession will be normal is $.94 \times .94 = .8836$. This gives as expected frequency of the CC genotypes of .8836.

What is the chance that one of the X chromosomes will be normal and the other abnormal? This can happen in either of two ways: either the first will be normal and the second abnormal $(.94 \times .06 = .0564)$ or the first will be abnormal and the second normal $(.06 \times .94 = .0564)$. The total probability, then, of one normal and one abnormal chromosome (that is, a heterozygote Cc) will be $2 \times .0564$, or .1128. Finally, the chance that both chromosomes will have the recessive c is $.06 \times .06 = .0036$, which is less than 1 percent. Because this condition is recessive (two alleles of type c being necessary in the female for color blindness to express itself), we have a very much reduced frequency of expectation for color-blind women than for similar men. Population surveys, in fact, confirm the low level of color-blind women, less than 1 percent, predicted by elementary arithmetic.

This type of approach to relating allele frequencies with genotype and phenotype frequencies in populations was first proposed explicitly by the British mathematician G.H. Hardy and the German physician W. Weinberg in 1908 and is often referred to as the *Hardy-Weinberg rule.* In the exercise we have carried out here, there is no difficulty in comprehending the basic principle. We will have more to say about this principle in Chapter 19.

Our rules for the expression of genes carried on the X chromosome obviously break down for some of the syndromes we have already discussed. Females who have Turner syndrome, with a single X chromosome, obviously have a probability of carrying X-linked defects equal to that of males. This theoretical expectation is borne out by observation of the high frequency of red/green color blindness in Turner syndrome females, which is approximately the same as the male frequency. This would not be the case, of course, if the nature of the defect put the female at a severe disadvantage (e.g., hemophilia). A medical practitioner who came upon the case of a female with a rare sex-linked disorder might well immediately look for signs of the Turner syndrome, particularly if the father was known not to have the disorder—for example, in the case of a female affected with muscular dystrophy of the Duchenne type.

Additional Reading

MacNichol, E. F., Jr. 1964. Three-pigment color vision. *Scientific American,* **211**:48–56.

McKusick, V. A. 1965. The royal hemophilia. *Scientific American,* **213**:88–95.

Marx, J. L. 1975. Hemophilia: New information about the "royal disease." *Science,* **188**:41–42.

Questions

1. Draw a pedigree showing the manner of inheritance of a genetic defect with its locus on the Y chromosome, that is, one with holandric inheritance.
2. Can you present any arguments supporting the thought that, except for genes important in male sex differentiation, the Y chromosome cannot have any other really important genes?
3. How did early work with *Drosophila* prove to make an unusual contribution to our understanding of inheritance in humans?
4. Brown enamel of the teeth is a well-known sex-linked dominant trait. Draw a pedigree showing how it will be inherited (a) if the father of a

family carries the defect, that is, is hemizygous for it; (b) if the mother is heterozygous for it; and (c) if the mother is homozygous for it.

5. Why do the terms *dominant, recessive, homozygous,* and *heterozygous* lose some of their original meaning when applied to X-linked or Y-linked alleles?

6. In what sense are red or green color-blind persons actually color-blind? How common are these defects? What is the genetic basis for such color-perception deficiencies?

7. What is the frequency of color blindness among females? Explain why it is so much different from the frequency in males. If the frequency of red-deficient color blindness is 36 percent in females in a certain population, you might calculate that more than half of all males would be similarly affected. How many more?

8. If a female should receive an X chromosome carrying an allele for red deficiency from her father and one for green deficiency from her mother, would her color vision be affected? Explain your answer.

9. What is hemophilia? Why is it so very well known compared with other sex-linked defects?

10. If a relatively uncommon disease such as muscular dystrophy of the sex-linked type occurs in a young girl, what might you suspect?

Abnormal Human Chromosomes: Structure and Transmission; Cancer

The range of variation in human chromosome structure is enormous; there is a good chance that any person picked at random will have at least one minor, but distinguishable, variant chromosome that will differentiate his or her karyotype from that of some other unrelated person also picked at random. An abbreviated listing of the symbols used in the nomenclature of abnormal karyotypes and chromosome variants is given in Table 8-1. Many of these variations are of no known genetic significance. In fact, most of them involve changes in the constitutive heterochromatin adjacent to the centromere, DNA that is thought to be largely inert genetically. Other types of changes, however, are responsible for clinical manifestations in the persons carrying them or may be responsible for genetic abnormalities in subsequent generations. How this comes about is our concern in this chapter.

Variant Autosomal Types

Such a morphologically deviant chromosome with no phenotypic effect is an extra-long chromosome 1, found in newborns with a frequency of about one percent (Figure 8-1). This chromosome is sometimes called the *uncoiler* because of an early hypothesis that the extra length was caused by a failure of a section near the centromere to coil tightly. It is now better explained by the duplication of material near the centromere. It is transmitted from a carrier to half the progeny, as would be expected.

Table 8-1 A condensed list of symbols used in designating abnormalities in karyotypes. (An International System for Human Cytogenetic Nomenclature (1978). Birth Defects: Original Article Series XIV, No. 8 The National Foundation–March of Dimes.)

A–G	the chromosome groups
1–22	the autosome numbers (Denver system)
X,Y	the sex chromosomes
Diagonal (/)	separates cell lines in describing mosaicism
Plus sign (+) or minus sign (−)	when placed immediately after the autosome number or group letter designation indicates that the particular chromosome is extra or missing; when placed immediately after the arm or structural designation indicates that the particular arm or structure is larger or smaller than normal
Question mark (?)	indicates questionable identification of chromosome or chromosome structure
Asterisk (*)	designates a chromosome or chromosome structure explained in text or footnote
ace	acentric
cen	centromere
del	deletion
dic	dicentric
h	secondary constriction or negatively staining region
i	isochromosome
inv	inversion
mat	maternal origin
p	short arm of chromosome
pat	paternal origin
q	long arm of chromosome
r	ring chromosome
rob	Robertsonian translocation
s	satellite
t	translocation
Repeated symbols	duplication of chromosome structure

In another interesting type of anomaly, a chromosome may have a "fragile" site. The first case to be discovered involved chromosome 16, the long arm of which appeared to undergo spontaneous breakage in many cells of persons in one kinship. Chromosomes 2, 10, 11, 16, 20, and the X have also been shown to have inheritable variants with fragile sites. The broken pieces may then undergo rapid asynchronous replication, producing cells with several small chromosome fragments.

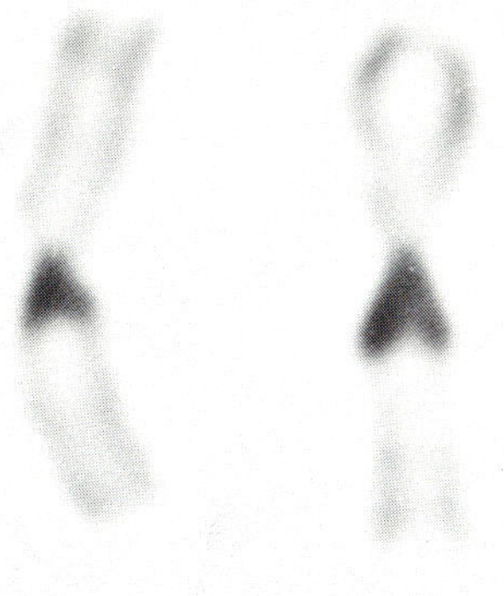

Figure 8-1 A comparison of a normal chromosome 1 and the variant known as *uncoiler 1*. The staining method used shows an unusual amount of heterochromatin adjacent to the centromere. (Courtesy of P. Pearson, University of Leiden.)

Many new variants have now been found with quinacrine staining. Chromosomes are particularly variable in the centromeric regions, where some stain brightly and other faintly. Two chromosomes that are different in structure are said to be *heteromorphic*. If different forms of chromosomes are found in a population with frequencies of greater than a few percent, that population is said to be *polymorphic* with respect to that chromosome.

Variations in the Y Chromosome

An argument was presented earlier that the Y chromosome must be relatively free of essential loci other than those that trigger male sex differentiation. It should not be surprising, then, to find that the Y is one of the most variable of the chromosomes, because changes in structure—additions or subtractions—would probably not involve the few genes essential to normal development.

Many cases have been found in which the Y is modified in structure or stainability (Figure 8-2). The Japanese have larger Y chromosomes with brighter quinacrine fluorescence; Australian aboriginals have smaller ones

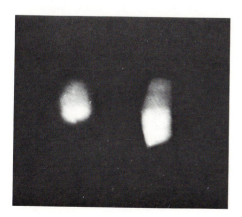

Figure 8-2 Polymorphism for the Y chromosome. At the left is a normal Y, at the right a large one with a brightly staining block. (Courtesy of P. Pearson, University of Leiden.)

with little or no fluorescence. There are some pedigrees of Caucasian males with very long Y chromosomes, others with very short ones, neither type having any particular effect on the phenotype of the males carrying them. Pedigrees have been described in which a long Y chromosome produces much the same symptoms as the extra Y in XYY males, as though some of the limited essential material in the Y was duplicated.

Origin of Chromosome Rearrangements

With a very low frequency, a chromosome may break spontaneously or may be broken by radiation passing through the cell or by noxious chemicals present in the environment. Such chromosome breaks may give rise to several categories of new chromosome types. Here we shall consider the various types produced, along with some of their genetic properties.

SINGLE CHROMOSOME BREAKS.   At the point of a chromosome break, the two newly formed free ends behave as though they were "sticky," or unsaturated, in the sense that they can now rejoin with other broken ends. In contrast, normal chromosome ends are stable and do not join with other ends—if they did, the fusion of two normal chromosomes, end to end, would give rise to serious problems in normal mitosis. In many cases, a broken chromosome may simply *restitute* to its original condition with no net damage, unless a gene located at or near the point of breakage has been damaged, in which case the chromosome may now carry a new mutation (Figure 8-3A).

At the next replication of the chromosome strand, newly broken ends of the sister chromatids may unite with each other in what has been called *sister strand union* (Figure 8-3B). As can be seen from the illustration, this union gives rise to serious problems at the following anaphase, for we now have two abnormal chromosomes, one with no centromere, or *acentric*, and a second with two centromeres, or *dicentric*. An acentric fragment, having no centromere to direct it to the poles of the division, will be lost during succeeding mitoses, leaving the daughter cells with a large deficiency for the genetic material carried in the acentric fragment. Figure 8-4 shows a mouse cell with a dicentric chromosome, along with acentric fragments.

The greater damage to the cell, however, may be done by the dicentric. At anaphase a *bridge* is formed, running between the two poles. In many cases, this bridge itself is sufficient to prevent the normal development of the two daughter cells. If the bridge should break, with the two opposing centromeres carrying fragmented chromosomes to their respective poles, at the next replication the broken chromosomes will rejoin upon themselves to produce additional dicentrics. Eventually such cell lines must die out or produce abnormal cell lines in which there is a broken chromosome or in which the broken chromosomes have been eliminated after successive anaphases with dicentric bridges. Thus the predominant chromosomal damage, that of the single break, generally is self-eliminating and for this reason makes little contribution to human abnormalities.

In rare instances, however, some chromosomes do survive as dicentrics

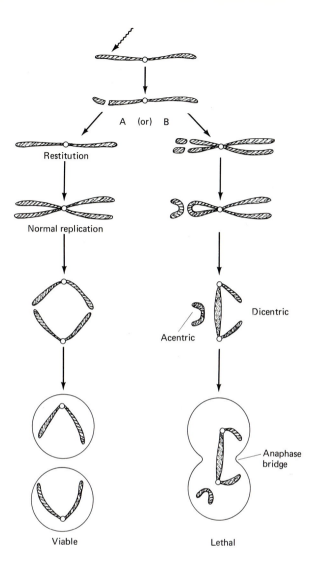

Figure 8-3 Normal restitution of a chromosome break versus the production of acentrics and dicentrics resulting in an anaphase bridge.

after breakage and manage to perpetuate themselves through a number of mitoses. Cells carrying such dicentrics are abnormal, because they lack the chromosome material of the acentric fragment and carry one extra chromosome arm.

Kinds of Chromosome Rearrangements

When two breaks occur in the same cell, several distinctly different kinds of chromosome rearrangements are possible. Unlike the dicentrics and acentrics produced by single breaks, two-break rearrangements may have serious consequences to the person carrying them, and to the progeny that carriers may produce. Fortunately the spontaneous incidence of new arrangements is only about 3 per 10,000 births.

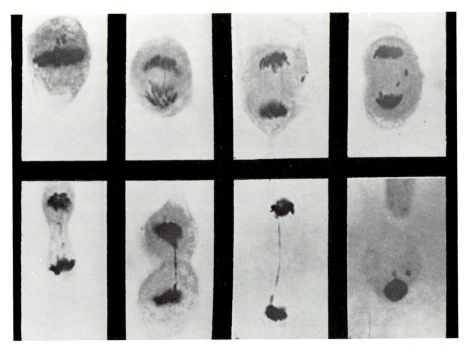

Figure 8-4 Cells of the mouse at telophase showing dicentric bridges formed at anaphase, along with acentric fragments. (Courtesy of T. H. Roderick, The Jackson Laboratory.)

TRANSLOCATIONS. If breaks occur in two different chromosomes, there are several possible consequences (besides restitution, which we shall disregard because it leads to nothing new). Figure 8-5*A* illustrates the result after two different chromosomes are broken and the acentric fragments of each unite with the centric fragment of the other chromosome to form a *reciprocal translocation*. Although some of the genetic material of the two chromosomes is interchanged, it should be noted that the cell still contains the same total amount of genetic material, and that the chromosomes involved in reciprocal translocations are structurally normal in containing one and only one centromere. Persons who carry such balanced translocations will appear to be quite normal. Note, however, that when such a translocation is found in a normal diploid cell, there still exists one normal homologue of each altered chromosome in that same cell. If the cell in question is a gonial cell, then problems may arise when the homologues attempt to pair at meiosis. The unusual consequences of the meiotic behavior of translocated chromosomes will be discussed later in this chapter.

DICENTRICS AND ACENTRICS. The rejoining may occur in a different way, shown in Figure 8-5*B*. In this case, dicentric chromosomes and acentric fragments are produced; their behavior is much like that of the dicentrics and acentrics produced by single breaks.

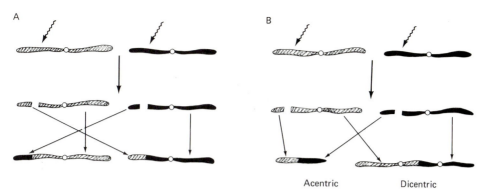

Figure 8-5 *A.* Breakage of two chromosomes resulting in a reciprocal translocation. *B.* Breakage of two chromosomes resulting in a translocation involving acentrics and dicentrics.

DEFICIENCIES. If two breaks occur in the same chromosome, the results depend on whether they both occur in one chromosome arm or one occurs on each side of the centromere. If the first is the case (Figure 8-6A), and if the breaks do not restitute, then the piece between the two breaks may be lost so that a *deletion* or *deficiency* is produced. A chromosome deficiency that involves a small number of genes may not have any obvious effect, because of the presence of the unaffected homologue in the normal diploid cell. Larger deficiencies, however, are more often detrimental to normal development.

The best-known deletion occurring with a considerable frequency is that of the cat cry (*cri du chat*) syndrome, first described in France (Figure 8-7). The afflicted infant is characterized by a cry strongly reminiscent of that of a cat (Figure 8-8), a feature which disappears later in life. Although the overall frequency may be as low as 1 per 100,000, more than 200 cases have now been reported. The affected individuals are invariably mentally retarded, with an array of abnormal clinical features. This syndrome is not lethal, as are most others involving the absence of chromosome material on one of the

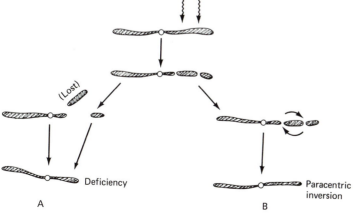

Figure 8-6 Double breakage of a chromosome resulting in a deficiency *(A)* or a paracentric inversion *(B)*.

Figure 8-7 A child with the cat's cry *(cri du chat)* syndrome, caused by a deletion for part of the short arm of chromosome 5. (Courtesy of E. Polani, Guy's Hospital Medical School, London.)

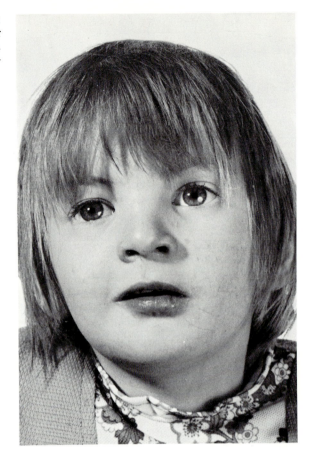

two homologues, and some of the known patients have survived beyond adolescence.

In the vast majority of cases, it can be shown that such individuals have a deficiency (−) for a part of the short arm (p) of chromosome 5 (Figure 8-9). This deficiency (5p−) is variable in size from one case to the next, ranging from less than half to almost all of the short arm.

INVERSIONS. In some cases, the broken piece of chromosome rotates and reunites with the other pieces of that chromosome in reversed order (Figure 8-6B) to form a *paracentric inversion*. The structure of the chromosome is now restored, except for the inverted segment and, in fact, is not usually detectable as having been altered. Without special staining techniques it would appear normal and, with its normal gene complement, would probably have no genetic effect. Paracentric inversions might be expected to occur in the population as one of the more common types of abnormal chromosome structure, although their detection is difficult.

Still another possibility exists. When the two breaks occur in the same chromosome, with one on each side of the centromere, other types of rearrangements are found (Figure 8-10). Obviously it is not possible to have a viable deletion of the middle segment because this would also remove the

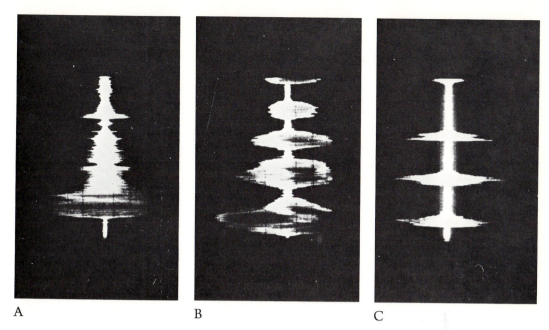

A   B   C

Figure 8-8 An electrical recording of the cry of a normal child *(A)*, a cat *(C)*, and a child afflicted with the cat's cry syndrome *(B)*, showing the strong resemblance of the sound of the cry to that of a cat. (J. Legros and C. Van Michel. Analyse de la voix dans un cas de "maladie du cri du chat," *Ann. Genet.* **11**:59–61, 1968.)

necessary centromere region. However, that segment could rotate, with the broken ends uniting to produce an inversion, in this case a *pericentric inversion* (Figure 8-10*A* and *D*). Depending on the relative distances of the breaks from the centromere, the new chromosome may have altered arm lengths, as indicated in the figure. However, the chromosome does have the same total number and kinds of genes that it had originally. Persons carrying inversions are usually normal, although they may produce abnormal offspring as we shall see later.

RINGS. Another very interesting anomaly that may arise from this type of breakage is the *ring* chromosome (Figure 8-10*B* and *C*), produced when the two acentric chromosome ends are lost and the two ends of the centric

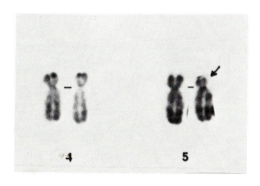

Figure 8-9 Two number 5 chromosomes (along with two number 4 chromosomes for comparison) from a karyotype of a child with *cri du chat,* showing one with a deletion in the short arm, responsible for the cat's cry syndrome. (J. Legros and C. Van Michel. Analyse de la voix dans un cas de "maladie du cri du chat," *Ann. Genet.* **11**:59–61, 1968.)

Figure 8-10 Double breaks in chromosomes resulting in pericentric inversions with unaltered *(A)* and altered arm lengths *(D)*, or in ring chromosomes *(B and C)*.

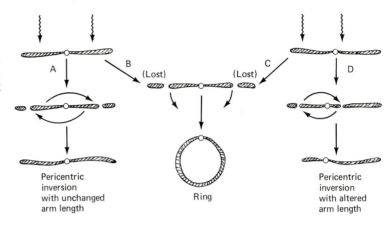

Pericentric inversion with unchanged arm length

Ring

Pericentric inversion with altered arm length

section join together. The capability of the two chromatids of a ring chromosome to separate at mitosis have always intrigued cytologists. Any simple model that we might construct simulating the replication of a ring of great size (molecularly speaking) leads to the expectation that the daughter rings will be interlocked and will be lost at anaphase. Human patients with ring chromosomes do tend to show a high degree of morphological variation of that chromosome; perhaps the more interesting question is why the loss does not always take place within a few mitotic divisions after formation.

Figure 8-11 shows two ring chromosomes separating at metaphase in a *Drosophilia,* with the rings lagging as though they had somewhat more difficulty in separating than the nonring chromosomes. This is not an atypical picture for ring chromosomes.

In the formation of a ring chromosome, some genetic material at the ends of the chromosome must be lost, and the ring-bearing person will have deficiencies of varying extent, depending on the specific ring. For this reason, the clinical features presented by ring-bearing individuals are of extremely variable sorts; these people show different degrees of mental deficiencies and other anomalies, and the life span can be quite variable from one individual to the next.

In one case a female with a ring chromosome 18 produced a daughter who also carried the ring, indicating that the ring was also able to manage to get through the meiotic, as well as the mitotic, divisions. A striking case of the effect of a ring for chromosome 1 is shown in Figure 8-12.

Figure 8-11 Two ring chromosomes in *Drosophila* separating at anaphase somewhat later than the others, suggesting some difficulty in separating from each other.

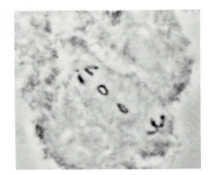

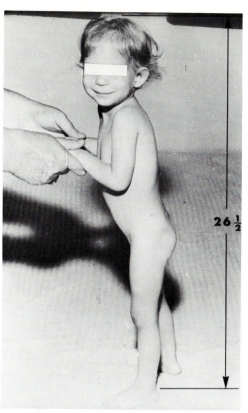

A

Figure 8-12 *A*. A young girl with a ring derived from chromosome 1. At twenty-eight months of age, her total height was slightly over 2 ft and her weight 10 lb 4½ oz. Even at age three her clothes were made from the pattern for a 26-inch doll. *B*. The ring chromosome found in the girl above. (C. B. Wolf, J. A. Peterson, G. A. LoGrippo, and L. Weiss. Ring 1 Chromosome and Dwarfism— A Possible Syndrome, *J. Pediatr.* **71**:719, 1967.)

$26\frac{1}{2}$

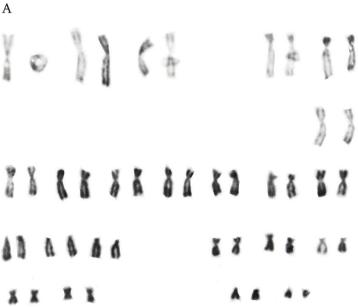

B

Cancer and Chromosomes

ORIGIN OF CANCER. The reason why normal cells occasionally revert to a primitive state of rapid mitosis, sometimes becoming malignant or cancerous, is not completely understood, although an enormous amount of information has been accumulated on the circumstances that lead to this state. There are two main theories proposed to account for this anomalous behavior: (1) that a cell becomes malignant after a virus penetrates that cell and (2) that it results from random mutational changes in the DNA of that cell.

Specific viruses have been implicated in a number of cases in both experimental animals and in humans. A tumor called the *Burkitt lymphoma* is found in areas of low altitude and high humidity in Africa. It affects mostly the young, or new migrants into areas of high risk, and its epidemic characteristics are generally those of an infectious disease. Examination of lymphocytes by electron microscopy has revealed a virus that has been named the *Epstein Barr virus* (EBV) for the discoverers. The Epstein Barr virus is also implicated in a nonmalignant infection of the lymph gland called *infectious mononucleosis*, a debilitating disease which is well known to many college students.

On the other hand, evidence for viral implication in most other cancers is not compelling, and the theory of spontaneous mutational change in the DNA may appear more attractive. While the frequency of spontaneous change can be estimated at about one per million (10^{-6}), several such changes occurring over a period of time might be necessary to convert a normal cell into a cancerous one. Such a view is consistent with the observation in other species that all chemicals that produce carcinomas (i.e., are *carcinogenic*) also produce mutations. Furthermore, this would account for the much greater incidence of most cancers in older people, who would have accumulated the required number of mutations over a period of time to convert some of their cells to cancerous types later in life.

CHARACTERISTICS OF CANCER CELLS. Regardless of how the initial event is induced, by mutation, by a virus, or by some combination of the two, a few characteristics of tumors have been well established. It is known, for instance, that tumors are derived from a single cell which then reproduces mitotically, that is, they are *monoclonal*. Thus, if Burkitt lymphoma cells from a woman who is heterozygous for two biochemically distinguishable X-linked alleles are tested, they all prove to carry only one, not both, of the two alleles. The enzyme *glucose-6-phosphate dehydrogenase* (G-6-PD) is ideal for this purpose. This strongly suggests that all the cells of the tumor came from a single cell which happened to have an active X chromosome with that particular allele. Similar results have been obtained with more than a dozen different tumors. Even the common wart, an ordinarily benign tumor known to be virus-caused, can be shown to be monoclonal in origin.

One characteristic of cancerous cells in general is their great variability in chromosome number (Figure 8-13). Mitosis in these rapidly dividing cells

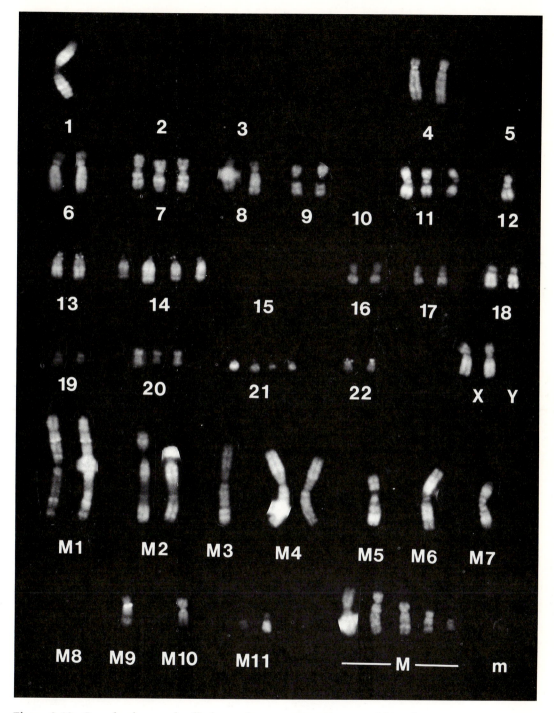

Figure 8-13 Grossly abnormal cells from a human prostate cancer. Note that some chromosomes are present in triplicate or quadruplicate, whereas others appear to be missing. The genetic material of the latter is presumably incorporated into the larger unidentifiable chromosomes, in the bottom two rows, indicated as M (Marker) chromosomes. (Courtesy of Y. Ohnuki et al., Pasadena Foundation for Medical Research.)

may be quite abnormal. Spindle formation may be defective, with three instead of two poles. The chromosomes may replicate asynchronously or move to the poles irregularly. The stage of early prophase is virtually absent in all cancers, and a higher proportion of cells than expected is found at metaphase. It is not surprising, then, that a common feature of tumor cells is an abnormal chromosome number, but this is to be regarded as a consequence of, rather than the cause of, the abnormal cell behavior.

In several cases, however, there does appear to be clear evidence that specific chromosome changes are associated with particular types of cancer.

THE PHILADELPHIA CHROMOSOME IN LEUKEMIA. The first clear association of a chromosome anomaly with a specific cancer was made in a patient in Philadelphia who had *chronic myelogenous leukemia.* It turns out that about five out of every six patients with this type of leukemia have a defective chromosome 22 (Figure 8-14). This chromosome has been named *the Philadelphia chromosome,* Ph[1]. It has been shown that part of the long arm of 22 has been translocated, in most cases, to the end of chromosome 9 (Figure 8-15), and occasionally to other chromosomes. The exchange of material between chromosomes 9 and 22 appears to involve no net change in the total amount, suggesting that all of the genetic material is still present in the cell. One explanation for this change in the behavior of the cell is that the genes on the altered chromosomes are now no longer able to function

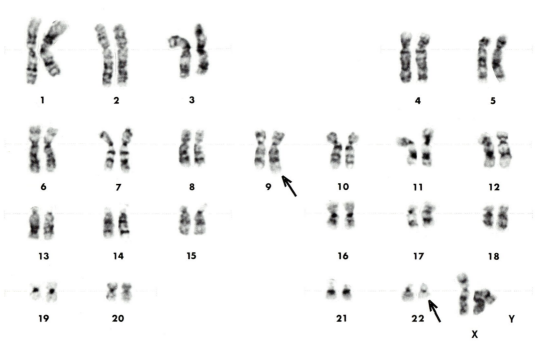

Figure 8-14 Karyotype of a patient with chronic myelogenous leukemia. The arrows point to an extra long chromosome 9 and a short chromosome 22. The latter is the *Philadelphia chromosome* which has lost a terminal segment by translocation to chromosome 9. (B. Kaiser-McCaw, The Genetics Center of the Southwest Biomedical Research Institute.)

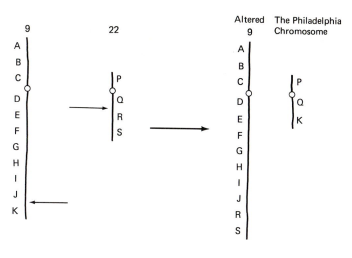

Figure 8-15 Hypothetical scheme showing the origin of the Philadelphia chromosome. The ends of the long arms of chromosomes 9 and 22 (*K and RN,* respectively) are interchanged to produce two new chromosome sequences. Chromosome 9 becomes slightly longer, and 22 is transformed into the smaller Philadelphia chromosome. The breakup of the original gene sequences and/or the new gene associations (either *J* with *R* or *K* with *Q*) may then be responsible for a position effect giving rise to the leukemia.

normally because, on the one hand, their normal genetic neighbors, with whom they ordinarily cooperate, have been removed to a distant position and, on the other, they are now closely associated with different loci that interact to produce new developmental effects. Such a changed behavior of genes, when they are moved to new locations, is called a *position effect* (Figure 8-15).

RETINOBLASTOMA.   About one child in 20,000 is afflicted with *retinoblastoma,* a malignant tumor of the eye. In many cases it can be treated by surgery. It appears about 60 percent of the time in pedigrees as an autosomal dominant; in the other 40 percent, it is sporadic. In the former cases, which appear to be genetic, it is usually bilateral; in the sporadic cases only one eye is affected.

Children with a deletion in chromosome 13 have an unusually high risk of retinoblastoma. Contrarily, some children with retinoblastoma have been shown to carry a deletion of band 14 of the long arm (q) of chromosome 13, or deletion 13q14. It has been suggested that this is the locus of the dominant allele responsible in part for familial retinoblastoma, and that the deletion, whether inherited or newly derived, acts like the dominant. In addition, however, another gene located in the retina of the eye must also mutate for retinoblastoma to develop. Even though the mutation rate at this second locus may be very low, there are millions of cells in the retina and the probability that at least one will suffer the second necessary mutation will be quite high. Thus, a person carrying the change at 13q14 (or the corresponding dominant allele) will stand a good chance of developing the tumor, even in both eyes. Any person without that change, however, would have to suffer two specific mutational events, one at 13q14 and the other at the second locus, for the tumor to develop. In this case, the tumor would occur sporadically, appear not to be genetic, and would usually affect only one eye.

WILMS TUMOR.   About one in 15,000 children get a bladder cancer called *Wilms Tumor.* The frequency of *aniridia,* the absence of an iris, is a thousand times more frequent in these children than in the general population. A deletion in the short arm of chromosome 11 is also often present. Since the tumor is sometimes curable by surgery or chemotherapy if detected early, it has been suggested that any child born with aniridia should be checked for the chromosome deletion in order to predict the likelihood that the tumor will develop.

ATAXIA TELANGIECTASIA (AT). Individuals homozygous for the recessive responsible for AT are extremely sensitive to high energy radiation; they may die if exposed to doses that are normally given to treat cancer. Although only one in 40,000 is so affected, about 1 percent of the population carry the allele in the heterozygous state. It appears that these heterozygotes may be five times more likely to die of cancer before the age of 45 than a nonheterozygote. AT patients tend to have lymphocyte clones that carry a balanced translocation in which one of the chromosomes involved is number 14.

OTHER CHROMOSOME INVOLVEMENTS. In one extensively studied pedigree, members with a certain translocation involving chromosomes number 3 and 8 have a seven in eight chance of developing cancer of the kidney before 60 years of age, compared to persons taken at random who have a chance of only one in a thousand of developing a similar cancer by that age. As the study of this family progressed, it developed that three persons had kidney cancer in its early stage, without their knowing it. When their karyotypes showed that they had the translocation, further investigation revealed their condition.

Abnormalities in the long arm of chromosome 14 have also been linked to Burkitt lymphoma. However, the chromosome most commonly found to be abnormal in blood diseases like leukemia is number 8, with 7 and 9 involved to a lesser degree.

In twelve out of thirteen cases of breast cancer investigated by one team, chromosome 1 was found to be translocated to one of five other chromosomes. The general applicability of this observation is not yet known.

OTHER GENETIC PREDISPOSITIONS TO CANCER. In a rare growth disorder characterized by dwarfism, *Bloom syndrome,* the affected individuals who are homozygous for a recessive show a high frequency of spontaneous chromosome breaks in their cells. Such persons have a high risk of developing cancer. *Fanconi anemia,* another hereditary disease involving defects not only of the blood but also of the skin pigmentation and malformations of the heart and other organs, shows chromosome instability and an increased probability of leukemia.

Xeroderma pigmentosum (Figure 13-5) is caused in most cases by a defect in the enzyme systems which normally repair lesions in DNA. In the early life of homozygotes, heavy freckling appears in those parts of the skin exposed to ultra-violet light and these develop into severe lesions, which eventually become cancerous and lead to death. Heterozygotes are also heavily freckled, although not so severely as the homozygotes. They have a normal life expectancy. The DNA lesion responsible for this disease is discussed in Chapter 13.

HYDATIDIFORM MOLE. A grossly abnormal pregnancy, without an embryo or placenta, may produce a tumor called a *hydatidiform mole.* These are usually benign but can develop into a malignancy. Studies of the chromosomes of twenty such moles showed that all had two X chromosomes. Some of the chromosomes of the couples involved showed heteromorphisms; these made it possible to show that the chromosomes of the mole were all paternally derived. One reasonable explanation for the origin of this type of teratoma is that a cell (presumably an egg cell) was penetrated by a sperm, the maternal chromosomes were excluded, and the paternal chromosomes underwent replication twice before commencing cell division to form the tumor. Thus the haploid sperm produced diploid cells. These purely paternal cells must all be XX, since a Y-bearing sperm going through the same would lack all X chromosome genes and could not produce a viable cell.

THE BENIGN OVARIAN CYST. When *benign ovarian cysts* are examined cytologically, they prove to be diploid with chromosomes only from the mother, indicating that they do not originate from abnormal embryos. However, when the mother's chromosomes carry heteromorphisms, staining differences at their centromere regions, the cells carry only one of the two types, that one type being present in duplicate. The simplest explanation for this is that the chromosomes from two of the products of the second meiotic division have combined to produce a diploid cell, and that cell has multiplied to produce the cyst. How do we know that a single haploid nucleus has not replicated itself, thereby achieving the diploid state prior to mitosis, as appears to be the case for the hydatidiform mole? When the woman is heterozygous for a pair of alleles, each of which yields products that are biochemically identifiable, the cyst will usually carry either one or the other in the homozygous state. This would be consistent with the doubling of a haploid nucleus. But occasionally the cyst will be heterozygous, carrying both alleles. The explanation for this is that a chiasma occurring between the centromere and the alleles has interchanged the association of the alleles and the centromeres (see Figure 3-1). Such an explanation makes sense only if the origin of the cyst is by fusion of two second division products of meiosis, presumably of the egg and one of the polar bodies.

CANCER IN GENERAL. Except for the few cases described above, simple inheritance of genes leading to a predisposition to cancer is very rare. Occasionally pedigrees do appear in which the same type has developed repeatedly in various family members. It is known that women who have close relatives with breast cancer will themselves stand a greater chance than average of developing a similar cancer. Nevertheless, as we shall see in Chapter 17, studies of twins indicate that the development of malignancies may be genetic in the sense that the genetic material of the affected cell is altered but is generally not transmitted from one generation to the next.

Transmission of Abnormal Chromosome Types

What kinds of offspring will a woman with Down syndrome produce? When a person has an abnormal chromosome, will it be passed on to the progeny? If a couple produces one child with an abnormal chromosome, what are the chances that the next child will be similarly affected? Is it possible for a normal couple to produce chromosomally abnormal children repeatedly?

The science of human cytogenetics has advanced to the stage at which most of these questions can be answered with reasonable assurance, although often the answer given in a genetic counseling session, for instance, would be phrased in terms of *empiric risks*—that is, the estimates of probability based on population studies of similar cases, rather than on theoretical probabilities from Mendelian principles.

So many different types of chromosome abnormalities have been described in humans—well into the thousands—that it would be useless to try to describe any appreciable fraction of them here. However, they do fall into the characteristic types already described, about which some general statements can be made. The transmission of these types has been studied in other organisms—corn and the fruit fly primarily, but in many other plants and animals as well. As a rule, human chromosome rearrangements behave in much the same way as similar anomalies in other species. This is the case because meiosis is much the same in all organisms. In fact, it would be difficult to find a meiotic phenomenon in humans that does not have a parallel in other organisms.

TRANSMISSION IN SIMPLE ABNORMALITIES.   If a person carries an altered chromosome—one with a deletion, with an inversion or with some distinctive staining property on one of the two homologues—it will be found in half of that person's gametes and, on the average, in half the progeny. Such distinctive chromosomes can be very valuable in determining the path of a chromosome as it comes down through a pedigree. Because they are "marked" by their differences from their homologues, such variants are also called *marker chromosomes.* Usually these differences are found in the amount or distribution of the constitutive heterochromatin found adjacent to the centromere.

Nondisjunction and Trisomy-21

An excellent example of the use of such heteromorphisms is given by the analysis of the chromosomes found in Down karyotypes, along with those of their parents, which enables us to specify the parent, and the meiotic division, in which the nondisjunctional event occurred.

A convincing demonstration that this nondisjunctional event usually occurs in the female comes from examination of the karyotypes in the mothers, the fathers, and their trisomy-21 offspring. As an illustration of the use of heteromorphic chromosomes in determining the origin of the extra chromosome in trisomy-21, we can take the uncommon case where all four chromosomes in the two parents are different from each other (Figure 8-16). Of the three chromosomes present in the trisomic offspring, each can be attributed to one or the other of the two parents; thus, if two of the three came from the mother, then we know that the nondisjunctional event occurred in her. Furthermore nondisjunction at the first division results in one

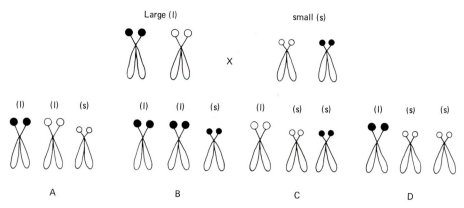

Figure 8-16 When two parents have structurally different chromosomes numbered 21, then it is possible to determine whether the nondisjunctional event responsible occurred in the male *(C and D)* or the female *(A and B)* parent as well as whether that event took place at the first meiotic division *(A and C)* or at the second division *(B and D)*.

of each of the two homologues going into the egg; therefore we recognize cases *A* and *C* as resulting from first meiotic nondisjunction. On the other hand, structurally identical chromatids separate at the second meiotic division, so that when we see that the two chromatids coming from one of the parents are the same, we conclude that the nondisjunction occurred at the second meiotic division (cases *B* and *D*).

In actual investigations where heteromorphisms in the parents of Down children made it possible to specify the origin of the extra division, the frequencies of the various possibilities are given in Figure 8-16. From this, it can be seen that the previous conclusion that trisomy-21 results mostly from nondisjunction in the female is supported. There is, however, a surprising amount of nondisjunction in the male, accounting for a bit more than 20 percent of all trisomy-21 births.

The conclusion that trisomy-21 is caused usually by nondisjunction at the first meiotic division in the female makes it possible to narrow down the possible causes of this syndrome. It has been suggested, for instance, that trisomy-21 might be the result of a larger average time between ovulation and fertilization of the egg in older as compared with younger women. However, we know that the egg is in the dictyotene stage up to the time of ovulation; then at ovulation it completes prophase and stops at metaphase of the *second* meiotic division until the egg is fertilized. Therefore a prolongation of the period between ovulation and fertilization can have no effect on first meiotic nondisjunction. The common and adequate explanation is that the oocytes, which have been stalled at the dictyotene stage since about the time of birth, deteriorate with age. Some forty years later the chromosomes may not go through meiosis properly; instead, they have a tendency to nondisjoin.

ORIGIN OF TRIPLOID EMBRYOS. A surprisingly high frequency (almost 20 percent) of chromosomally abnormal abortuses have three sets of chromosomes. Heteromorphic chromosomes in the parents make it possible to figure out which of the parents contributed two haploid sets to the triploid embryo and which contributed one. In two thirds of the cases analyzed, it appears that two different sperm fertilized the same egg. In almost a quarter of the cases, the egg was apparently fertilized by a diploid sperm. In the remaining 10 percent, the egg was diploid, presumably because of the failure of the first meiotic division. These determinations can be made on the basis of heteromorphisms present in the parents; the detailed argument for each case, which will not be presented here, is left to the ingenuity of the enterprising student. It may be helpful to note that the X and Y chromosomes constitute a regularly occurring heteromorphic pair in the male sex.

Transmission in Trisomies

DOWN SYNDROME. In rare instances females with the Down syndrome have produced children. In the slightly more than a dozen known cases in which this has occurred, slightly more than half of the progeny were normal and the rest were affected as the mother was affected. The expectation based on a simple model in which the gamete has an equal chance of carrying the extra chromosome, or not, would be equal numbers of each. In this case the ideal expectations may be modified; both the Down and normal children tend to have additional developmental defects. These appear because such offspring are often the result of an incestuous relationship, which increases greatly the chance that hidden recessives in the family will become homozygous.

There is no recorded case of progeny from Down syndrome males. Although it is known that they can produce motile sperm, the number is so small that it may be responsible for effective sterility. In addition, trisomy-21 males have reduced sex drive.

Individuals with Down syndrome have germ cells with three chromosomes 21. We have earlier seen how, when only the normal two chromosomes are present, those two homologues pair with each other, locus by locus, along their entire length, during the first meiotic division. What happens when there are three homologues, instead of two? Our best information comes from cytological studies of meiosis in trisomies of experimental organisms like wheat or corn. A helpful analogy is that of the zipper. Once pairing is initiated at certain homologous loci, the adjacent loci proceed to pair until both homologues are locked together. A third homologue could not pair with either one, any more than a third zipper strand could fasten itself to the two already joined. However, if two homologues are paired in one region but not in another, a third homologue may pair with one of them at the unpaired region to form a combination of three chromosomes (Figure 8-17). Two paired chromosomes and a third unpaired chromosome consist of a *bivalent* and a *univalent*; three chromosomes paired two at a time at different regions form a *trivalent*. A few isolated observations of first-division spermatocytes of Down syndrome males show both kinds of pairing; bivalents along with univalents in some, trivalents in others.

Because there are three homologues present but only two directions in

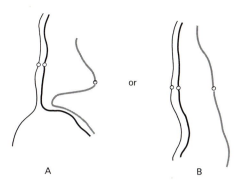

Figure 8-17 The two different kinds of behavior of three homologues at the first meiotic division. *A.* In some cases, two of the three chromosomes pair for some distance, and then one of the two pairs for the rest of the length with the third chromosome, to form a trivalent. *B.* In other cases, two of the homologues pair completely in a bivalent, and the third homologue remains unpaired as a univalent. In both cases, half of the gametes get one of the homologues and the other half get two.

which they can go at first anaphase, the separation will be of two of the homologues to one pole and one to the other. In Chapter 3 we discussed the production of gametes with two homologues by accident of meiosis from normal germ cells; this event is called *primary nondisjunction.* When gametes with two homologues are formed from germ cells with three, the event is called *secondary nondisjunction.* In such a case, the expectation of persons with trisomy-21 is that half of their progeny will be normal and half will have trisomy-21.

XXX FEMALES. Triplo-X females would be expected to produce 50 percent eggs with two Xs and 50 percent with one, an example of secondary nondisjunction. One might expect, then, that half of the eggs would be normal and the other half would have an extra X. The XX eggs lead to either XXX or XXY progeny, depending on whether the XX egg was fertilized by an X- or a Y-bearing sperm. Curiously, this does not appear to be the case; the children such women have are usually normal. It is possible that the women tend not to produce diplo-X eggs because the XXX germ cells (oogonia) do not readily go through the early germinal divisions. Perhaps a few of these oogonia suffer occasional mitotic nondisjunction or the anaphase loss of the extra X chromosome and produce XX oogonial clones that then become eggs. Even if her oogonia fail to become eggs, an XXX woman could produce enough normal XX oocytes by occasional, rare nondisjunction to last her entire reproductive life, as the female infant is born with well over a million reproductive cells.

Still another possibility is that the three chromosomes form a trivalent at the prophase of meiosis, and that at the anaphase of the first meiotic division the trivalent orients itself in such a way that two of the Xs are lost in the polar bodies and one ends up in the ovum. On the other hand, if a bivalent and univalent are regularly formed, the univalent might be lost by lagging at anaphase, so that each meiotic product would get one chromosome from the separation of the bivalent. Possibly some of these fertile XXX females are in fact mosaics, with some XX oogonia present to provide the functional eggs.

XYY MALES. It might be expected that the combination of the three homologues, (one X chromosome and two Ys) during meiosis would give rise to abnormal progeny. The outcome, of course, would depend on the way the chromosomes pair at the prophase and separate at the anaphase of the first division.

Analysis of spermatogenesis in XYY males gives conflicting and inconsistent results. Although a few cases are known where such males have produced XXY or XYY sons, presumably coming from their XY or YY sperm, their progeny are almost always normal XX or XY. Furthermore the cytology of some XYY men shows meiotic configurations involving a single X and a single Y rather than an X and two Ys. The best assumption consistent with most of the data is that there is strong selection against XYY spermatogonial cells (paralleling that presumed to operate against XXX oogonia) and that the progeny produced by XYY males come from the occasional XY cells that appear in the XYY germ line. This assumption would be consistent with the observation that there is wide variability in the fertility of these males, which ranges from complete sterility to apparently normal fertility.

Deletion Chromosomes and Sperm Activity

A chromosome that has had a section deleted lacks loci, the number depending on the size of the deletion. The normal homologue present in a heterozygote may allow deletion-bearing individuals to survive, often with little phenotypic effect, if the deletion is small. At first sight it would appear that the expected progeny from such a heterozygote would consist of half with the normal chromosome and half with the deletion chromosome.

However, another question arises here. Does a woman heterozygous for a deletion produce half of her gametes with that deletion, which are then at a disadvantage because some genes normally present in the gamete are completely missing? This is apparently not the case. As far as the egg is concerned, the diploid condition persists until the egg is ovulated, at which time meiosis (which had been stalled at the dictyotene stage since the fetal period) starts again. The haploid stage of the egg lasts for only a short time, for if fertilization is to take place, it will happen within a few days. In any case, all the synthetic activity necessary for the first few cleavage divisions has already been completed in the egg, and the occurrence of a transient haploid phase is of minor importance. On fertilization any genetic deficiency in the egg may be compensated for by normal genes for that region in the sperm. Thus there is no selection against a deficiency-bearing egg from a heterozygous female, and such females will produce progeny half of whom are similarly affected. However, if the mother herself has two normal chromosomes, with a deletion occurring anew in her germ line, subsequent heterozygous zygotes with that new deficiency may or may not develop to produce viable progeny, depending on the developmental effect of the new deficiency.

On the other hand, the sperm spend an appreciable period of time (at least several days) as independent functioning haploid cells, highly motile and therefore with unusual energy requirements. Do their survival and functioning depend on a normal haploid complement? We cannot answer this question unequivocally for human sperm. However, it seems likely that a human sperm is able to function normally, independent of the alleles it does (or does not) carry in the haploid state. We know, in the first place, that half of the sperm lack the X chromosome, having a Y instead, and these sperm must be as functional as those with an X because the ratio of males to females at birth is about 1:1. There is no mutant, no deletion, or no rearrangement in human chromosomes that has been proved to suffer from low sperm transmissibility because of a deletion in the haploid complement. In fact, in *Drosophila* it is possible to produce gametes that lack all of the major chromosomes (i.e., that have virtually no genes at all). These sperm are viable and can fertilize eggs.

The weight of the argument, from both logic and observation, is in favor of the view that the ability of a sperm to function normally in fertilization is determined during its earlier existence as a diploid cell, and that this ability is independent of the sperm's genetic content. One argument for the absence of gene activity in sperm is that the chromosomes are compressed into the tiny sperm head without the elaborate cellular machinery for converting raw material into gene products. The function of the sperm is merely to carry the haploid set of chromosomes from the male to the egg.

Transmission of Chromosome Rearrangements

There is an almost unlimited number of different ways in which the forty-six chromosomes of the human complement can be "broken" and the segments reattached. The specific instances described in humans are now in the thousands and the number continues to grow.

When rearrangements are formed, they appear as isolated cases and slowly disappear from the population. However, while they persist, they may cause relatively little damage to most of their carriers, but they are capable, as a result of meiosis, of causing grossly aneuploid progeny. Furthermore some (though not all) of the normal children carrying the rearrangement may also produce affected children. In the case of trisomy-21 resulting from chromosome translocation, two thirds represent new rearrangements and one third have received abnormal chromosome complements from a phenotypically normal parent who was heterozygons for a translocation.

As we have seen earlier, a translocation is a chromosome rearrangement in which two nonhomologous chromosomes are broken. This rearrangement may happen in somatic cells. Unless such a new arrangement of chromosomal segments has some rare and unusual property, such as the ability to transform a normal cell into a cancerous one, it will have no effect on the individual carrying it and will disappear from the population upon the death of the person carrying it.

Our attention here is focused on those chromosome changes that occur in the germ line and are transmitted from one generation to the next.

Translocations. Let us suppose that two breaks in different chromosomes of a germ cell have given rise to a translocation. When the descendants of that cell, heterozygous for that new translocation, reach the prophase of the first meiotic division, homologous chromosome regions will attempt to pair. In order to visualize the configurations produced by the pairing of homologues in these complex configurations, it is customary to number or letter each chromosome so that homologous loci have the same numbers. Then by matching the corresponding numbers, we can see what sort of configuration would be achieved by the chromosomes of a new rearrangement as they pair with their normal homologues. It should be noted also that a more accurate representation would have all chromosomes made up of two chromatids, as they are at prophase of the first meiotic division. However, when nothing is gained conceptually by this more elegant (and accurate) representation, single lines are used for the sake of simplicity.

The steps in the formation of a new translocation, and its pairing configuration, are shown in Figure 8-18. Whether they succeed in pairing completely may depend on the relative lengths of the homologous segments, with the longest uninterrupted sections having the greatest chance of getting together. Whether such a paired configuration, once achieved, will remain that way after the chromosomes have shortened or condensed during the prophase of meiosis may depend on whether or not a chiasma, of the sort that regularly occurs between all homologues, occurs between these new pairing segments. If the segment is long, as it is in the case illustrated, the chance of a chiasma is great; if it is small, the chance is considerably reduced. In any event, in an ideal case of the sort we have illustrated, an exchange might occur in all four arms, with the result that these chromosomes would be associated up to the time of metaphase.

The simplest rule governing the manner of segregation at the first division is that segments that have been involved in a chiasma, or exchange, usually separate from each other at the first meiotic division. It is obvious (Figure 8-19) that if there are chiasmata in all four arms of the configuration and the chromosome segments paired in those arms separate from each other, then diagonally opposite centromeres will go to the same poles. This kind of

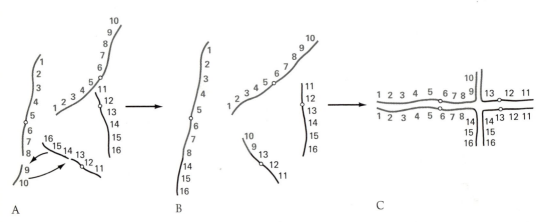

A   B   C

Figure 8-18 Two pairs of homologues, with homologous segments numbered identically, are located at random within the nucleus of a cell. Two breaks occur, at the positions marked with arrows *(A)*, and the broken segments reattach to produce a reciprocal translocation *(B)*. When homologous regions attempt to pair, the two normal chromosomes pair with the translocated ones in a cross-shaped configuration *(C)*.

A. Alternate Segregation

B. Adjacent Segregation (1)

C. Adjacent Segregation (2)

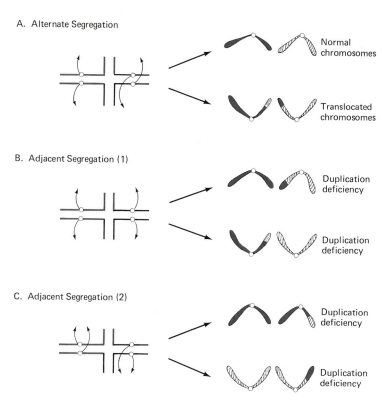

Normal chromosomes

Translocated chromosomes

Duplication deficiency

Duplication deficiency

Duplication deficiency

Duplication deficiency

Figure 8-19 The different types of segregation from a translocation heterozygote. *A.* Alternate members of the configuration go to the same pole, resulting in two gamete types, one with the two normal chromosomes and the other with the two translocated chromosomes. *B.* Adjacent segregation, of type 1, where homologous centromere regions go to opposite poles, producing duplication-deficiency gametes. *C.* Adjacent segregation, of type 2, in which homologous centromere regions go to the same poles, resulting in duplication-deficiency gametes.

segregation is referred to as *alternate segregation.* From such a configuration, two kinds of products may be formed, one with the two normal chromosomes and the second with the two translocated chromosomes. The first will give rise to normal progeny with normal chromosomes, the second to normal progeny heterozygous for the translocation.

It is easily seen, however, that this regular alternate result might not be achieved. Depending on the position of the breaks and on the configurations formed as a result, the translocation heterozygote may not be paired in all arms, or chiasmata may not hold all four arms together, with the result that segregation would not be of alternate members of the configuration but of two adjacent ones (Figure 8-19*B* and *C*) going to the same pole.

The gametes resulting from adjacent segregation are unbalanced; that is, they are deficient for some loci and duplicated for others. If the amounts of genetic material involved in the duplications and deficiencies are large, the embryo formed from a gamete of this sort, along with a normal gamete from the other sex, is aborted at a very early stage in embryonic or fetal life. If the duplications and deficiencies are not extensive, an unbalanced fetus may be born. When translocations are detected in humans, it is usually after a still-birth or the birth of an abnormal offspring who proves to have an abnormal chromosome makeup that came from the products of adjacent segregation in a normal heterozygous parent. Furthermore examination of the normal progeny in the sibship usually shows that about half of them have normal chromosomes whereas half carry the same balanced translocation as one of the parents (Figure 8-20).

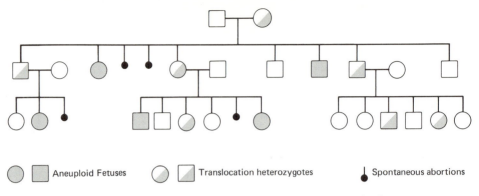

Figure caption legend: Aneuploid Fetuses — Translocation heterozygotes — Spontaneous abortions

Figure 8-20 A pedigree of a family with a 2-21 translocation. The heterozygotes are phenotypically normal but tend to have a high spontaneous abortion rate because of the grossly abnormal gametes they produce. Half of the phenotypically normal children carry the heterozygous translocation.

Generally speaking, more than half of the gametes from a translocation heterozygote of the sort we have described here have either normal chromosomes or a balanced translocation. These zygotes should develop normally. The less frequent embryos with grossly unbalanced chromosome complements from adjacent segregation are aborted early or, less frequently, come to term as quite abnormal children. When there is embryonic or fetal loss that limits the reproductivity of parents, one of whom carries a translocation, we would expect the translocations to be eliminated from that kinship slowly over several generations. Perhaps most of those that are now being observed have been in existence for only a few generations. Curiously it can be argued that translocations with more drastic aneuploid products causing early embryonic loss may, as a group, persist in the population for a longer period of time than those with lesser effects, because the early (often unnoticed) embryonic losses are more likely to be followed by additional children, half of whom would carry the translocation, whereas births of defective children tend to limit family size.

COMPOUND CHROMOSOMES. In 1916, W. R. B. Robertson described some chromosome rearrangements in grasshoppers in which he noted that two rod-shaped chromosomes (acrocentrics), appeared to be attached to each other at the centromeres to form V-shaped, or metacentric, chromosomes. He called these "compound chromosomes," and this name can be applied to translocations that result from breaks near the centromere regions of the two chromsomes involved so that entire chromosome arms are attached to each other. However, such whole-arm translocations are referred to by human cytologists as *centric fusions, fusion chromosomes,* or, more commonly, as *Robertsonian translocations,* and when the chromosome arms so attached appear to be either two long or two short arms of the same chromosome number, they are called *isochromosomes* (iso = "same").

TRANSLOCATIONS OF 14 AND 21. One rearrangement found in a large number of pedigrees involves the attachment of the long arms of chromosomes

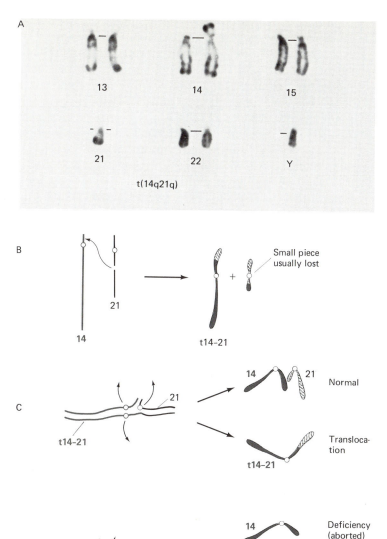

Figure 8-21 *A.* Chromosomes taken from the karyotype of a person heterozygous for a 14–21 translocation. The other unaffected chromosomes of the D and G groups, and the Y, are included to verify the point that the translocation does in fact involve 14 and 21. Note that the long arms of both chromosomes are attached and that the very small product made up of the two short arms of each chromosome is missing. (Courtesy of B. Dutrillaux, University of Paris). *B.* The formation of a compound chromosome, a Robertsonian translocation, with most of the material of chromosomes 14 and 21 found on the large piece. The smaller segment may be lost without having any phenotypic effect on the carrier. *C.* Alternate segregation from the heterozygote, yielding one half of the gametes with normal chromosomes, the other half with the translocation. *D.* Adjacent segregation giving rise to deficiency and duplication gametes. Although the zygotes with the deficiency for chromosome 21 are lethal, the complementary class with the duplication will produce a trisomy-21 zygote.

14 and 21 (Figure 8-21*A*). When such compounds are formed (Figure 8-21*B*), there is undoubtedly initially a small fragment that can be lost without seriously affecting development. When such a compound in a normal cell pairs with a normal 14 and 21, the expectations are relatively simple. In this sort of configuration, the homologues will pair at prophase and separate from each other at anaphase to give a high proportion of balanced alternate gametes producing normal progeny (Figure 8-21*C*) and a very small proportion of unbalanced gametes (Figure 8-21*D*).

Of all cases of trisomy-21, about 95 percent arise by primary nondisjunction and are sporadic in nature, being found primarily among children born to older women. Another 3 percent are translocation heterozygotes. Of these almost three quarters result from a newly arisen translocation, that is, one not present in either parent, and the remainder result from adjacent segregation in a translocation heterozygote in one of the parents. The latter are clustered in kinships (i.e., are familial) and occur independently of the age of the mother. Although the trisomy as such is not inherited, the translocation responsible is, and the risks are very much greater for the birth of other children with Down syndrome to the same mother or to her relatives, both male and female, heterozygous for the translocation. The risk that a heterozygous female will produce a Down child is 10 percent; the risk for a male is 2 percent. The differences in risk for the two heterozygous parents probably reflect the different frequencies of adjacent segregation in meiosis of the two sexes (Figure 8-18D). Note that other kinds of adjacent segregation will also produce inviable types (sometimes three of the elements go to one pole and one to the other), so that a pedigree involving such a translocation will show more fetal loss than usual.

One very unusual rearrangement is the rare 21–21 isochromosome. In this case, the two homologues are attached, and the person carrying them is phenotypically normal but can produce only two kinds of gametes: those with the 21–21 combination and those with no chromosome 21 at all. Those zygotes having only one chromosome 21 (coming from the normal parent) will die *in utero*. All of the viable offspring of such a person will receive the 21–21 isochromosome, which, along with the single 21 from the other parent, will result in Down syndrome. Here the risk of an affected offspring is 100 percent!

INVERSIONS. The essential difference between the two kinds of inversions, paracentric and pericentric, is shown in Figure 8-22. They are given different names because they behave quite differently at meiosis after an exchange has occurred within the inverted region.

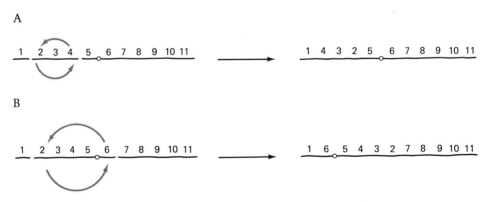

Figure 8-22 The events that lead to inversion formation. *A.* The paracentric inversion produced when both breaks occur in the same chromosome arm. The relative lengths of the two arms remain unchanged, so that the inversion is not cytologically conspicuous. *B.* The pericentric inversion, produced when one break occurs in each of the arms. The centromere region may be shifted in position, giving a new configuration with changes in the relative lengths of the two arms.

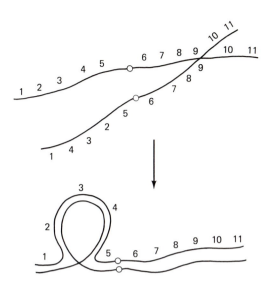

Figure 8-23 Two homologues, different in that one carries a paracentric inversion. In order to synapse completely, the homologues have to throw themselves into an "inversion loop."

The difficulty that two homologues have in pairing when one has an inversion is quite different from the problem with a translocation heterozygote. In this case one chromosome has to throw itself into a loop to pair with the homologous loci (Figure 8-23). The chromosomes often do not succeed in doing this. In these cases the frequency of chiasmata (and therefore crossing over) may be reduced. In fact, early experimental work with such rearrangements in *Drosophila* described their characteristics as "crossover suppressors" rather than as specific chromosome rearrangements.

Crossing over within the inversion, however, has very interesting consequences that can be illustrated best by figures showing the four-strand configuration of the synapsed homologues during the first meiotic division. For the case of the paracentric inversion, shown in Figure 8-24A, a single exchange within the inversion causes the formation of a dicentric as well as an acentric chromatid. Both of these are lost immediately because they cannot go through cell division. The two normal chromatids, however, can.

Crossing over within a pericentric inversion produces two kinds of chromatids (Figure 8-24B), both of which are monocentric but differ by carrying duplications and deficiencies. At the first anaphase division, these new chromosome types go to opposite poles, and at the second division they can get into the gametes just as well as the two normal chromatids can. Hence we may expect, from exchange in pericentric inversion heterozygotes, the production of abnormal offspring. When abnormal chromosome types are manufactured anew by crossing over within a balanced heterozygote, as in the preceding case, the process is referred to as *aneusomy by recombination*.

An example of this phenomenon is given in Figure 8-25, in which a pericentric inversion (A) present in a normal female has, by crossing over with the normal homologue, produced an abnormal chromosome in a son (B). The pericentric inversion in Figure 8-25A is obvious because of the shifted centromere positon, and the crossover product in Figure 8-25B has an abnormally long arm. The child with the duplication-deficiency product is shown in Figure 8-26.

Figure 8-24 Crossing over within a heterozygous inversion. The pairing configuration at prophase of meiosis, showing the completely different consequences depending on the type of inversion. In each case, the chromosome is represented as having replicated, so that the two homologues consist of four strands. At first anaphase, the two chromatids making up each chromosome are attached at the centromere regions and so go to the same pole. Separation of the centromere region occurs at the second meiotic division. *A.* When crossing over occurs within the inverted region of a paracentric inversion, a dicentric is formed as well as an acentric fragment. The dicentric and acentric are lost immediately, so that no abnormal progeny will be produced. *B.* Crossing over in a pericentric inversion loop gives duplication-deficiency products. At mitosis, they can replicate and move to opposite poles, as normal chromosomes do, but because of their abnormal gene content, they may be responsible for defective embryos.

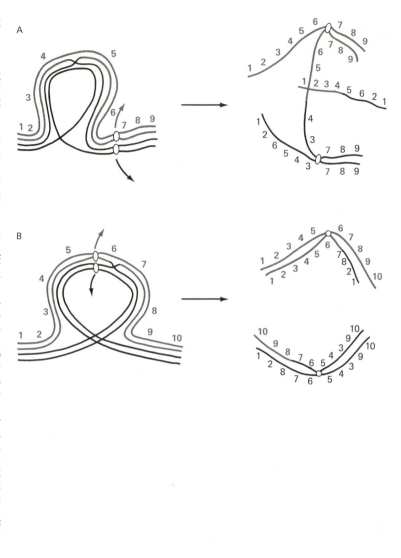

MOSAICISM. When nondisjunction, a new chromosome rearrangement, or any genetic change appears during the lifetime of an individual, that individual becomes a *mosaic*. Considering the tremendous number of cells in the human body, some must be genetically different, and from that point of view all humans are mosaics, at least to a small degree.

When the new genetic cell type appears early, as, for instance, during the first few cleavage divisions of the zygote, a large fraction of the cells may be involved and the embryo may be affected developmentally. Mosaics have been described for every common syndrome. A few individuals with some of the characteristics of Down syndrome who, at the same time, appear at the higher end of the intelligence scale have proved to be mosaic. Some of their cells are trisomy-21 and others are normal. Deviations from the normal

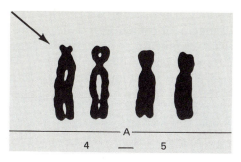
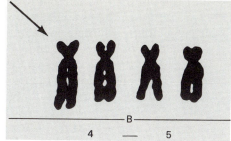

Figure 8-25 *A.* The B-group chromosomes of a phenotypically normal mother heterozygous for a pericentric inversion. *B.* A similar set of chromosomes of her son, showing the abnormal chromosome resulting from crossing over between the normal and inverted chromosomes of the mother. (M. Wilson et al. Inherited Pericentric Inversion of Chromosome No. 4, *Am. J. Hum. Genet.* **22:**679–90, 1970. © 1970 by American Society of Human Genetics. Published by The University of Chicago Press.)

chromosome number is such a common and persistent phenomenon that human cytologists usually karyotype a number of cells before coming to any conclusion about a person's chromosome constitution.

For a genetic alteration to be transmitted, it must occur in the germ

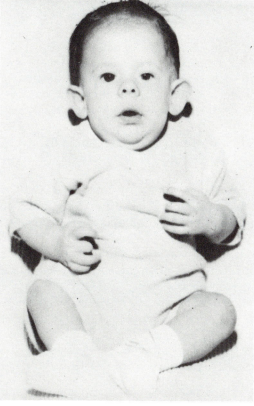

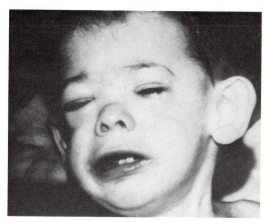

Figure 8-26 The child with the crossover chromosome illustrated in Figure 8-25B, shown at three months *(A)* and in a deteriorated condition at six months *(B)*. (M. Wilson et al. Inherited Pericentric Inversion of Chromosome No. 4, *Am. J. Hum. Genet.* **22:**679–90, 1970. © 1970 by American Society of Human Genetics. Published by The University of Chicago Press.)

A   B

cells. It is not uncommon to find persons all of whose cells carry a rearrangement, as though they inherited the rearrangement from one of their parents, although none of the cells available for karyotyping from these parents show any evidence of the genetic change. On rare occasions several sibs carry the same change, proving that one of the two parents must have been a gonadal mosaic. Any prediction about the possible consitution of the offspring of such mosaics, becomes very risky.

ORIGIN OF CHROMOSOMAL ABERRATIONS. Many of the aberrations observed in the human populations studied seem to involve breaks that are randomly located. These breaks cannot usually be attributed to any specific cause. There is, however, a category of regularly recurring rearrangements that deserves special mention. Compounds involving chromosome 21 with members of the D group are found with a very high frequency, and it appears that chromosome 14 is preferentially involved. One explanation is that chromosomes 21 and 14 are more similar to each other than to other members of the complement, particularly in that region adjacent to the centromere. During the course of meiosis or mitosis, these partly homologous regions may engage in a kind of pairing activity. Very often slides of human cells show a nonrandom association of the satellite regions of these chromosomes. Such a physical association of the chromosomes would greatly enhance the frequency with which the initial event giving rise to translocation could take place, or the actual event might be an "illegitimate" meiotic crossover having something of the nature of a crossover occurring in the mitotic cells (a somatic crossover).

RECURRENCE RISKS. Once a translocation heterozygote has been identified, a question invariably arises about the recurrence risk. Because each translocation is usually quite different from the next, with breakpoints at random along the length of the chromosomes, it is difficult to make specific predictions. Because those elements of the translocation heterozygote which have shorter pairing segments are more likely to nondisjoin (having fewer or no chiasmata), a simple meiotic diagram of the translocation could be helpful in predicting the relative likelihood of the different disjunctional possibilities.

Furthermore the extent of the unbalance deduced from such a diagram may indicate, roughly, the probable viability of the offspring to be produced. To be conservative, the general rule might be adopted that the translocation heterozygote will produce up to 50 percent unbalanced gametes, with the possible saving feature (from the standpoint of the trauma to the parents) that if the duplications and deficiencies are sufficiently great, the embryo will probably be lost spontaneously very early.

On the other hand, for those translocations involving whole chromosome arms, the Robertsonian translocations, the large homologous segments synapse, and exchange occurs within these segments which tends to ensure their segregation to the opposite poles at the first meiotic division. One might expect to find fewer nondisjunctional types from whole arm translocations than from other types with breaks randomly occurring within the arms. It must be kept in mind in this connection also that because male and

	Translocation Carrier	
	Father	Mother
Type of Offspring	Risk (percentage)	Risk (percentage)
Down syndrome	2	10
Translocation carrier	49	45
Normal chromosomes	49	45

Table 8-2 Empiric risks in D-21 translocations: probability of an offspring of one of three different types when the father, or the mother, is a translocation heterozygote. (F. Hecht, Autosome Chromosome Abnormalities, in *Metabolic, Endocrine and Genetic Disorders of Children,* Harper & Row, New York, 1974; M. Mikkelsen, *Humangenetik* **12:**1, 1971.)

female meiosis are grossly different, with the prolonged dictyotene stage in the female, there is no reason that the frequency of the different segregating types should be the same in both sexes.

As an illustration of the above, Table 8-2 gives the observed recurrence risks for the production of offspring with normal chromosomes, translocations, and Down syndrome from parents having the D-21 translocation. The frequency of observed individuals with Down syndrome may be somewhat depressed because of their greater inviability prior to birth than normal individuals. However, such tables, based on observation rather than theory, serve as the most realistic basis on which to predict the results at birth of translocation segregation.

Additional Reading

DIASIO, R. B., and R. H. GLASS. 1970. The Y chromosome in sperm of an XYY male. *Lancet,* **19:**1318.

EVANS, E. P., et al. 1970. XY spermatocytes in an XYY male. *Lancet,* **4:**719–720.

FELLOUS, M., and J. DAUSSET. 1970. Probable haploid expression of HL-A antigens on human spermatozoan. *Nature,* **225:**191–193.

DE GROUCHY, J., and C. TURLEAU. 1977. *The Clinical Atlas of Human Chromosomes.* New York: Wiley.

GERMAN, JAMES, ed. 1974. *Chromosomes and Cancer.* New York, London Sydney and Toronto: Wiley.

HALIM, K., and H. PESTENSTEIN. 1975. HLA-D on sperm is haploid, enabling use of sperm for HLA-D typing. *Lancet,* **2,** pt 2:1255–1256.

HECHT, F., et al. 1968. Nonrandomness of translocations in man: Preferential entry of chromosomes into 13–15/21 translocations. *Science,* **161:**371–372.

SCHIMKE, R. NEIL. 1978. *Genetics and Cancer in Man.* Edinburgh, London and New York: Churchill Livingstone.

SUMNER, A. T., J. A. ROBINSON, and H. J. EVANS. 1971. Distinguishing between X, Y and YY-bearing human spermatozoa by fluorescence and DNA content. *Nature (New Biology),* **229:**231–233.

Questions

1. Why do many variant chromosome types have no effect on the individual carrying them?
2. If a chromosome of unusual structure were found in a single karyotype, would it be reasonable to conclude that it was a chromosome variant? What additional information might be desired?
3. What is the distinction between heteromorphic and polymorphic chromosomes?
4. Why would it not be surprising to find that the Y chromosome is particularly variable?
5. What happens after a chromosome is broken? In your answer, use the terms *restitute, sister strand union, acentric, dicentric,* and *anaphase bridge.*
6. When two different chromosomes each undergo a single break and rejoin, what are the two completely different consequences?
7. What is a deletion, or a deficiency, in a chromosome? Can you describe the curious phenotype of the deletion for a segment in the short arm of chromosome 5?
8. If two breaks occur within the same chromosome and the middle segment is twisted through 180 degrees, two different kinds of rearrangement may result. Why are they given different names?
9. Why would we expect practically all persons with ring chromosomes to show some type of physical or mental abnormality?
10. What are the two main theories on the origin of cancer? What specific kinds of cancer appear to arise in the two different ways?
11. What is meant by the monoclonal origin of cancer? How can this be demonstrated?
12. What is the *Philadelphia chromosome?*
13. Retinoblastoma and Wilms cancer appear to have a genetic basis that can be localized to a specific chromosome segment. Where are these chromosome segments located?
14. Name some additional genetic diseases that predispose a person to cancer.
15. Why are the hydatidiform mole and the benign ovarian cyst of unusual interest?
16. What is meant by an *empiric risk?* If two parents heterozygous for a recessive produce an offspring, would you say that the empiric risk that the child will be homozygous recessive is one fourth?
17. How can we be sure that the nondisjunctional event that gives rise to trisomy-21 occurs mostly in the female (80 percent of the time) rather than in the male? How is it possible to distinguish between nondisjunctions occurring at the first and the second meiotic division?
18. What is a triploid? Are triploid fetuses viable? Where does the extra set of chromosomes come from? How can the origin of this extra set be determined?
19. Using the three chromosomes in a Down patient, show what is meant by (a) a *univalent,* (b) a *bivalent,* and (c) a *trivalent* at meiosis. What is the distinction between primary and secondary nondisjunction?

20. In what way are the progeny from XXX and XYY individuals unusual?
21. A missing chromosome can have disastrous consequences to a developing embryo. Is the effect of a missing chromosome similar in an ovum or a sperm?
22. It can be argued that from the long-range point of view, chromosome rearrangements that are responsible for grossly aneuploid embryos do less damage to the species than those that have less drastic effects. Explain.
23. What is the difference between alternate and adjacent segregation in translocation heterozygotes? Which type is more likely to be responsible for an abnormal embryo?
24. How does a "Robertsonian translocation" differ from an ordinary translocation? From an isochromosome?
25. A person with a 21–21 isochromosome can produce only trisomy-21 viable offspring. Imagine, however, that two such persons marry. Could they produce normal offspring?
26. What is *aneusomy by recombination?*
27. The two types of inversions produce distinctly different results when crossing over occurs within the inverted segment in a heterozygous person. Can you explain why one type is likely to produce seriously defective progeny and the other not?
28. What is meant by a *recurrence risk?* Is it likely to be greater for a couple who have produced a Down child when one has a 14–21 translocation than when neither has any rearrangement? Explain.

Genetic Counseling

The Aim of Genetic Counseling

When a family is faced with a problem that they have reason to believe is genetic in origin, their best course of action is to obtain more information. They must understand the genetic medical implications of the disorder and be prepared to make any decisions about alternative courses of action rationally. The condition must be accurately diagnosed and explained to the family. The genetic basis should be determined so that the likelihood of recurrence can be estimated. The individuals must be made fully aware of the means by which future problems may be handled or avoided, consistent with their own philosophical, ethical, religious, or moral constraints. In brief, counseling amounts to an intensive course, lasting a few hours at most, in the elements of medicine and genetics as they relate to a specific problem.

Families seek genetic counseling for a wide variety of reasons. According to one study, approximately half of the cases prove to involve genetic defects determined by a single gene. Another 20 percent of cases concern chromosome anomalies, and of these more than 80 percent involve trisomy-21. Another 20 percent concern congenital defects with a polygenic or unknown genetic cause. On rare occasions a family may seek advice because of anticipated problems related to consanguinity or to exposure to chemicals or radiation.

The Background of the Counselor

In the early days of human genetics, genetic counseling was handled to some extent by medical practitioners whose knowledge of genetics was minimal and whose main interests may have been focused in other directions, but who tried to communicate to the patient the limited knowledge available at that time. More often, perhaps, counseling was done by academically trained geneticists with an expert knowledge of the basic principles of genetics but with little or no orientation or training or opportunity to deal in depth with the clinical aspects of counseling problems. In both instances the results were sometimes far from satisfactory.

Now, overall evaluation is carried out more often than not by a group of highly qualified experts, usually functioning in a clinical setting. When such persons are M.D.s, they are trained primarily in pediatrics, but also in pathology, hematology, internal medicine, obstetrics, gynecology, and neurology. Those with Ph.D. degrees include specialists in genetics, zoology, cytology, cell biology, biophysics, psychology, anthropology, and statistics. In some cases counselors have joint degrees in both medicine and genetics. There are programs now in operation at a number of colleges to train both medical and nonmedical personnel to provide specialized services to group counseling clinics.

The Procedures in Genetic Counseling

Medical Diagnosis.   The first step in counseling must be an accurate medical diagnosis of the condition, for if the problem should prove to have no genetic basis, as happens occasionally, genetic counseling is unnecessary. For instance, more than half of the patients referred to one center for muscular dystrophy proved to have something else. Some genetic disorders can be mimicked by environmental disorders, as is shown by the similarity of genetic phocomelia and the thalidomide-induced phenocopy, both reducing arm length to a fraction of the normal size. Achondroplasia, which is caused by a dominant gene, is similar to about half a dozen other physical anomalies, which in some cases are caused by a recessive gene. During the diagnostic period, if a chromosome abnormality may be involved in any way, it is essential to karyotype the affected individual and frequently also the parents and sibs. The precise diagnosis may—and probably does—depend on a medical examination, laboratory tests, and sometimes X-ray examinations, and for this reason the participation of medically qualified personnel at this stage is mandatory.

The Pedigree.   Making a detailed pedigree is absolutely essential, including as many relatives as possible, with their ages, and their reproductive history, including stillbirths, abortions, and deaths. In some cases a simple inquiry about the country of origin of the ancestors of the two parents, and their surnames, may suggest consanguinity. Questions about religion and ethnic origin may be valuable because certain diseases are more prevalent in one group than in another. Sometimes the pedigree shows the method of inheritance unambiguously and, as a result, makes a distinction between two similar disorders with different modes of inheritance. Both parents should participate in these sessions from the beginning, when the contribution of family history information from each is important, to the final counseling periods, when both parents must understand the nature of the problem and the range of possible solutions and must agree on a final decision.

INFORMATION ABOUT THE DEFECT. There exists a vast body of medical literature on genetic defects. A catalog, *Mendelian Inheritance in Man,* issued by V. McKusick of Johns Hopkins University and brought up to date every few years, is useful in providing the basic information about all known genetic disorders. Another useful source is *Birth Defects: Original Articles Series,* edited by D. Bergsma.

ESTIMATING THE RISK. For simple dominant and recessive genetic defects, there is no difficulty in making a precise statement of the probability of recurrence after the birth of an affected child. If one parent is heterozygous for a dominant defect and the other is homozygous normal, the chance of an affected child is one half. If two normal parents have produced a child homozygous for a recessive, it can be assumed that both are heterozygous and that the chance of an affected child is one fourth. An affected child may carry a dominant trait not found in either of his parents; such a new mutation has little chance of being found in any additional children.

If a normal woman has produced a male child with an X-linked defect such as muscular dystrophy, the simplest assumption is that she is heterozygous and that half of her subsequent sons will be affected and half of her daughters will be carriers. However, it will be shown (Chapter 13) that roughly a third of all such cases of affected males must result from new mutations. The counseling in the two cases would, of course, be quite different. It is possible to identify those heterozygous mothers, with some degree of confidence, when they carry alleles for X-linked hemophilia and several other X-linked diseases. It would be very important for the mother of a severely affected child with an X-linked disorder to know that she is not heterozygous and so will not be likely to produce any more defective sons or heterozygous daughters.

These are the simple cases. When the problem is caused by a chromosome anomaly such as a translocation, or a developmental defect such as cleft palate, there are no specific algebraic expectations for the occurrence of an affected individual, and it is necessary to rely instead on *empiric risks.* These are probabilities based on statistics of similar past cases (Table 9-1).

EDUCATION OF THE PATIENTS IN PROBABILITY. Probably the greatest difficulty in counseling is educating the individual in the principles of probability. A common misunderstanding concerns the independence of the probability of successive events. It is not unusual for a couple who are both heterozygous for a simple recessive to be told that there is a one-in-four chance of having a child homozygous for that recessive defect, and for them to assume that their already having had one affected child now ensures that the next three children will be completely normal. Sometimes the use of

Table 9-1 Empiric risks, for counseling purposes, of cleft lip with or without cleft palate (CL ± CP) and cleft palate (CP) in various family situations. (F. Clarke Fraser, reprinted from the *American Journal of Diseases of Children* **102**:853, 1961. Copyright 1961, American Medical Association.)

	Proband has	
	CL ± CP	CP
	(per hundred)	
I. Frequency of defect in the general population	.01	.04
II. My spouse and I are unaffected		
A. We have one affected child		
1. What is the probability that our next baby will have the same condition if:		
a. We have no affected relatives?	4	2
b. There is an affected relative?	4	7
c. Our affected child also has another malformation?	2	2
d. My spouse and I are related?	4	2
2. What is the probability that our next baby will have some other sort of malformation?	same as general population	
B. We have two affected children		
1. What is the probability that our next baby will have the same condition?	9	1
III. I am affected (or my spouse is)		
A. We have no affected children		
1. What is the probability that our next baby will be affected?	4	6
B. We have an affected child		
1. What is the probability that our next baby will be affected?	17	15

simple examples like tossing a coin may seem to get the point across, but if many beginning college students cannot understand probability, even after hours of patient classroom explanation by skillful teachers, how can parents taken at random from the population be expected to master it? Explaining probability to prospective parents is all the more difficult because the issue is not hypothetical, but one vitally important to them personally. They must make a serious decision on the vagaries of chance for which they will have to accept the consequences for the rest of their lives.

DECIDING ON A COURSE OF ACTION. Every family with a genetic problem experiences some emotional trauma. The parents may feel guilty about being responsible and angry about being the ones to suffer the misfortune, and they may even deny the legitimacy of the child or the existence of the problem. Parents may recall any of a large number of factors that they feel might have been responsible: tobacco, alcohol, drugs, weight loss (or gain), coffee, tea, and so on. What the parents need is an experienced counselor with an understanding of the intensity of the trauma and sufficient time and patience to try to educate them to the genetic facts of life—that all persons carry their share of deleterious genes, independent of status, class, wealth, or ethnic group, and that no one should assume any personal blame for genetically defective children.

After a rational perspective has been assumed, plans can then be made for a future course of action, if any, depending on the nature of the problem and the magnitude of the risk. The chance of one in ten seems to be a cutoff point that many persons unconsciously adopt. When told that the risk of a serious malformation or disease is more than ten percent, two thirds of parents decide not to take that risk. On the other hand, when told that the risk is less than ten percent, three quarters of such parents decide that they will take the chance.

In some cases, if the parents are so inclined, future pregnancies can be monitored so that the status of a fetus can be determined. In other cases, parents may decide not to have any additional offspring. This is a decision that is left to the parents and is highly individual.

SUBSEQUENT REVIEWS. Many counseling centers attempt to maintain contact with the subjects. This contact can be helpful in a number of ways. The situation may change with time so that a reevaluation of the original decision may be called for. New clinical methods may develop that make the earlier decision obsolete. Or the initial information given to the parents, as well as their decisions, may become garbled with the passage of time. In fact, studies have shown that many people who appeared to understand the entire situation during the first counseling sessions actually did not, and follow-up studies made it possible to correct any misunderstandings. Furthermore, when the counseling was performed by a family physician (who passed on to the patient the information provided by a genetics clinic), this physician often had a limited and somewhat distorted view both of the genetic nature of the disease and of the consequences to the patient. This problem will no doubt disappear as the newer generations of genetically sophisticated medical practitioners replace the old.

Reviews give clinics and counselors an opportunity to assess their effectiveness. Subsequent studies of counseled parents revealed the following: that the family's decision was based more on their assessment of the emotional and financial trauma to the family than on the risk of having an affected child, that the amount of the genetic information they retained was minimal, and that the principal barriers to family limitation were religious principles prohibiting contraception, sterilization, or abortion.

Problems of Counseling

Most counselors make a strong effort not to influence the parents with respect to the course of action to be taken, but to allow the parents maximum freedom of choice in their decisions after supplying them with all available information. However, it may be very difficult to avoid a facial expression, a manner of speech, or other behavior that expresses a point of view. The classic example is the choice forced on a counselor who must inform the parents that they are both heterozygous for some crippling recessive condition such as Tay-Sachs disease. The parents could be told that "There is a chance of three out of four that your child will be perfectly normal," or, equally well, that "There is a chance of one in four that the next child will be hopelessly doomed to an early death." Each statement is essentially correct, yet the reaction of the parents may be quite different in the two cases. Either statement would betray the counselor's views of the matter and subtly suggest a course to be followed. The precise way in which this information is communicated may be determined largely by the counselor's intuitive assessment of the psychology of the parents and their likely reaction to the various ways of presentation.

In any case the background of a counselor may very well determine the attitude with which he or she views human problems, tempering the advice likely to be given. A doctor of medicine whose code of ethics and many years of training have centered almost entirely on the preservation of life is quite likely to give an opinion based primarily on the immediate well-being of the patient, whereas a doctor of philosophy, trained as a population geneticist, may feel that some anticipated future needs of society should take precedence over the welfare of the patient. It would be desirable for counselors to identify unconscious biases and, if possible, to compensate for them when appropriate. This, however, may be an impossible ideal, because most people refuse to admit having biases. Perhaps the most rational decisions are reached by several counselors with different backgrounds acting as a team. Such decisions can then be presented to the patient by an individual counselor.

HUNTINGTON DISEASE. Some of the problems of genetic counseling are illustrated by the situation that has been described for Huntington disease, also called *Huntington chorea*. This is a degenerative disease of the brain, determined by a single dominant gene that expresses itself, on the average, at about age thirty-five. Close to 100,000 persons in the United States either now have, or are threatened by, Huntington disease. It is characterized by mental deterioration and by purposeless involuntary movements suggesting the waving motions of a dance (hence the name, from the Greek *choros*, meaning "dance"). On the average, patients with this disease survive about ten years after the disease has become evident.

Because of the late onset of the disease, most afflicted persons have married and often have completed their families before becoming aware of their condition. For this reason this gene continues to be perpetuated generation after generation. Its overall incidence is 1 per 10,000 persons, but it tends to be concentrated in a limited number of kinships. On the other hand, as a simple dominant with high penetrance, it would appear, in principle, to be a relatively simple disease to eliminate. All that is required is that all persons in families in which the disease has appeared not have children.

But is there any justification for suggesting a limitation on the reproductivity of normal persons because they stand a 50 percent chance of carrying an abnormal allele (when there is an equal chance that they do not carry that allele)? Even if it were possible to detect the heterozygotes early in life, would it be acceptable to discourage their reproduction because they stand a 50 percent chance of having a heterozygous child at each birth?

Studies of the attitudes of persons in kinships segregating for Huntington disease show that the desire for family size limitation was greatest among the spouses of the affected and among those relatives at low risk (not likely to have the allele), but that those actually affected or young adults with a 50 percent chance of being heterozygotes were less enthusiastic about curtailing family size. This restraint on the part of the unaffected, or those not likely to be affected, and the lack thereof on the part of those affected are shown in one study indicating a 27 percent decrease in reproductivity of unaffected sibs compared with a 22 percent increase (over an average sample of the population) for those affected. In fact, a quarter of those in the high-risk category, the possible heterozygotes, indicated that they would probably refuse to take a test, if one were available, to determine whether they actually were heterozygotes.

The understandable human desire of persons at risk, or those with relatives at risk, not to be reminded of this serious problem may be shown in the extent of the response to a study undertaken in New Jersey, where questionnaires requesting participation were sent out to 125 families on a mailing list of those with Huntington disease among the relatives. Only nine positive responses, with consent to be interviewed, were obtained.

Prenatal Diagnosis (by Amniocentesis)

Because the possibility of a deformed or mentally defective child haunts virtually all parents at some time, reassurance about the probable well-being of a future child is greatly appreciated by prospective parents. On the other hand, if a fetus can be diagnosed as afflicted with a grossly debilitating disease, some parents may prefer to know and, if their religious and ethical feelings allow, prevent the fetus from coming to term. The decision to decrease the number of births of severely abnormal children may become more prevalent as more and more married couples decide to limit family size anyway, for example, to two offspring.

Amniocentesis (Figure 9-1), the puncturing of the amniotic membrane to obtain amniotic fluid and fetal cells, was developed originally in the diagnosis and treatment of *hemolytic disease of the newborn (HDN)*. Cytological examination of the cells reveals the sex of the fetus and possibly any chromosome abnormalities carried by it; in addition, about forty genetic disorders may be detected biochemically (Figure 9-2).

COMPOSITION OF AMNIOTIC FLUID. The amniotic fluid is interchanged quite rapidly between the fetus and the mother, being completely replaced about once every three hours. The total amount of amniotic fluid is variable, but it averages between 500 and 1,000 milliliters at the time of birth. The cellular components of the fluid are epithelial cells from the skin, the alimentary tract, and the respiratory tract, as well as cells from the amnion and other tissues. In addition, there is a concentration of protein of about 1 per-

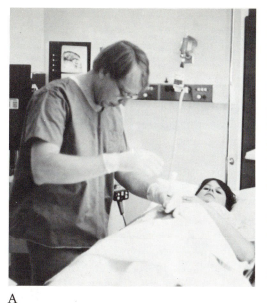
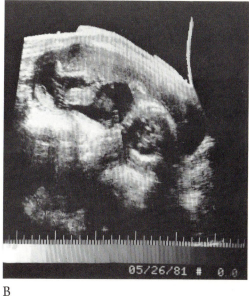

A   B

Figure 9-1 *A.* A modern set-up for amniocentesis. At the upper left an ultrasound apparatus displays the fetus and its position in the womb. *B.* A twenty week old fetus as seen by ultrasound equipment. The head is at the right and the legs are at the left. (Courtesy of G. Prescott, University of Oregon Health Sciences Center.)

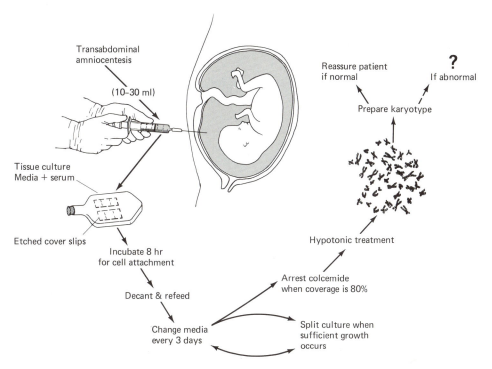

Figure 9-2 Prenatal diagnosis by amniocentesis. The cells present in the amniotic fluid can be cultured and tested for cytological and biochemical properties. (From C. B. Jacobson. Cytogenetic Techniques and Their Clinical Uses, in *Genetics in Medical Practice,* Lippincott, Philadelphia, 1968.)

cent. There is a high concentration of some free amino acids in early pregnancy; other chemicals include sugars and fatty acids and a number of enzymes and steroids. Some of these components are probably maternal in origin; others are of fetal origin.

THE TECHNIQUE OF AMNIOCENTESIS. The amniocentesis operation consists of inserting a needle into the uterine cavity through the abdominal wall and removing about twenty ml of fluid and associated free cells. The procedure is best carried out between fourteen and sixteen weeks of pregnancy. Amniocentesis cannot be accomplished very early because the volume of amniotic fluid is insufficient. On the other hand, if either biochemical or chromosomal cell studies are to be made, it cannot be delayed too long because amniotic cells become more difficult to culture successfully as pregnancy progresses. Culturing the cells takes time—sometimes as long as a month— and the biochemical and cytological studies add another week or two. In the event that serious defects are found in the fetus, or there is some other compelling reason for considering a therapeutic abortion, the parents must be able to make this decision prior to the legal limit if it is to be done in one of those states in which it is legal to perform an abortion to prevent the birth of an abnormal child. The law varies from one state to the next, and the limit may be twenty-two to twenty-four weeks. If the attempt to culture cells for analysis should fail, or if they should prove refractory to either biochemcial or cytological analysis, then a second amniocentesis, with the time necessary for analysis, may extend the required time beyond that allowed for legal abortion.

PROBLEMS OF AMNIOCENTESIS. Amniocentesis is usually carried out for specific women at risk (Table 9-2), and then only to the extent demanded by the nature of the problem. As a general procedure undertaken for the general population in all pregnancies, it would simply present an intolerable expense.

Several problems arise in performing amniocentesis. In the first place, although the risk to the mother is low, the approximate proportion of pregnancies in which complications develop after amniocentesis—including spontaneous abortion, fetal death, or excessive maternal bleeding—is of the order of 1 percent. The use of ultrasonic equipment (Figure 9-1B) that presents a view of the amniotic cavity, the fetus, the placenta, and other tissues makes it possible to position the needle more judiciously, obviously helping to reduce possible error. And, of course, as physicians have become more skillful in this procedure, the operation has become safer.

Second, the attempt to obtain amniotic fluid may be unsuccessful.

Third, the procedures for producing chromosome preparations may fail even in the most skillful hands.

Still another problem is that the sample of cells taken may include maternal cells. Of course, if the cultured cells have a Y fluorescing body, which usually indicates a Y chromosome, the danger of confusing maternal with male fetal cells is minimal. Nevertheless, in a small number of XX female cells (2–3 percent), "Y fluorescence" is present, but it is usually caused by a fluorescent polymorphism of an autosome. Errors resulting from the con-

Table 9-2 Responses to questionnaires from a number of U.S. and foreign medical laboratories indicating the criteria used by them for doing amniocentesis. (From *Birth Defects: Orig. Art. Ser.,* ed. D. Bergsma. Published by Williams & Wilkins Co., Baltimore, for the March of Dimes Birth Defects Foundation, White Plains, N.Y.)

Reason	United States		Foreign Countries	
	Yes	No	Yes	No
Family history of gene-transmitted or chromosomal abnormality	77	6	36	7
Previous birth of retarded or malformed child	43	33	19	23
Parent with balanced chromosomal translocation	76	7	37	6
Viral infection first trimester	18	58	8	32
Poor previous pregnancy history	24	52	12	30
History of exposure to mutagen(s)	24	52	12	27
Rh factor	59	19	24	19
Advanced age of mother	69	10	28	14
Sex determination (for X-linked disorders)	51	28	30	10
Research control data	24	44	13	22

fusion of maternal for fetal cells can be minimized by routinely karyotyping the mother, as well as the child, so that chromosome polymorphisms present in one set but not the other will serve to distinguish the two.

Twin pregnancies present an unusual problem in amniocenteses. Ultrasound pictures are very helpful in determining or verifying such pregnancies, and in suggesting the best procedure for obtaining amniotic fluid from each of the cavities. A harmless dye may be injected into one of the cavities, in order to make sure that a sample of fluid is taken from each. However, parents may face a particularly painful dilemma if one of the twins should prove to be genetically defective and the other not.

In general, amniocentesis is not performed unless there is a substantial medical reason for it. At the present time, for instance, most physicians would object to performing amniocentesis for the sake of ascertaining the sex of the forthcoming child in the absence of any compelling medical justification.

ALPHA FETOPROTEIN. Malformations of the central nervous system—such as spina bifida, in which the spine is incompletely closed in its lower regions, and anencephaly, where the head and brain are poorly (or not at all) developed—occur with a frequency of about 1 in 500 births in the U.S. *Alpha fetoprotein (AFP)* may be found at a high level in the amniotic fluid in these central nervous system malformations and in other congenital malformations as well.

It has also been found to reach high levels in twin pregnancies and when the fetus is a trisomy-13. AFP may be assayed after amniocentesis, making it possible to detect ninety-eight percent of the cases of spina bifida and anencephaly. The false positive rate (i.e., high AFP with no unusual birth condition) is about half of 1 percent.

In the serum of most women carrying fetuses with central nervous system defects, the concentration of AFP may be more than twice as high as in women with normal fetuses. This high concentration is found in only a small percentage of normal pregnancies. Thus it is possible to screen for these defects, with some degree of accuracy, by testing the serum of the mother's blood, with a view to a subsequent ultrasound study and possibly amniocentesis.

RISK CATEGORIES.   One proposed scheme for classifying patients eligible for amniocentesis is based on the relative risk involved:

1. A *high-risk* group. When the mother is known to be heterozygous for a serious X-linked recessive disorder, half of the male progeny can be expected to carry this disorder. If a woman has had a child with muscular dystrophy of the Duchenne type, she might prefer not to undergo a similar experience a second time. Of course, if an accurate prenatal diagnosis is possible, as it is for more than forty sex-linked diseases, she need be seriously concerned only if the fetus is an affected male. In those cases where prenatal diagnosis is not possible, she might choose not to have any male offspring. Techniques for the accurate prenatal diagnosis of the better-known sex-linked disorders like muscular dystrophy and hemophilia are being developed rapidly.

When the parents are known to be heterozygous for a recessive, as is the case when they have already produced a homozygous child, then the risk that their next child will also be homozygous is one in four. Tay-Sachs homozygotes can be identified *in utero;* here amniocentesis would identify the abnormal fetus. On the other hand, the risk for a PKU child would also be one fourth if both parents are heterozygous, but PKU is not recognizable until after birth, and amniocentesis cannot yet identify the affected fetus.

Another group in the high-risk category includes those in which one parent is heterozygous for a chromosomal rearrangement as a result of which there is a sizable chance (5 percent or more) that the fetus will carry an abnormal chromosome complement. Usually this condition can be detected by karyotyping fetal cells.

2. A *moderate-risk* group. Here we include women who are of age thirty-eight or older, as the probability of a child with a chromosome aberration, such as one of the trisomies, may run as high as 2 percent.

3. A *low-risk* group. In this group are women aged thirty-five to 38, and those who have normal chromosomes but who have previously had a chromosomally abnormal offspring. In these cases the likelihood of an affected child is greater than that for the population at large, but it is less than that for the moderate-risk category.

Of course, these are only rough guidelines for determining the desirability of a course of action. In practice the final decision would have to depend to a great extent on the patient's perception of the magnitude of the risk, the

Indication	Pregnancies Studied	*Affected Fetuses Found* *
Chromosomal		
Translocation carriers	290	29 (10.0%)
Maternal age ⩾35 years	3,012	79 (2.6%)
>40 years	864	36 (4.2%)
35–39 years	979	16 (1.6%)
Previous trisomy-21	1,887	23 (1.2%)
Miscellaneous	1,284	26 (2.0%)
X-linked diseases	429	199 (46.4%)†
Biochemical defects	438	106 (24.2%)
Neural-tube defects	1,571‡	86 (5.6%)**
Total	10,754	610 (5.7%)

Table 9-3 Reasons for, and results of, prenatal diagnosis. (Charles J. Epstein and Mitchell S. Golbus, Prenatal Diagnosis of Genetic Diseases. *American Scientist* **65**:703–711, 1977.)

* For X-linked diseases, all males, whether actually known to be affected or not, are listed as affected.
† One trisomy-21 fetus is included.
‡ Since individuals from a family with a history of neural-tube defects are at a 4–5 percent risk of having children with such a defect, they are the ones referred for prenatal diagnosis.
** Another ten fetuses with neural-tube defects found when amniocentesis was performed for other genetic indications are not included.

quality of life if the fetus has an abnormality, and the likelihood that any positive action would follow the discovery of an abnormal fetus.

Table 9-3 shows the outcome of pregnancy in more than 10,000 cases in which amniocentesis was performed; 610 affected fetuses were detected.

Screening Programs

With our advanced genetic knowledge and sophisticated clinical methods, what better use is there to put them to than in the prevention of abnormal births? Screening programs are those in which large numbers of individuals are routinely checked for the presence of some defective allele or chromosome. These have fallen into two broad groups: those in which all newborns in a given area are checked for defects, and those in which as many adults as can be persuaded in an ethnic group at risk are tested, usually for deleterious alleles in the heterozygous (or carrier) state.

The best-known screen of newborns is that for phenylketonuria (PKU), now required by law in many states. In addition, there are about a dozen other defects that can be detected shortly after birth, with the possibility of ameliorating the condition by treatment in early childhood. Such diseases include cretinism (or hypothyroidism) and maple syrup urine disease (in which the urine smells like maple syrup).

One might argue that the more information a person has about his or her genetic constitution, the more intelligent can be the choice made about the production of offspring and the better off the individual and society will be. To accomplish this ideal, a screening program that will identify those with genetic defects for which we now have tests is a first step.

If the preceding paragraph seems idealistic, self-evident, and noncontroversial, it is only superficially so. In fact, when screening programs have been instituted, they have created a great deal of dissension, and although there is no doubt that they have accomplished much good, they have also done some harm. In fact, in some cases the harm may outweigh the good. To illustrate this point, we will examine four different screening programs: for Tay-Sachs and sickle-cell anemia in adults and for phenylketonuria and chromosome abnormalities in newborns.

Tay-Sachs Disease

Tay-Sachs disease is a simple recessive disease in which the nervous system degenerates. It is obvious after a few months of age and always leads to death in early childhood. Usually the child does not learn to walk; muscular control degenerates rapidly. At age eighteen months, the affected child is usually deaf and blind, with complete rigidity of the muscles, and will die before the fourth or fifth year. The affected child lacks an enzyme, hexosaminidase A, and this lack results in the accumulation of lipid bodies in the cells. Accumulation in the brain cells, in particular, leads to the loss of coordination, seizures, blindness, and finally death.

This disease occurs in approximately 1 in 5,000 Ashkenazic Jewish births, but in only 2 out of 1 million non-Jewish births. From these figures we can calculate that about 1 in 35 Ashkenazic Jews and about 1 in 400 non-Jews are carriers for this recessive.

The absence of the enzyme hexosaminidase A in the homozygote can be detected by amniocentesis, thus making it possible to monitor pregnancies at risk if the parents wish, with the choice of abortion being left to the parents. Furthermore the enzyme is found to be present in only about 50 percent of the normal amount in known heterozygotes, so that heterozygotes can be detected (Figure 9-3), but parents are ordinarily not aware of their heterozygosity until the birth of the first affected child. Thus amniocentesis might be expected to decrease the frequency with which second affected children appear. However, on the average, one homozygote is expected in every four children, and family size is usually less than that number. In fact, 82 percent of all homozygotes are the first affected children, alerting the parents to their heterozygous condition, and only 18 percent are second affected children detectable by amniocentesis. Under these conditions, amniocenteses after the birth of an affected child do not decrease the frequency of homozygous children very much. Detection of heterozygous parents in a screening program would be much more effective.

A screening program was set up in 1971 in the Baltimore–Washington area, where it was estimated that about 8,000 persons are heterozygotes. In most cases heterozygotes have married normal homozygotes, but both husband and wife of several hundred couples are heterozygotes, and their prog-

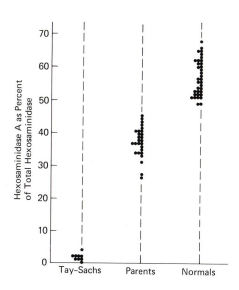

Figure 9-3 The detection of heterozygotes for the recessive allele causing Tay-Sachs disease. The percentage of hexosaminidase A is very low in affected children, and lower in their parents (who must be heterozygotes) than in normal individuals. (Reprinted by permission of J. S. O'Brien and the *New England Journal of Medicine,* Tay-Sachs Disease—detection of heterozygotes by serum hexosaminidase assay, **283:**15–20, 1970.)

eny are therefore at risk. The propriety of therapeutic abortion of homozygotes detected *in utero* is a matter of considerable disagreement among Jewish religious leaders. If elective abortion were decided on by all carriers of this disease when the fetus appears to be a homozygote, then clearly the incidence of affected newborns would diminish rapidly. On the other hand, because such early losses of the homozygous fetus would probably be compensated for by the parents' deciding to have an additional offspring—a phenotypically normal offspring, but heterozygous two thirds of the time—the net effect of the therapeutic abortion would be to increase the frequency of the defective gene in that population.

Sickle-cell Anemia

Sickle-cell anemia, caused by homozygosity for the hemoglobin *S* allele, affects 1 in every 500 black children in the United States (about 50,000 of this nation's 22 million blacks). Homozygotes suffer from considerable pain as well as chronic anemia, strokes, and other disasters of the circulatory system. Such homozygotes very often die before ten years of age and the majority of them by thirty. More than 2 million blacks are heterozygous, described as having the *sickle-cell trait.* Under ordinary conditions, heterozygotes with the sickle-cell trait are indistinguishable from normals, although they may at times be slightly anemic, and a drastic reduction in oxygen pressure may cause their otherwise normal cells to sickle. There have been reports of deaths of individuals with the sickle-cell trait under anesthesia, during severe physical stress, and after blood transfusions, but these cases are most unusual, and heterozygotes should be considered quite normal. It is possible to detect heterozygotes by a relatively simple test (one of the least expensive and best-tested systems for detecting them costs only 2 cents per test!), so mass screening programs are technically feasible.

ETHICAL PROBLEMS. Few issues of genetic concern have given rise to such widespread controversy as the sickle-cell screening program, and some of its severest critics have come from the black population itself. Some criticisms are directed toward the manner of carrying out the programs, sometimes by untrained and insensitive personnel. Others are directed at the racial overtones of this kind of screening program, particularly when it is made compulsory by law, as it has been at one time or another in ten states.

It is certainly legitimate to wonder if any ethnic group should be subjected to a screening program that is aimed specifically at them as a population group. Still another question that arises is whether individuals who do not wish to be exposed to a medical (or quasi-medical) examination should be forced to do so.

From the standpoint of the blacks, there has been some skepticism as to whether the millions of dollars that have been legislated for the alleviation of this disease were given not so much for humanitarian purposes as for political gain. Counseling has been interpreted as pressure for the heterozygous black person to have fewer children, and the word *genocide* has been used to describe counseling that might limit productivity. Some have questioned the motives of those in the medical profession and in hospitals who have accepted large sums of money from the government to do research in this area.

There has been some confusion of the homozygote with the heterozygote in the public eye: the first is afflicted with sickle-cell anemia, whereas the second is a carrier and is said to have the sickle-cell trait. The distinction may not be clear. The original Massachusetts law refers to "the disease known as sickle-cell trait or sickle-cell anemia." The legislation passed by the Congress of the United States includes the statement that "the Congress finds and declares (1) that sickle-cell anemia is a debilitating inheritable disease that afflicts approximately two million American citizens." Surely this statement refers to the number of heterozygotes with the trait, not to the homozygotes with the anemia. The New York State law requiring testing prior to marriage (along with tests for veneral disease, a most unfortunate association) states that it is for the purpose of "discovering the existence of sickle-cell anemia," when it is clearly meant to detect the heterozygote. In many cases even medical personnel have confused the sickle-cell trait and sickle-cell anemia, and this unfortunate terminology is the source of a great many of the problems that arise. In some cases individuals who have the sickle-cell trait (i.e., who are heterozygous) have been fired from their jobs or otherwise discriminated against on the grounds that they have a serious disease, which is simply not the case.

Phenylketonuria

Large numbers of studies have been made on phenylketonuria (PKU) incidence in different populations, and it appears to occur with a frequency ranging from 3 to 10 per 100,000 in different populations of Northern European origin (Table 9-4). It has also been found in Japan and in the Middle East, but it is very uncommon among both Jews and blacks. Most of the

Country	Estimated Number Screened	Cases Found
Australia	479	0
Belgium	20,000	5
Canada	65,898	6
Denmark	12,500	0
Germany	67,309	10
Ireland	25,000	4
Israel	65,000	5
New Zealand	1,840	0
Poland	130,912	21
Scotland	16,500	2
Sweden	21,505	1
Switzerland	7,002	1
United States	2,238,634	205
Wales	1,536	0
Yugoslavia	23,690	3
	2,697,805	263

Table 9-4 The results of screening newborn infants for phenylketonuria in fifteen countries. (From R. Guthrie, Screening of Inborn Errors of Metabolism in the Newborn Infant. In *Birth Defects: Orig. Art. Ser.*, ed. D. Bergsma. Published by Williams & Wilkins Co., Baltimore, for the March of Dimes Birth Defects Foundation **4[6]**:93, 1968.)

homozygotes are found in mental institutions, where they constitute about 1 percent of the mentally defective population. The age distribution of these patients shows a less-than-expected frequency under five years and over thirty-five years. The former is undoubtedly a reflection of a late diagnosis or a late development of the phenotype. The lack of affected individuals at older ages results from the relatively early deaths of homozygotes, usually from the infectious diseases that are common in institutions.

DIAGNOSIS. The homozygotes can be detected by the presence of phenylpyruvic acid (one of the breakdown products of phenylalanine when not converted into tyrosine) with the use of ferric chloride ($FeCl_3$). When phenylpyruvic acid is present in a test sample of urine, the ferric chloride turns the color to olive green. This test, however, depends upon the activity of a liver enzyme which may not become active until some time after birth and so is not a completely dependable test for PKU in newborns.

The most common and practical test is called the Guthrie test, in which a drop of the infant's blood is checked to see if it promotes the growth of a special mutant strain of bacteria. This happens when the blood carries minute quantities of the phenylalanine present in higher quantities in PKU children than are normal. It is a very sensitive test and may be used shortly after birth. Most states in the U.S. and many countries now require that a test for PKU be performed on newborn infants.

It is interesting that the original diagnosis was made because the mother of two of the first patients insisted that biochemical tests be made because she was surprised by the physical similarities of her two offspring, although they were of different sex, and by their characteristically smelly urine. It is from the presence of the metabolites of phenylpyruvic acid, the *phenylke*-tones in the *u*rine, that the abbreviation *PKU* is derived.

TREATMENT. The basic principle in treating this disease is to feed the child a diet with a low concentration of phenylalanine. Such a diet was attempted originally by replacement of the protein in a normal diet by a mixture of pure amino acids. This procedure, however, is very costly. A more practical method is to remove the phenylalanine from a diet containing otherwise normal proteins. The problem that arises is including sufficient protein along with phenylalanine to maintain body protein levels adequate for normal growth, at the same time not exceeding the phenylalanine requirement. A diet too deficient in phenylalanine may cause severe complications, including anemia, with eventual fatal consequences. After a proper phenylalanine treatment, the biochemical abnormalities disappear. The peculiar odor of the urine vanishes, skin blemishes clear up, hair and skin pigmentation darken, and the patient becomes more manageable, with an increase in attention span and improvement of motor performance. Even PKU adults who are beyond the period when specific nerve development can be improved by a low-phenylalanine diet can have their behavior patterns improved considerably by such a diet.

The homozygotes are apparently born normal, and degeneration begins within the first few weeks after birth. The enzyme responsible, phenylalanine hydroxylase, does not, however, become fully functional in the liver until shortly after birth. Because the treatment must be applied within a few weeks of birth, before intellectual impairment sets in, it becomes a major problem to diagnose the disease in time to start treatment. The more the treatment is delayed, the less the improvement, as a rule.

INFLUENCE OF MATERNAL PKU ON OFFSPRING. One of the curious aspects of PKU retardation is that when homozygous females marry and have children (Figure 9-4), which happens fairly frequently for those without severe mental deficiency, their non-PKU children may nevertheless be abnormal. In an international survey covering five hundred and twenty-four pregnancies in one hundred and fifty-five PKU women, 95 percent of the women with high phenylalanine levels had at least one mentally retarded child and 81 percent had only retarded children. In fact, the degree of mental retardation was greater in the children than in the mothers. In addition, the children had a low average birth weight, and sometimes microcephaly. It is reasonable to conclude that elevated levels of phenylalanine (or its products) in the mother cross the placenta and reach similar high levels in the blood of the fetus. This problem may be compounded by the fact that the circulating phenylalanine may reach levels much higher than normal during pregnancy, as well as during fever and infection. These compounds are then harmful to the brain development of the fetus. These cases can be handled to some extent if the mothers follow a low-phenylalanine diet during pregnancy. One difficulty, however, is that the pharmaceutical dietary preparations with a low phenylalanine content may be unpalatable or even nauseating to the patient. Another problem is that the phenylalanine concentration in the blood must be constantly monitored so that the mother gets neither too much nor too little.

As the treatment of PKU infants becomes more common, an increasing

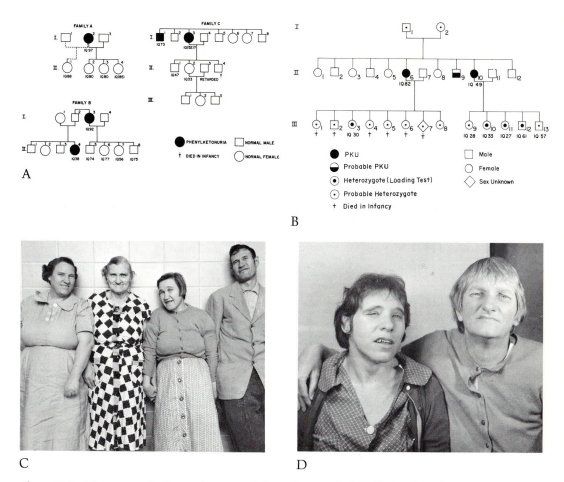

Figure 9-4 The unusual effects of maternal phenylketonuria (PKU). *A.* Three families in which the mother was homozygous and all children (heterozygotes) suffered as a result. *B.* A pedigree with high mortality in one sibship, and striking effects on all progeny, four of whom *(III3, 10, 11, 12)* could be shown to be heterozygotes. *C.* Individuals *III10, II10, III11,* and *III12* of pedigree *B. D.* Individual *III3* and her mother, *II6* of pedigree *B.* (Reprinted by permission of C. C. Mabry and the *New England Journal of Medicine,* Maternal phenylketonuria: Cause of Mental Retardation in Children Without Metabolic Defect, **269**:1404–1408, 1963.)

number of homozygous women will appear normal, get married, and have children. Reagents have been developed for widespread screening of pregnant women to uncover any unsuspected homozygotes.

Chromosome Abnormalities

Screening programs that center on adults have provided important cytological information about the distribution of chromosome defects in the population. Not only have relative frequencies been determined, both inside and outside of institutions, but information on the range of physical and mental normality provides a basis for predicting the likely course of new cases as they arise. There can be no doubt that without the kind of information that comes from population surveys, there would be a tendency to exaggerate both the frequency and the effects of chromosome abnormalities found in adult populations.

Several large-scale studies have been made of the chromosomes of newborn infants (in Boston, Denver, New Haven and Edinburgh) and of young children (Baltimore). In principle, this seems like a desirable procedure, making it possible to determine those cases in which a behavioral condition may stem from some previously undetected chromosome abnormality. Some of the problems that arise in this connection are those common to all screening programs. A unique one is whether such determinations ever become self-fulfilling prophecies. In the case of an XYY boy, who, according to the law of averages, would probably be indistinguishable from an XY boy in height, intelligence, and behavior, will his parents and teachers be able to treat him as a normal boy, or will their attitudes of solicitude or apprehension divert his behavior to an abnormal direction? Is there any likelihood of his developing the feared traits as a consequence of his being informed that he is XYY and that such persons often behave in an antisocial way? Questions such as these have forced the suspension of well-established and medically approved programs of screening the chromosome compositions of newborns.

Problems Arising During Counseling and Screening

MISTAKEN PATERNITY. With some small frequency, children have mistaken paternity; that is, their mother's husband (and their legal father) is not their biological father, a fact not generally known to either the husband or the child. It is understandable that the mother might be unwilling to transmit this information during a counseling session, particularly if her husband is present. Mistaken paternity may be revealed if a child is homozygous for an allele that only the mother carries (Tay-Sachs disease or PKU) or if the child carries an allele that neither the mother nor her husband has (sickle-cell trait). It can also be uncovered accidentally in the course of routine blood-group testing of the family. When evidence of mistaken paternity is uncovered, the confidentiality of this information must be maintained with utmost discretion.

THE PROBLEM OF CONFIDENTIALITY. Who is to have access to the information revealed during counseling procedures? Suppose that a pedigree analysis uncovers some serious defect. Who, if anyone, is to inform relatives who might be similarly affected, or who might be carriers? It is possible that the patients themselves will prefer not to discuss this problem with close relatives—and may even decide that this essential information should not be communicated to them by anyone. The ethics of confidentiality are involved in a case like this, because physicians cannot take it upon themselves to discuss the case or the nature of the defect with other people. This ethical stance is reinforced in thirty-four states of the United States, which have laws governing the confidentiality of privileged medical communication.

ETHICAL ISSUES. Counseling and screening programs bring up many fundamental ethical questions. Do parents have a fundamental right to produce children under any circumstances? Does this right exist even if it is known that there is a high probability, or even a certainty, of defective offspring? Does the woman have an absolute right over the embryo and fetus developing within her? If the fetus is considered a living being, does it have the inalienable right to be born free of physical and/or mental defects? Does the fetus have rights in society? If so, when in gestation do they begin? If it were possible to examine all fetuses for physical and mental defects prior to birth, should this be done? If a woman has had one trisomic offspring, should she consider not having any additional offspring? Should a woman known to be heterozygous for a serious sex-linked recessive gene by the birth of an affected son feel free to abort all subsequent male fetuses?

If parents refuse to abort a defective fetus, can the child later sue the parents for having been allowed to come to term? Should the possibility of any defect, even a minor one, that distresses the parents be considered an adequate justification for amniocentesis and possible abortion?

Does society have the right or the responsibility to lay down rules for parents at risk of producing abnormal offspring? Or "too many" offspring? Is it fair to develop a screening program in which the focus is entirely on "bad genes" found predominantly in one ethnic group? Is it within the jurisdiction of the religions to stipulate the conditions for having (or not having) offspring? If a physician fails to inform a woman of a high probability, or a certainty, of having a defective child, and the alternative courses of action available to the parents, is that physician legally responsible? Does a human have a right to a certain tranquility of spirit, peace of mind, without being badgered by unsolicited, unwanted, and sometimes depressing genetic information about which she or he can do nothing? Would it be desirable to have a national repository of information containing what is known about each person's genetic constitution?

These are only a few of the many questions that have become acute during the past few years. Because there can be no hard-and-fast answers to these questions, the controversy surrounding them can be expected to continue for years to come.

Additional Reading

BODMER, W., and A. JONES. 1974. Genetic screening—The social dilemma. *New Scientist*, **63**:596–597.

COHEN, B. H., A. M. LILLIENFELD, and P. C. HUANG. 1978. *Genetic Issues in Public Health and Medicine.* Springfield, Ill.: Thomas.

CULLITON, B. J. 1975. Genetic screening: N.A.S. recommends proceeding with caution. *Science*, **189**:119–120.

Cutting the risk of childbirth after 35. 1979. *Consumer Reports,* **44**:302–306.

EPSTEIN, C. J., and M. S. GOLBUS. 1977. Prenatal diagnosis of genetic diseases. *American Scientist,* **65**:703–711.

HILTON, B. and D. CALLAHAN, eds. 1973. *Ethical Issues in Human Genetics. Fogarty Institute Proceedings,* **13**, New York/London: Plenum.

KOLATA, G. B. 1980. Prenatal diagnosis of neural tube defects. *Science,* **209**:1216–1218.

LENKE, ROGER R. and HARVEY I. LEVY. 1980. *The New England Journal of Medicine,* **303**:1202–1208.

MILUNSKY, A. 1979. *Genetic Disorders and the Fetus.* New York, N. Y.: Plenum.

OMENN, G. S. 1978. Prenatal diagnosis of genetic disorders. *Science,* **200**:952–958.

POWLEDGE, T. 1976. Prenatal diagnosis—Now the problems. *New Scientist,* **69**:332–334.

REID, R. 1974. Birth defects and society's obligation. *New Scientist,* **64**:746–747.

RISTOW, W. 1979. Genetic counselling: The third decade. *New Scientist,* **83**:976–978.

WERTZ, R. W., ed. 1973. *Readings on Ethical and Social Issues in Biomedicine.* Englewood Cliffs, N. J.: Prentice-Hall.

Questions

1. For what reasons do families seek genetic counseling? What does the genetic counselor try to accomplish?
2. Explain why the participation of medically qualified personnel is essential in the process of counseling.
3. What kinds of questions are usually asked of the family in making out a pedigree?
4. Can you explain the difficulties that those being counseled may have in understanding the simple rules of probability?
5. How successful has genetic counseling proved to be, based on later studies of counseled parents?
6. Can you imagine a situation in which counseling by an untrained person could lead to a harmful result?
7. If you were to act as a genetic counselor, what personal biases or convictions that you have would inevitably show up in the kind of advice that you would give? How could such an expression of personal bias be avoided?
8. One might imagine that individuals who definitely know that they carry a seriously debilitating genetic defect might produce fewer children than average. Does the experience with Huntington disease bear this theory out?
9. What is amniocentesis? What factors determine the time at which the procedure is to be carried out? How often do complications develop afterward?

10. What does a high level of AFP indicate? Where is it found?
11. Which general classes of genetic disorders belong to the various risk categories?
12. What is meant by *screening programs?* For which diseases have screening programs been instituted in one part of the country or another?
13. What is meant by *sickle-cell trait?* How has this terminology given rise to unnecessary problems?
14. Do you think that screening programs serve a useful purpose?
15. How is PKU treated? What problems may result from these treatments? When do heterozygotes for PKU stand a higher probability of being abnormal?
16. Describe some of the problems that arise when large segments of the population are screened for chromosomal abnormalities.
17. Do you think that a genetic counselor has an obligation to inform a patient of the availability of amniocentesis, if the condition under consideration can be detected *in utero?* Should the counselor inform the patient of all courses of action available (such as abortion) even though they may be contrary to his or her own moral principles?
18. Should a genetically defective child be allowed to sue its parents for "wrongful life" in those cases where the defect might have been detected prior to birth?

10

Locating Genes

One of the aims of human genetics is to pinpoint, as accurately as possible, the position of specific loci on the twenty-three human chromosomes. In pursuit of this effort, there are two general types of investigations:

1. Assigning a genetic factor to a specific chromosome, as, for instance, the loci of the Rh blood-group genes to chromosome 1 or the locus of the Tay-Sachs alleles to chromosome 15. Sometimes it is possible to limit the locus to a certain region of the chromosome, such as the narrowing of the ABO blood-group locus to the long arm of chromosome 9.
2. Demonstrating that two loci are both found on the same numbered chromosome (i.e., are syntenic on a specific one of the twenty-three different chromosomes) without necessarily knowing specifically which of the twenty-three carries those loci. We shall use only one or two examples to illustrate each of these methods.

Assignment of Loci to Chromosomes

X- AND Y-LINKED FACTORS. There is no difficulty in assigning a chromosome to characteristics that show X-linked inheritance; their loci must all be found on the X chromosome. That is, all X-linked loci are syntenic. Because this assignment can be made simply by knowledge of the sex-linked mode of inheritance of that characteristic (such as color blindness, hemophilia, and Duchenne muscular dystrophy), it is not surprising that more loci have been unambiguously assigned to the X chromosome than to any of the autosomes. Similarly, all loci that show holandric inheritance (and we have seen that there are very few of them that do) are syntenic and must necessarily be assigned to the Y chromosome.

AUTOSOMAL ASSIGNMENT.   Obviously determining the exact autosome on which an autosomal locus is to be found presents a greater problem, because there are twenty-two different autosomes, any of which could carry any particular locus. However, considerable progress has been made, and more than a hundred loci have now been assigned, with considerable certainty, to specific autosomes. There are several methods by which this has been done.

Cell Hybridization

A most spectacular recent advance has been the development of somatic *cell hybridization,* a process by which somatic cells, grown in culture, are induced to combine with somatic cells from another source to produce a so-called somatic cell hybrid. The properties of these hybrid cells and their descendents can give valuable information about the nature of the parental types.

Like so many other important observations in genetics, cell hybridization was first noticed in the course of experiments set up to check on an entirely different cell property. In 1960 several researchers in Paris mixed together two different mouse cancer lines that differed not only in their morphology but also in their chromosome number. During these tests to see whether the two kinds of cancer cells had an effect on each other, a third type was found in the culture. On investigation, it turned out to be carrying the chromosomes of both the original types. These cells were hybrids that had arisen by the fusion of cells of the two "parental" types. It was found that these hybrid cells could be perpetuated as distinct cell lines with some of the characteristics of both parents. Initially they had the total chromosome count of both parents together, but as mitosis continued over a period of time, the cells gradually lost some of the chromosomes and developed into individual lines with chromosomes of varying number.

Ordinarily cell fusion occurs spontaneously with a very low frequency, about one in a million. These particular hybrids were detectable because they had an advantage in rapid mitosis over the two parental types, eventually becoming a significant fraction of the total cell population. Since that time a number of treatments have been discovered that appreciably increase the frequency of hybridization. The method used initially was the addition of *Sendai virus,* related to those viruses that cause influenza. After the inactivation of this virus by ultraviolet light or chemical agents, it is added to a cell culture. It makes the cell surfaces "sticky," with the result that fusion occurs with a frequency from 100 to 1,000 times greater than in untreated cells. Now there are more convenient chemical treatments to promote cell fusion, such as the addition of polyethylene glycol.

SELECTION SYSTEMS. Even with this high fusion frequency, it is necessary to have some kind of selection system that will automatically differentiate between the very large number of parental cells and the relatively few hybrids. In principle this selection is very simple (Figure 10-1). This scheme is an adaptation of a method—perfected decades ago in genetic studies of molds and bacteria—for collecting rare combinations of new cell genotypes. One mutant cell line is unable to synthesize substance A, which is necessary for its survival and must be supplied as part of the basic medium in which the cells grow. Another strain is unable to manufacture substance B and survives and multiplies only when B is provided in the medium. We shall call the first strain A^-B^+ and the second A^+B^-.

These two cell types can be mixed together, and both will thrive as long as the medium contains both substances A and B. At this point a fusion-inducing material such as *Sendai* virus is added, and after an opportunity for fusion, the cells are removed from their former enriched medium and put into a new medium lacking both A and B. Under these conditions, cell types A^-B^+ and A^+B^- no longer flourish and eventually die. A hybrid cell, however, can now manage perfectly well because it carries the normal alleles for the A deficiency from its A^+B^- parent and the normal alleles for the B deficiency from its A^-B^+ parent, in a manner analogous to an F_1 heterozygote.

In this way, even though the incidence of fusion may be very small, hybrid cells are produced and detected very efficiently and are well differentiated from the parental stocks. In actual practice the number of deficient cell types available from humans is quite limited; most genotypes characterized by this kind of deficiency would also be lethal to the organism and therefore not recoverable. One of the few enzyme-deficient types comes from males

Figure 10-1 A simplified version of the procedure for obtaining cell hybrids. Two lines of cells, each with synthetic disability requiring the addition of a substance to the medium, are mixed in the presence of a virus or other hybrid-promoting agent. The two extraneous growth requirements are removed, killing cells of the original lines and allowing the survival only of cells with both synthetic properties. Unless new cell types are produced by a rather rare event as mutation, the survivors will be hybrids.

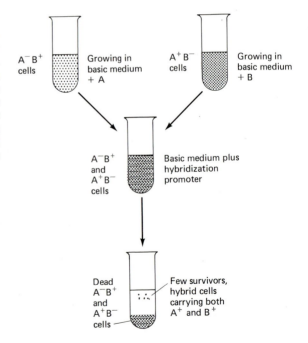

with the Lesch-Nyhan syndrome, a serious X-linked recessive defect characterized by mental and physical maldevelopment, self-mutilation, and short life span.

INTERSPECIES HYBRIDS.   Generally speaking, two grossly different species cannot mate and produce offspring. This restriction, however, does not apply to cell hybrids, and cells from quite unrelated organisms can be induced to fuse. Hybrids between cells of the mouse, the rat, the hamster, and humans are only a few of a large number that have been produced, and many other hybrid types will undoubtedly be possible when systems have been developed to select among the potential parent lines and their hybrids. In fact, the production of hybrid types among mammals is restricted only by the effort necessary to develop selection techniques versus the probability that the new hybrid will provide worthwhile information to the researcher.

It might be imagined that such an F_1 hybrid cell, once produced, would constitute a dead end, of minor value only as a scientific curiosity. These hybrids, however, exhibit a peculiarity of great importance when the parental lines derive from two different species. When the cells hybridize, their nuclear membranes also fuse so that all of the chromosomes of both species are found in a single nucleus. Thereafter the chromosomes become synchronous and progress together through the mitotic cycle, except that, for reasons that are not at all clear, the chromosomes of one species or the other may be lost. In this way, hybrid cells that start out with a complete diploid set of chromosomes from each of the parental cells lines gradually lose chromosomes so that after many generations the cell lines have a variable number of chromosomes.

In hybrids between mouse cells and human cells, the mouse chromosomes usually persist and the human cells are eliminated. This phenomenon might suggest that mouse chromosomes are better able to go through mitosis in hybrid cells, but in the cross of mouse and Syrian hamster cells, it is the hamster cells that preferentially survive. The loss of chromosomes makes it possible to perform a kind of segregation analysis by isolating clones of cells with varying number of chromosomes still present from one of the species.

MAPPING THE LOCUS OF THYMIDINE KINASE.   Let us take the case of a deficiency of the enzyme thymidine kinase (TK), which is known in mouse cell lines but has never been observed in humans. (This does not necessarily mean that the corresponding deficiency has never occurred in humans. It is quite likely that homozygous individuals with this defect would be lost very early during embryonic development.) From a mixed culture containing affected mouse cells, we obtain, using the procedures described above, an F_1 hybrid cell line. This hybrid line is normal with respect to the defect in question, so we have good reason for postulating that the human parental line has contributed genes that are capable of carrying out that function normally. As the hybrid cells undergo mitosis and the human chromosomes are preferentially lost, we can derive clones that have a small number of human chromosomes more-or-less randomly selected from the original

Table 10-1 Tests of clones derived from mouse-human hybrids, in which the original mouse line was thymidine kinase deficient (TK⁻). Note that whenever the clone is TK⁺ it carries the human chromosome 17.

Human Chromosomes Persisting in Clone	TK?
5, 9, 12, 21	−
3, 4, 17, 21	+
5, 6, 14, 17, 22	+
3, 4, 9, 18, 22	−
1, 2, 6, 7, 20	−
1, 9, 17, 18, 20	+

forty-six. After determining pecisely which human chromosomes are present in each clone, we test the clones for the original mouse enzyme defect. In the case in question we discover that every clone containing chromosome 17 from the original human complement is capable of synthesizing the enzyme TK (Table 10-1). In those cases in which chromosome 17 is missing, irrespective of which other chromosomes still remain from the human contribution, the clone is defective and TK activity is absent. It is a reasonable conclusion that the normal allele responsible for the synthesis of that enzyme in humans is located on chromosome 17.

A little thought tells us that we have performed a most remarkable experiment; we have used the mouse cells, which can be manipulated experimentally quite freely, to study the genetics of humans. By way of interspecific cell hybridization, we have made a significant contribution to our knowledge of the inheritance of various characteristics without the direct involvement of affected individuals. In the case of the TK deficiency, we can reasonably postulate that if it should occur in humans, it would be a disease with a simple genetic basis. We have even been able to state with some assurance that the locus responsible is carried on chromosome 17, a conclusion that would have been very difficult to come to even if this were a well-known characteristic in humans.

Cell hybrids may be used to advantage in still another way. Superoxide dismutase, an enzyme that mops up excess reactive oxygen in the cell, has a slightly different form in humans and mice and can be distinguished biochemically. If an array of cell hybrids with varying numbers of human chromosomes is examined, it is found that the human type of superoxide dismutase is present only when chromosome 21 is in the hybrid cell. Hence the locus must be on chromosome 21. The technique makes it possible, in principle, to draw on a store of mutant genes available in experimental animals, or to induce new ones in those animals using radiation or other mutagenic agents—procedures that would be scientifically as well as ethically prohibitive in humans. Furthermore we have been able to accomplish in a very short period of time an experiment that, were it not for the cell fusion technique, would have required many generations of human time. In fact, although this technique is fairly recent, it has been responsible for most of the information we now have on the allocation of loci to specific autosomes in humans.

It should be emphasized, however, that cell hybridization techniques can be applied to only a small fraction of genetic changes in humans, those that are identifiable in cell culture. At present are excluded a large number of important morphological characteristics, developmental and neurological anomalies that are found only in differentiated cells in the human organism.

MORE PRECISE LOCALIZATION BY CELL HYBRIDIZATION. Abnormal chromosomes, such as translocations and fragments, are found with a higher-than-normal frequency in cell hybrids, or abnormal chromosomes may be introduced by the human cell line at the time of hybridization. This procedure adds another possible refinement to the localization. Thus, if the long arm of chromosome 17 appears in a culture independent of the short arm of that chromosome, and if hybrid cells are TK$^+$ when the long arm alone is present, and not when the short arm alone is present, it follows that the human TK locus must be on the long arm.

Use of Unusual Chromosome Variants

The first unambiguous autosomal assignment was made in 1968 with the discovery of an unusual chromosome that had a long arm somewhat longer than normal. It was called *uncoiler 1* from its appearance (see Chapter 8) and is now known to be found in about one out of every several hundred persons (Figure 8-1).

This chromosome not only had the same abnormal appearance in the cells of the propositus but also appeared in some of his relatives. It was clear that the abnormal chromosome was transmitted from one generation to the next as a simple Mendelian characteristic. In several different studies of the inheritance of various blood groups in these kindreds, it was shown that for one blood group, called *Duffy (Fy)*, a specific allele present in an individual with the abnormal chromosome always segregated in the progeny with that chromosome. The conclusion from these observations is that the locus for the Duffy blood group is found on chromosome 1.

It is possible to get valuable information of a negative sort from trisomies and monosomies. Thus the common blood groups—ABO, MNS, and Rh—have more than three alleles or closely linked loci that can be identified by blood typing. If Down syndrome patients occasionally showed that they carried three alleles at any of these loci, instead of the normal two, this would constitute good evidence that the locus of the blood group in question was carried on chromosome 21. Evidence to the contrary indicated that these loci are on some chromosome other than 21, a fact that is borne out by the assignments that have now been made of these blood groups to chromosomes other than 21.

GENE DOSAGE IN DUPLICATIONS AND DEFICIENCIES. Deficiencies of chromosome material give an opportunity to pinpoint a locus. Five persons with such chromosome variations involving chromosome 13 were analyzed for the amount of esterase D activity. The results indicated the locus to be 13q14 (chromosome 13, long arm, and band 14) because there was reduced activity whenever the band 14 was absent. Although esterase D is an enzyme of unknown function, it may serve a useful purpose here because the same band, 13q14, is found to be deleted in many patients who develop retinoblastoma, the cancerous, and often fatal, condition of the eye. Thus it may be possible to identify with considerable certainty individuals at risk for retinoblastoma on the basis of the esterase D alleles they carry, after it has been determined which allele was carried by the parent who transmitted the allele for this condition.

DNA–RNA Homologies

In some instances it is possible not only to identify the chromosome on which a specific locus is found but even to approximate its location on the chromosome by chemical means. The two complementary strands that make up the DNA of the chromosome may be separated from each other by gentle heating. These two strands then tend to reassociate as they cool because of the complementarity of their structures. If two complementary DNA strands are completely separated and subsequently mixed, they tend, with time, to reassociate precisely with each other. RNA, which has a similar structure, also reassociates with its complementary DNA. In a few cases it is possible to obtain in quantity relatively pure RNA for specific loci. One such example is the RNA of hemoglobin. This RNA may be tagged so that it can be later identified; tagging is accomplished either by making the RNA radioactive or by attaching a fluorescent compound to it so that it may later be detected visually. If the DNA strands of the chromosome of the cell are separated and the tagged RNA is added to the cell, this RNA tends to associate with the homologous DNA at the position of the locus itself. In this way it has been shown that the two loci responsible for synthesis of the two different polypeptides of the hemoglobin molecule are on chromosomes 11 and 16.

CLONING. A powerful tool for locating genes on chromosomes depends on *cloning* a section of chromosome containing an identifiable gene. The locus for human growth hormone (GH) has been allocated to chromosome 17 in this way. The DNA from white blood cells was broken up into large fragments by enzymatic digestion. These fragments were then combined with the DNA from small viruslike plasmids, structures which reproduce themselves inside a bacterial cell. Many single bacterial cells are isolated that have a plasmid carrying a random chromosome segment attached. Each cell is then allowed to grow into a colony of many millions of identical cells, or a clone.

The clones thus derived can now be tested for whether they carry the segment of the chromosome with the locus for GH. When a clone is obtained that does, it can be grown in a medium containing radioactive phosphorous, ^{32}P, so that the DNA of the plasmid is radioactive. If the DNA is now added to human cells, the homologous DNA of the GH locus on the plasmid anneals with the DNA of the GH locus on the chromosome. If a photographic film is placed over the chromosomes, it will show dark spots at the region of the disintegration of phosphorous and in this way will mark the chromosome that carries the GH locus.

In practice it is somewhat more efficient to add the plasmid DNA to hybrid cells (such as man–mouse) that contain different human chromosomes. When the plasmid is added to a hybrid cell line that carries the chromosome with the locus, it anneals with that segment, but when the cell line does not, it will not. Therefore, by measuring the amount of radioactivity retained by various hybrid cell lines with different human chromosomes, it can be deduced, if one knows the human chromosomes present, which carries the cell locus. In this instance all the cell lines that showed high radioactivity had chromosome 17 and those showing low values did not. The reasonable conclusion is that the locus for GH is on chromosome 17.

Recombination Between Syntenic Loci

The method of determining the genetic distances between two loci on the same chromosome is best illustrated by the use of X-linked characters, for the simple reason that the X-chromosome genotype of a male can be unambiguously determined from his phenotype, as there is only one X chromosome in a male. Thus, if a female is heterozygous at two different loci—for instance, color blindness and muscular dystrophy (Becker type)—she can produce four types of male offspring: a type unaffected by either recessive, a type affected by both, and a type affected by one or the other singly. Our task is to make sense out of the relative frequency with which these four types occur.

Let us take the case shown in Figure 10-2. A woman with one X chromosome carrying the recessive for muscular dystrophy, all other alleles being the normal dominant, marries a male who is color-blind and who has the normal, dominant, allele M for muscular dystrophy. Among the children they produce, half the daughters will be doubly heterozygous, having re-

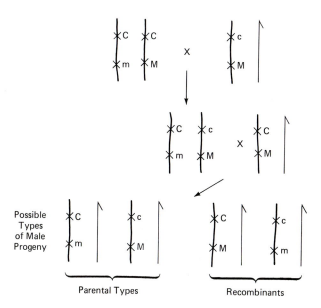

Possible Types of Male Progeny

Parental Types | Recombinants

Figure 10-2 A series of matings showing how genetic distances can be readily determined for X-linked loci. If two parents happen each to carry different alleles at two loci, they can produce a daughter who is doubly heterozygous. The male offspring of such a daughter can then be classified according to whether they carry the original combinations, the parental types, or recombinants.

ceived the recessive *m* allele from the mother on one X chromosome, and the recessive *c* allele on the other X chromosome, from the father. When such a female reproduces, we can immediately classify her sons according to the alleles they have received from their mother.

These sons may receive chromosomes, carrying the same alleles that were present in the mother, that is *mC* and *Mc*. These are *parental* combinations. However, with some frequency a crossover will occur between the two loci so that two new combinations, *MC* and *mc*, will be found in the male progeny. These are *recombinants*. The percentage of males showing recombinants, *mc* and *MC* in the total, is the frequency of recombination.

Of course, any one female will produce at most only a few male offspring, so the frequency of recombination may appear to be quite variable. There is, however, no reason that we cannot add together a number of similar pedigrees. Thus we might find several dozen females who are similarly heterozygous at these loci. On classifying all of their male progeny, we find the following distribution:

36 not dystrophic but color-blind   = *Mc*
40 dystrophic, not color-blind   = *mC*
10 neither dystrophic nor color-blind = *MC*
14 both dystrophic and color-blind   = *mc*

The first two categories, *Mc* and *mC*, are the parental combinations, non-recombinants, and the latter two, *MC* and *mc*, are recombinants. Because there are 24 recombinants out of a total of 100, the frequency of recombination is 24/100 or 24 percent. This percentage is usually referred to as 24 units.

The frequency of recombination is regarded also as being the genetic distance between the loci, and treated as distances, the frequencies of recombination between several loci can be used to construct a *chromosome map*. Before we consider the rationale behind mapping, however, it is necessary to review the events at meiosis that are responsible for genetic recombination.

THE MEIOTIC BASIS FOR CROSSING OVER. In the earlier discussion on meiosis in Chapter 3, it was noted that several unusual aspects of chromosome behavior are found at the prophase of the first division. In the first place, the forty-six chromosomes, which had each replicated into two identical chromatids at the preceding interphase, now pair with their respective homologues to produce twenty-three bivalents. The pairing is quite precise; *synapsis*, as it is called, occurs locus by locus along the entire length of the two homologues. The X and Y chromosomes, which lack such homology, pair only at the ends of their short arms. The chromatids of the homologous chromosomes exchange segments, or cross over. In the region of such an exchange, the chromosomes form a chiasma (plural, *chiasmata*), which can be seen as an X-shaped configuration between two chromatids of the homologous chromosomes (Figure 10-3). As a result of crossing over, two segments of chromatids are exchanged, with the result that of the four products of meiosis, each carrying its own chromatid, two have recombinants, or crossovers, and two have the original parental types (Figure 10-4).

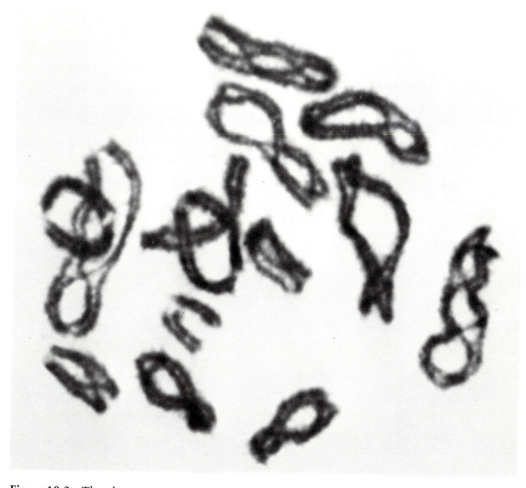

Figure 10-3 The chromosomes at late prophase of meiosis, showing the X-shaped chiasmata where homologous chromatids have undergone exchange.

Several points are suggested by Figure 10-4. The first is that to be detected genetically, an exchange must occur between two marked loci. In fact, exchange occurs regularly between homologous chromosomes in virtually all plants and animals, with between one and four exchanges occurring per bivalent. In the second place, if the positions of exchanges occur randomly along the chromosome length, then the chance that one will occur between two loci is greater the further apart the two loci are. We are justified, then, in concluding that two loci with 1 percent exchange between them are genetically closer than two with 10 percent exchange.

It is also clear that given several loci with recombination data, we can construct a chromosome map. Thus, if we find that the distance from locus A to locus B is 5 percent and from B to C is 15 percent, we can construct a map as in the lower left of Figure 10-5. However, an equally consistent map with such limited data is shown in the lower right of Figure 10-5. Whether

Figure 10-4 A diagrammatic representation showing exchange between two homologous chromosomes. The upper exchange is genetically undetectable because it does not occur between heterozygous loci. The lower exchange yields four genetically distinct types, two like the original parental chromosomes and two recombinants.

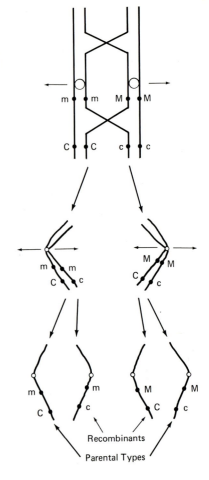

the first or the second is the correct map would be determined by data from a cross involving double heterozygotes at A and C. If the frequency of recombination between A and C is about 20 percent, then the map shown in A is the correct one. However, if it is about 10 percent, the map shown in B is correct. It would be helpful if we could obtain data from pedigrees in which *A, B,* and *C* are all segregating simultaneously, as from the triple heterozygote *ABC/abc*. Although this sort of information is routinely obtained in experimental plants and animals, it is virtually unknown in humans. Genetic chromosome maps in humans must be arduously built up from pedigree data, using two loci at a time.

Another aspect of recombination obvious in Figure 10-3 is that a single crossover event involves only two of the four chromatids. For each exchange, there are two crossover chromatids, and also two noncrossovers. Because it is a matter of chance which of the four chromatids will be recovered in a progeny, we can say that there is a 50 percent chance of recovering a crossover after an exchange, or, equally, that the frequency of recovered crossovers must be multiplied by two to give the number of exchanges that gave rise to them.

Figure 10-5 The principle of chromosome mapping, using crossover data, showing the additivity of crossover values. In one experiment, *A* and *B* prove to be 5 units apart, and in another, *B* and *C* are found to be 15 units apart. The quick conclusion is that *A* and *C* are 20 units apart (possibility I). However, there exists an equally likely possibility (II) that *A*, not *B*, is the middle locus. In that case, the expected distance between *A* and *C* would be 10 units. Additional information, such as data involving *A* and *C* specifically, would be needed to distinguish between these possibilities.

Can two exchanges occur between the two marked loci? As the distance becomes greater, the chance that two exchanges will occur between them also becomes greater. The effect of a second exchange might appear, at first sight, to reverse that of the first, as in Figure 10-6. However, this does not hold because any second exchange will involve two strands of the homologous chromosomes randomly with respect to those involved in the first exchange. Thus there are four possibilities for the second exchange with respect to the first exchange (Figure 10-7). If we examine the consequences of the combinations of two exchanges, we find that the doubles appear to be noncrossovers because the parental combination of the two alleles at different loci is restored. Thus the double exchanges produce 50 percent observed crossing over, just as the singles do. This exercise can be extended to triple exchanges as well as even higher orders; 50 percent is always the maximum amount of crossing over that can be observed between two loci, regardless of how far apart they are and how many exchanges may occur between them. It should be noted also that 50 percent recombination gives any four gametic types, *AB*, *Ab*, *aB*, and *ab*, equally frequently; that is, it is the equivalent of

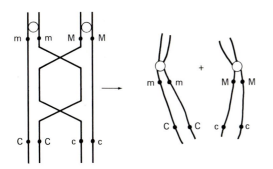

Figure 10-6 The occurrence of two exchanges between two loci such that the second restores the original condition changed by the first.

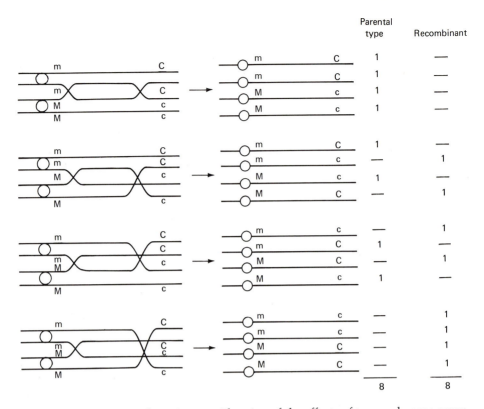

Figure 10-7 A more exhaustive consideration of the effects of two exchanges occurring between two marked loci. If the exchange at the left involves a certain two of the four strands, then the one at the right can involve any two of the other four strands at random, except that two sister strands do not exchange with each other. When all possibilities are taken into account, as they are in the drawing, then inspection of the consequences shows that half of the recovered chromosomes will be parental types and the other half will be recombinants, similar to the effects of a single exchange.

the expectation from an independent assortment of pairs of alleles with loci on different chromosomes.

It comes as no surprise, then, that cases are known where syntenic loci do not appear to be linked. Thus we can be positive that both red and green color blindness and the Xg blood group have loci on the X chromosome. However, from abundant data available because both loci have a high frequency of heterozygosity, recombinant types are as frequent as parental types. This finding would ordinarily suggest independent assortment resulting from the random segregation of alleles of two loci on different chromosomes. In this case, we can be sure that the high frequency of recombination can be attributed to a great genetic distance between the loci, so that one or more exchanges regularly occur between the loci.

As an aside, it may be noted that Gregor Mendel found only independent assortment when he tested all seven of his pairs of characters with each other. Later work has shown that the garden pea has only seven chromosomes. Actually several of the allelic pairs he used were syntenic, but he did not happen to use any close enough to show any deviation from random assortment. With many more loci he should eventually have discovered linkage, but this premature observation could not have fit into his conception of the rules of inheritance. As a result he might not have published his results, or if he had, his presentation would have been even less convincing to his contemporaries than it was.

Mapping of the Human Autosomes

The use of the X-linked characters of color blindness and muscular dystrophy as an illustration of the procedure of mapping was chosen because X-chromosome alleles, whether dominant or recessive, express themselves in every male offspring. Thus all parental combinations and recombinants can be unambiguously identified from every doubly heterozygous mother. Historically, more effort has been devoted to the study of pairs of autosomal loci in an effort to uncover any evidence for linkage. The problems of detecting linkage for autosomal loci are considerable but not insurmountable. Ideally, one would look for pedigrees in which one parent was *AB/ab* (or *Ab/aB*) and the other *ab/ab*. Then, in the absence of linkage, the alleles at the A and B loci would assort independently, giving parental combinations and recombinants in the progeny with equal frequencies. On the other hand, if there were complete linkage, only the parental combinations *AB* and *ab* (or *Ab* and *aB*, for the second case) would be found in the progeny. Any intermediate case would indicate synteny with some amount of crossing over.

But how often is a family found in which one parent has two loci heterozygous and the other is homozygous recessive at both loci? In the case of the loci for albinism and PKU, for instance, the mating of one person heterozygous at both loci to another homozygous recessive at both has never been observed and, in fact, has probably never occurred. This problem can be circumvented to some extent by using loci which possess several readily detectable (i.e., dominant) alleles that occur with a high frequency in the population. The alleles of the various human blood groups have these characteristics. This is why the first three cases of synteny to be well established (in the early 1950s) all involved human blood groups (Lutheran blood group and Secretor, six units apart; Rh blood groups and elliptocytosis, two units apart; ABO blood group and nail-patella syndrome, ten units apart).

Data on possible cases of linkage continue to accumulate, particularly in medical genetic laboratories which can classify kinships for a large number of blood groups, enzymes and other loci characterized by a number of readily detectable alleles. Progress is impeded on the one hand by steadily decreasing family size but promoted on the other by the use of advanced biochemical methods for detecting genetic differences and by the use of computerized techniques for analyzing extensive data.

A chart showing the relative positions of various loci on chromosome 1 is shown in Figure 10-8. The combination of this type of information, along with the localizations that come from the procedures described earlier, make the mapping of the human genome (Figure 10-9) one of the most rapidly developing areas of human genetics.

Figure 10-8 Maps of chromosome 1. *A.* A physical map, showing the approximate location of some loci. *B.* A genetic map. The percentage of recombination between adjacent loci is approximately equal to the numerical difference between the numbers marking the position of the loci. (D. C. Rao, et al. *American Journal of Human Genetics* **31**:680–696, 1979; reprinted by permission of The University of Chicago Press.)

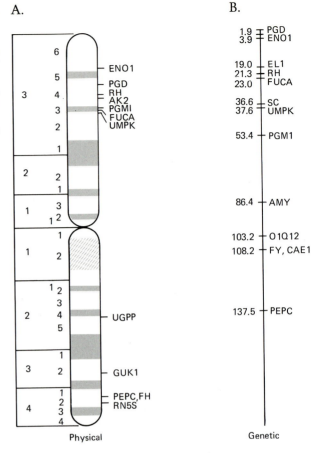

A.

B.

Physical   Genetic

Additional Reading

DEISSEROTH, A., et al. 1977. Localization of the human α-globin structural gene to chromosome 16 in somatic cell hybrids by molecular hybridization assay. *Cell,* **12**:205–218.

DEISSEROTH, A., et al. 1978. Chromosomal localization of human β-globin gene on human chromosome 11 in somatic cell hybrids. *Proceedings National Academy of Sciences,* **75**:1456–1460.

DONAHUE, R. P., et al. 1968. Probable assignment of the Duffy blood group locus to chromosome 1 in man. *Genetics,* **61**:949–955.

EPHRUSSI, B., and M. C. WEISS. 1969. Hybrid somatic cells. *Scientific American,* **220**:2–11.

FRANTS, R. R., et al. 1975. Superoxide dismutase in Down syndrome. *Lancet,* **2**, pt 1:42

GEORGE, D. L., and U. FRANCKE. 1976. Gene dose effect: Regional mapping

Figure 10-9 The location of genes on the human chromosome map. Note that many of the locations are tentative and not confirmed, that the loci are usually those of enzymatic properties, and that many common defects, such as albinism or cystic fibrosis, are conspicuously absent, unless their chromosome allocation is unambiguous because of *X*-linkage. (Courtesy of P. Meera Khan, University of Leiden.)

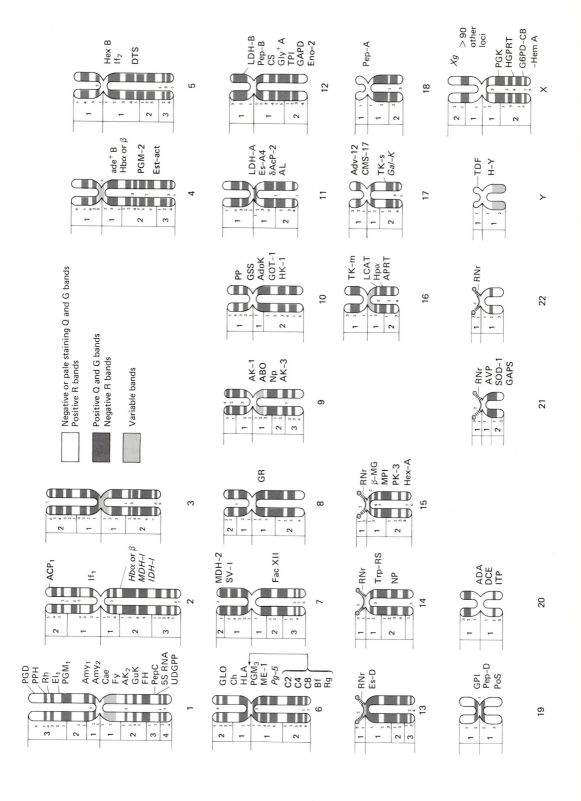

of human nucleoside phosphorylase on chromosome 14. *Science,* **194**:851–852.

HARRIS, H., et al. 1965. Mitosis in hybrid cells derived from mouse and man. *Nature,* **207**:606–608.

KLEBE, R. J., C. TCHAW-REN, and F. H. RUDDLE. 1970. Mapping of a human genetic regulator element by somatic cell genetic analysis. *Proceedings National Academy of Sciences,* **66**:1220–1227.

MAGENIS, R. E., et al. 1975. Gene dosage: Evidence for assignment of erythrocyte acid phosphatase locus to chromosome 2. *Proceeding National Academy of Sciences,* **72**:4526–4530.

MILLER, O. J., P. W. ALLDERDICE, and D. A. MILLER. 1971. Human thymidine kinase gene locus assignment to chromosome 17 in a hybrid of man and mouse cells. *Science,* **173**:244–245.

NABHOLZ, M., V. MIGGLANO, and W. BODMER. 1969. Genetic analysis with human-mouse somatic cell hybrids. *Nature,* **223**:358–363.

OWERBACH, D., et al. 1980. Genes for growth hormone, chorionic somatomammotropin and growth hormone-like gene on chromosome 17 in humans. *Science,* **209**:289–292.

SCHWARTZ, A. G., P. R. COOK, and H. HARRIS. 1971. Correction of a genetic defect in a mammalian cell. *Nature (New Biology),* **230**:5–8.

SPARKES, R. S. 1980. Regional assignment of genes for human esterase D and retinoblastoma to chromosome band 13q14. *Science,* **208**:1042–1044.

WEISS, M. C., and H. GREEN. 1967. Human-mouse hybrid cell lines containing partial complements of human chromosomes and functioning human genes. *Proceedings National Academy of Sciences,* **58**:1104–1111.

WESTERVELD, A., et al. 1971. Loss of human genetic markers in man-Chinese hamster somatic cell hybrids. *Nature (New Biology),* **234**:20–24.

Questions

1. Why is it relatively simple to show that some characteristic (brown enamel of the teeth, for instance) is determined by alleles on the X chromosome?
2. How was the phenomenon of cell hybridization discovered? How is it carried out in practice?
3. What peculiar behavior of the chromosomes in all hybrids makes it possible to localize genes to certain chromosomes?
4. List some of the advantages of the cell hybridization technique over standard human genetics methods, like pedigree studies. What are the limitations of this technique?
5. How was it possible to show that the locus for a blood group known as *Duffy* was on chromosome 1?
6. How might enzyme levels in trisomics and monosomics provide some information about the location of genes? What are the limitations of this method?
7. Do you think that you might come to a valid conclusion about the position of the ABO locus with respect to chromosome 21 simply by collect-

ing information on the frequency of group 0 (homozygous 00) in Down patients?

8. How is it possible to locate genes by making use of the fact that DNA in the single-stranded state tends to reassociate with homologous RNA? What are the limitations of this method? For which loci has it been particularly successful?

9. How can cloning segments of DNA in bacterial cells be used as a method for determining gene locations?

10. What is meant by *recombination* between syntenic loci? If a woman carries two dominant alleles on one of her X chromosomes and two recessives on the other, what recombinant types of progeny might she produce?

11. Does the frequency of recombination have any physical relation to the chromosome, or is it simply an abstract measure?

12. When does recombination occur during meiosis? What is the relationship of recombination to the chiasmata?

13. How can recombination data involving three or more loci be combined into a chromosome map?

14. Will a double crossover act in the same way as two single crossovers occurring simultaneously, so that the second crossover "undoes" the first?

15. If two loci are syntenic, but very far apart, so that several—even many—crossovers occur between them, will there ever be more than 50 percent crossing over observed in the progeny?

16. Gregor Mendel worked with seven pairs of allelic factors, and the garden pea has seven chromosomes. He observed only independent assortment, yet we now know that several of his pairs were syntenic. Can you explain why he did not observe linkage?

17. Give several reasons why mapping autosomal genes is much more difficult than mapping genes on the X chromosome.

11

The Blood Groups

The periodic bouts that we have with viral and bacterial diseases, illnesses of several days with gradually increasing severity, are usually followed by an apparently miraculous recovery, often unaided by any medication, antibiotics, or medical care. These experiences serve as recurring reminders of our dependence on our own *immune system* to combat external invaders. Any person who has wandered into a patch of poison ivy or poison oak may have become painfully aware of the existence of the body's immune system. Those who suffer from hay fever dramatically exhibit an *immune reaction*. The case of the unfortunate person who has become sensitized to as common an event as a bee or wasp sting, and who subsequently dies after another single sting, is not an uncommon news item.

Infectious Disease

Humans have always been at the mercy of infectious disease; throughout time it has been the most common cause of death. During the Middle Ages, the Plague decimated entire populations, paralleled by the more recent introduction of diseases by white men into areas occupied by peoples not previously exposed. Just as the Indians were wiped out in Yucatán by the first epidemic of yellow fever, introduced by Europeans in about 1650, so were previously unexposed Europeans who later migrated to the West Indies and West Africa similarly stricken with yellow fever. The deaths from these epidemics were not limited to children and the weak; in fact, sometimes these diseases seemed to favor robust, mature, and apparently quite healthy adults. Even in recent times epidemics have occurred: it is estimated that in 1918 and 1919 influenza killed a total of 20 million people worldwide, 12 million in India alone.

As time goes on, more and more of these scourges are being eliminated: poliomyelitis, typhoid fever, cholera, smallpox, diphtheria and scarlet fever, to name a few. The colored poster nailed to the entrances of homes warning of the presence of a patient with a serious communicable disease, common a few years ago, is now virtually unknown in the Western world. To be sure, knowledge of the biological nature of the causes of disease, the use of vastly improved sanitation, and the development of effective medicines have all been important in the struggle, but there can be no doubt that the application of most elementary techniques involving the immune reaction has played a leading role.

Some of the characteristics of the immune system are technically complex, but the most important are not. Every person is familiar with the widespread occurrence of childhood diseases such as chickenpox, measles, and mumps, and with the fact that once a person has had such an illness it will in most cases never return. One classic example of the long-range memory of the immune system comes from the measles epidemics that ran through the Faeroe Islands in 1781, 1846, and again in 1875. In each instance large numbers of people were affected, but in the later epidemics persons who had been alive during the previous one and had contracted the disease at that time were not again affected, despite the intervening decades.

On a more personal level we are reminded of our own immunity characteristics when our blood is typed and we are designated as O-positive or A-negative, for instance, and are vaguely aware that this information is of great importance if our blood is to be transfused into another person (or, less fortunately, if we ourselves require a transfusion). The adverse effect of the *Rh factors* in newborn infants has received considerable attention in the daily press, and women who have been tested and found to be Rh-negative are forewarned that their progeny might suffer as a result. Finally, the press reports case after case of organ transplantation from one human into another, many of them successful and many not; in the latter cases, we know that the failure is often due to *rejection* of the transplanted organ, a property of the immune system.

The ABO Blood Groups

The art of the blood transfusion, which has been instrumental in saving literally millions of lives over the past sixty years (developed just in time to have a sizable impact on the chances of survival of the wounded during World War I), evolved slowly from a historical standpoint. The reasons for its late arrival include a prior lack of information about the circulatory system and about the necessary sterile techniques, and an absence of the tools (e.g., hollow needles) necessary to perform the operation properly. The major obstacle, however, was the lack of any understanding of the blood groups.

HISTORICAL DEVELOPMENT. Possibly the earliest attempt to "transfuse" blood took place in 1492 when Pope Innocent VIII, in a coma, was administered the blood of three young men. He died, along with the three donors. (It seems likely that he was given the blood to drink rather than as a true transfusion.) In the 1660s the physician to Louis XIV of France performed a successful transfusion of about half a pint of lamb's blood into a boy who was weak and feverish from excessive bloodletting, a common practice then thought to have curative effects. However, subsequent transfusions were unsuccessful and that physician was charged with murder. Blood transfusions, which had such a high fatality rate at that time, were banned in France, England, and Italy.

At the beginning of the nineteenth century there was a resurgence of interest in this technique after it was shown to be effective in some cases of anemia. Almost 500 transfusions were performed during that century throughout Europe, the majority with human blood and the rest with blood from animals. Again, however, there were severe unexpected reactions in such a large proportion of the patients that the technique was abandoned once more.

LANDSTEINER'S INITIAL DISCOVERY. The primary cause for this incompatibility of the blood of different persons was discovered by the immunologist Karl Landsteiner, who showed in 1900 that human blood can be classified into several distinct groups. Subsequent work quickly established the number of such groups to be four.

We can demonstrate the existence of these four groups quite easily. Let us take samples of blood from 100 persons, number the samples carefully, and mix a drop of each sample with a drop from every other sample. After a few minutes we will see that in about a third of the mixtures clumping of the red cells, or *agglutination,* occurs (Figure 11-1), but the rest remain fluid. At first, it may appear that agglutination is taking place at random between different samples. However, a closer inspection (and some thought) reveals something more: that the samples fall into groups, and that samples do not agglutinate when mixed with others of the same group but do when mixed with those of other groups. Furthermore the number of such "self-compatible" groups is four.

THE BASIS FOR AGGLUTINATION. To understand why the cells in some mixtures agglutinate and others do not, we must consider the way by which the body protects itself against foreign invaders. When certain substances, usually large complex molecules, enter the bloodstream, the body is able to recognize them as alien and reacts by producing highly specific protein molecules, called *antibodies,* against them. Any substance that triggers antibody formation is called an *antigen.* The particular antibody produced as a result of the presence of a given antigen is able to form a chemical union with that antigen. When the antigen is located on a red blood cell, the antibody joins two cells together, one at each end of the molecule, causing the

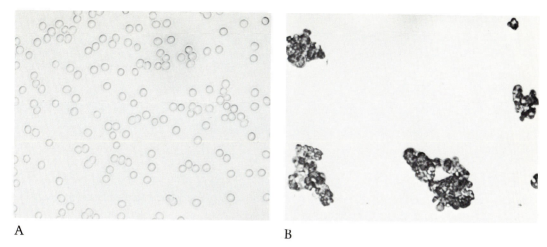

A   B

Figure 11-1 A comparison of normal red cells *(A)* of group A with cells that have been agglutinated *(B)* by the addition of the anti-A antibody. Note that the clumps are formed as cells associate with each other. This phenomenon is not to be confused with clotting, which is caused by the formation of fibrin in the blood fluid. (Courtesy of J. Puro, University of Turku.)

cells to agglutinate (Figure 11-2). Each of us possesses a large number—thousands—of compounds with antigenic properties as part of our natural makeup; others may be introduced into our system from outside the body. Similarly the body normally carries hundreds of different kinds of antibodies, most of them having been previously induced in response to exposures earlier in life to foreign substances, and the body is able, on call, to produce antibodies to new antigenic stimuli.

THE PLASMA AND CELL COMPONENTS. Clearly our first experiment was inadequate to indicate the role of the antigens and antibodies in the agglutination reaction. For this we must refine our technique by separating the red blood cells, or *erythrocytes*, from the fluid, or *plasma* of the blood. This is

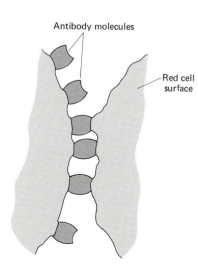

Antibody molecules

Red cell surface

Figure 11-2 Detail showing how a specific kind of antibody molecule, with reactive sites at each end, can link two red cells together. If the antibody molecules attach themselves over the entire surface of the cell, large numbers can be held together in single clumps. (Adapted from F. Haurowitz, *Immunochemistry and the Biosynthesis of Antibodies,* Interscience Publishers, New York, 1968.)

	Erythrocytes			
	1	**2**	**3**	**4**
1	−	+	+	+
2	−	−	+	+
3	−	+	−	+
4	−	−	−	−

Table 11-1 The occurrence of agglutination (+) or not (−) when plasma of persons from the four different blood groups is added to the four types of erythrocytes.

Plasma (row label for the left side of the table)

easily accomplished by putting tubes with the blood samples into a centrifuge and spinning the cells to the bottom. Now we have two components, the cells and the plasma, for each of the four blood groups, and we can test all four classes of erythrocytes against all four classes of plasma. The results from such combinations are shown in Table 11-1; a positive (+) sign means that an agglutination reaction has occurred, and a negative (−) sign represents no reaction. It appears that this table does not make any sense because the + and − signs are somewhat mixed together. We shall see, however, that there is a simple explanation for the pattern of + and − signs.

Human erythrocytes may carry either one of the two antigens, A or B, or both together, or neither. The four groups into which we have classified our blood samples are named according to whether their erythrocytes carry neither antigen (group O), antigen A (group A), antigen B (group B), or both antigens (group AB). The plasma may contain antibodies against the A antigen, against the B antigen, against both together, or against neither. When the erythrocytes of a person of each of the four groups are added to antibodies of either A or B, the results are seen in Figure 11-3.

Each person possesses, as part of the *ABO* antibody constitution, those antibodies for which he or she does not have the corresponding antigens. Why humans carry these antibodies is not immediately obvious; one reasonable explanation is that bacteria and other microorganisms have antigens similar to the human A and B antigens, and these induce the formation of the corresponding antibodies in individuals to whom these antigens are foreign. The antigen-antibody compositions of persons of each of the four groups are listed in Table 11-2 and schematically represented in Figure 11-4. We can now reinterpret the results shown in Table 11-1 on the basis of the antigen–antibody components found in each of the four blood groups based on Table 11-2 and Figure 11-3. These are given in Table 11-3. A positive agglutination reaction occurs only when the plasma carries the antibody corresponding to the antigen on the red cell.

TRANSFUSION PROBLEMS. When the erythrocytes of a person carrying either of the two antigens are transfused into a person with the corresponding antibody, the transfused cells are subject to agglutination. The agglutinated red blood cells no longer flow freely to carry oxygen to the other cells of the body, and agglutinated blood masses lodge in the smaller blood vessels, cutting off the flow of the nonagglutinated blood often with fatal consequences.

Anti A

Anti B

Group O

Group A

Group B

Group AB

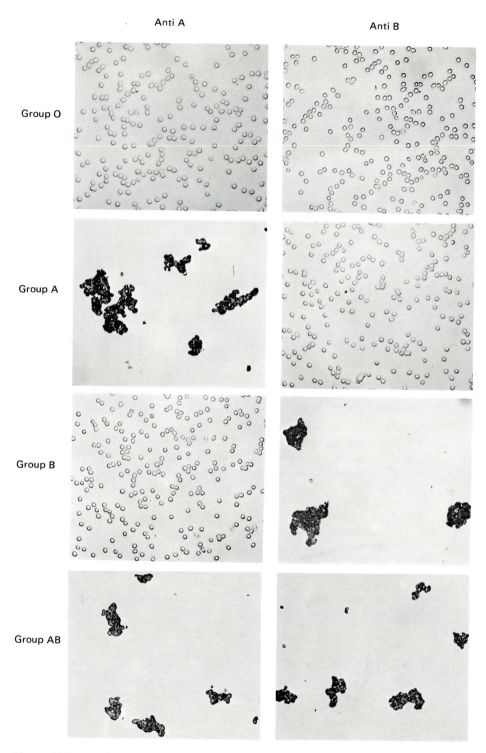

Figure 11-3 Agglutination response when erythrocytes of all four groups are mixed with antiserum containing either anti-A or anti-B antibodies. (Courtesy of J. Puro, University of Turku.)

Blood Group	Antigen on the Red Cells	Antibody in the Plasma
O	Neither	Both A and B
A	A	Anti-B
B	B	Anti-A
AB	Both A and B	Neither A nor B

Table 11-2 The presence (or absence) of the A and B antigens on the erythrocytes of individuals of the four blood groups, and the presence (or absence) of the antibodies to those antigens in the plasma. Note that no person simultaneously carries both a given antigen and the antibody against it but does carry the antibodies for any lacking antigen.

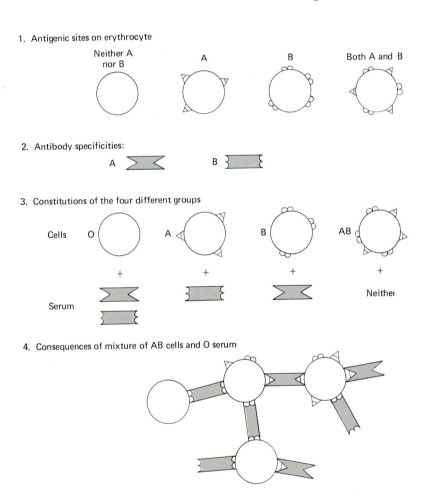

1. Antigenic sites on erythrocyte

2. Antibody specificities:

3. Constitutions of the four different groups

4. Consequences of mixture of AB cells and O serum

Figure 11-4 The complementary relationship between the specific antigenic binding sites on the cell and the corresponding antibody sites. The cells from the four groups have different surface configurations *(1)* corresponding to the antigenic structure *(2)*. Each of the four groups has the antibodies for which cells have no reactive sites *(3)*. When cells of one group are mixed with the plasma of another, agglutination occurs if the antibodies of the mixture find the complementary site on the cells *(4)*; in the case illustrated, the cells have both A and B sites, and both antibodies are present.

Table 11-3 Reinterpretation of Table 11-1 in view of the actual antigen–antibody compositions of each blood group. When a given antigen and its corresponding antibody are found in the same square, a positive agglutination reaction results.

	Erythrocytes			
Plasma	*Group O (Neither Antigen)*	*Group A (Antigen A)*	*Group B (Antigen B)*	*Group AB (Both A and B Antigens)*
Group O (both anti-A and anti-B antibodies)	–	+	+	+
Group A (anti-B antibody)	–	–	+	+
Group B (anti-A antibody)	–	+	–	+
Group AB (neither)	–	–	–	–

As an illustration of the cases where difficulties may be anticipated, consider the results of transfusing blood from a person of group A into four recipients, each of a different blood group. These results are given in column 2 of Table 11-3. When the recipient is of group A, the erythrocytes have the same antigens as the donor's, and the blood plasma the same antibodies, so that no complications will result. The bloodstream of the group AB recipient possesses no antibodies of either group, so again no reaction will occur. On the other hand, both group B and group O plasma contain anti-A antibodies, so the incoming group A erythrocytes will be agglutinated.

One might think that the erythrocytes with a given antigen (A, for instance) in the recipient would be agglutinated by the corresponding antibodies (as anti-A) from the donor's blood plasma but in practice this reaction is found to be relatively unimportant because the transfused antibodies are quickly diluted and absorbed by other tissues in the recipient. For this reason, compatible transfusions may not be reciprocal: O blood transfused into an A, B, or AB recipient ordinarily presents no problem, whereas A, B, or AB blood transfused into an O recipient does. Because it is the *erythrocyte composition of the donor* and the *plasma composition of the recipient* that determine the compatibility of different combinations in transfusions, Table 11-3 can be used to indicate likely transfusion reactions if the word *erythrocyte* is replaced by *donor,* and *plasma* by *recipient.* In medical practice, however, it would be unwise to rely on presumed compatibility based on such a chart. The customary procedure is to make a laboratory test of the donor's and the recipient's blood to make sure that no agglutination will occur, either because of ABO blood-group incompatibility or for any other reason, such as an unsuspected antigen-antibody incompatibility involving other factors.

Phenotype	Genotype
O	OO
A	AA, AO
B	BB, BO
AB	AB

Table 11-4 The phenotypes and corresponding genotypes of individuals of the four different ABO blood groups.

INHERITANCE OF THE ABO BLOOD GROUPS. Up to this point in this book we have been concerned with those loci at each of which is one "normal" allele, vaguely defined as that allele commonly found in the average person, which has an alternative allele, either dominant or recessive, considered abnormal because it may be responsible for a phenotype that deviates from normality. For the ABO blood groups there are three major alleles, whose locus is on chromosome 9. These alleles were originally designated by superscripts A, B, and O following the letter I. However, the symbolism I^A, I^B and I^O soon degenerated into simply A, B and O. Following common practise, we too shall designate the three alleles by the italicized letters *A*, *B* and *O*. Of course only one allele is found on each chromosome 9 and, because each person has two number 9 chromosomes, only two of the three alleles are found in each cell. Thus the possible genotypes are *AB*, *AA*, *AO*, *BB*, *BO* and *OO*.

A person who has the *A* allele produces antigen A, and a person who has the *B* allele produces antigen B. Thus the homozygotes *AA* and *BB* are in groups A and B, and because *O* can be considered recessive, *AO*, *BO*, and *OO* are in groups A, B, and O, respectively. The heterozygote *AB* is in group AB. Because the *A* and *B* alleles express themselves completely in heterozygotes, they are referred to as *codominant* alleles, and such sets of more than two allelic possibilities are referred to as *multiple alleles*.

The list of possible genotypes in the normal diploid individual, along with their corresponding phenotypes, is given in Table 11-4.

From this listing we can immediately see what progeny might result from the mating of any two parents. If they are both group O (*OO*), for example, all of their children must also be O, because there is no possibility of their acquiring any other allele. On the other hand, if one parent is A (of heterozygous genotype *AO*) and the other B (of genotype *BO*), then all four blood groups are possible among the progeny: *AB*, *AO*, *BO*, and *OO*. A complete listing is found in Table 11-5.

The MN System

The ABO groups were discovered before any of the other blood groups because of the adverse and frequently fatal reaction after the transfusion of the blood of one person into another. This agglutination is immediate because each person naturally carries the antibodies for those antigens of the AB system not present. This is the case, however, only for the AB antigens;

Table 11-5 Progeny expected from matings of each of the four groups in all combinations. For clarity the possible genotypes are included.

Father		Mother O OO	A AA, AO	B BB, BO	AB AB	Mother not specified
O	OO	O	O, A	O, B	A, B	O, A, B
A	AA AO	O, A	O, A	A, B, AB, O	A, B, AB	O, A, B, AB
B	BB BO	O, B	O, A, B, AB	O, B	A, B, AB	O, A, B, AB
AB	AB	A, B	A, B, AB	A, B, AB	A, B, AB	A, B, AB
Father not specified		O, A, B	O, A, B, AB	O, A, B, AB	A, B, AB	

for other blood groups, antibodies are not naturally found to the antigens not present, and in many cases they can be induced in humans only with difficulty.

THE DISCOVERY OF THE MN GROUPS. More than a quarter of a century (1900 to 1927) elapsed between the discovery of the ABO groups and the description of a new independent system, the MN groups. Because humans do not usually carry antibodies to the MN antigens, the corresponding antibodies must be induced in experimental animals, in the following way. The erythrocytes of a person who has been completely characterized with respect to all known blood groups are injected into an experimental animal—a rabbit or a guinea pig (Figure 11-5). Several injections at intervals of a week or more increase antibody production in the animal. Plasma is then removed from the animal; it contains large numbers—undoubtedly many dozens—of antibodies newly induced by the human tissue. To the serum

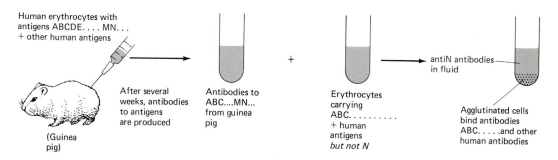

Figure 11-5 The technique for testing for additional erythrocyte antigens, leading to the discovery of the MN groups.

Table 11-6 Types of progeny produced by parents with the three different phenotypes of the MN system.

		Phenotype of Mother		
		M	MN	N
	M	M	M,MN	MN
Phenotype of Father	MN	M,MN	M,MN,N	MN,N
	N	MN	MN,N	N

containing all of these antibodies are added erythrocytes from another person identical to the first in all known groups. Practically all of the newly formed antibodies are absorbed at the reaction sites on the erythrocytes; these include antibodies to many antigens peculiar to humans and not rabbits. There remain in the plasma however, antibodies corresponding to any antigens possessed by the person who provided the erythrocytes for the initial injection, but not by the person providing the erythrocytes for the absorption. These antibodies are evident because the apparently completely absorbed plasma is still capable of agglutinating the original erythrocytes. Tests can then be made to determine the possible genetic basis of the new antigen responsible for the unexpected antibody.

In this way, it was found that all persons could be categorized into three antigenic classes, quite independent of the ABO groups both serologically and genetically. These classes were called M, N, and MN: the first class carried the new antigen M, the second N, and the third carried both.

GENETICS OF THE MN GROUPS.   On chromosome 4 is found the locus of two alleles, *M* and *N*. Individuals of group M are homozygous *MM*, those of group N are homozygous *NN*, and those of group MN are heterozygous *MN*. The kinds of progeny expected from each possible combination of matings is easily deduced from the fact that group M produces only *M* gametes, N produces only *N*, and MN produces half *M* and half *N* gametes. The possible combinations are given in Table 11-6.

The Rh Factors

Prior to 1940 there were several mysterious diseases of newborns, apparently unrelated, with severe symptoms, sometimes ending in death. In some cases the child, or more often the stillborn infant or aborted fetus, was characterized by an extreme excess of fluid that made it appear swollen. In others, the skin turned yellowish (technically, *jaundiced*) shortly after birth. Another congenital disorder was called *erythroblastosis fetalis,* because of the large number of erythroblasts (immature red blood cells) in the fetus's blood. Such children had an excess of blood-forming tissue, which spread over the entire body, causing swelling of the liver, the spleen, and other organs. We now know that most of these cases have a similar cause and accordingly are grouped under the common name *hemolytic disease of the newborn (HDN)*. About 1 child in 200 was born with HDN before 1940.

A number of medical workers in the 1930s made notable advances in attributing the cause of these maladies to an antigen-antibody reaction, but the solution to the problem came from two directions. Karl Landsteiner and A. S. Wiener demonstrated in 1940 that when the blood of a rhesus monkey was injected into rabbits, the rabbit blood developed antibodies that could agglutinate not only rhesus blood but most human blood as well. From a sampling of New York City residents, Landsteiner and Wiener found that 85 percent had an antigen on their erythrocytes that would react to this anti-rhesus antibody. The antigen, found in both humans and monkeys, was referred to as the *Rhesus antigen*.

A year earlier, Levine and a co-worker had pinpointed a specific antigen ("X") as the cause of HDN in the successive fetuses of a patient and as the cause of a severe agglutination reaction in that patient after a blood transfusion from her husband. The Rhesus antigen was subsequently shown to be identical to "X." This blood group is now referred to as the *Rh group*; individuals carrying the antigen are called *Rh-positive*.

THE PATTERN OF INHERITANCE. Extensive research by immunologists, geneticists, and medical practitioners has shown that the genetic basis for Rh inheritance is a very complex allelic system—one of the most complex known in humans. These will be discussed later. However, it is possible to characterize the system simply, by assuming the existence of two alleles, *R* and *r*, located on chromosome 1. A person who is *Rh-negative* (Rh$^-$)—that is, who lacks the Rh antigen—is homozygous for the recessive allele *r*. *Rh-positive* (Rh$^+$) persons are either homozygous *RR* or heterozygous *Rr*.

THE CAUSE OF HDN. Unlike the ABO groups, in which each person carries the antibodies for those A or B antigens that are absent, Rh antibodies are not found in humans unless they are specifically induced. However, Rh$^-$ people who lack the Rh antigen may be good producers of anti-Rh antibodies if Rh$^+$ cells are introduced (one drop of blood may be sufficient!) into their bloodstreams. In fact, about 70 percent of all Rh$^-$ persons injected with Rh$^+$ red blood cells develop a good level (or *titer*) of anti-Rh antibodies.

If a pregnant Rh$^-$ woman has anti-Rh antibodies in her bloodstream, they may pass through the placenta into the circulatory system of the fetus. If the fetus is Rh$^+$ (having inherited an *R* allele from the father), the antibodies going from the mother into the child's circulatory system can destroy the fetal erythrocytes, causing extensive damage. Figure 11-6 illustrates the course of events leading to HDN caused by the Rh factors. Of all Rh$^+$ children born to Rh$^-$, but sensitized, mothers, about one third die if not treated.

The mother may have developed the anti-Rh antibodies in response to an earlier transfusion. Prior to that transfusion, the ABO compatibility would have been checked and a crossmatch would have been made to test for the likelihood of an immediate antigen-antibody reaction, but a crossmatch is not sufficient to eliminate the possibility that the donor is Rh$^+$ and the

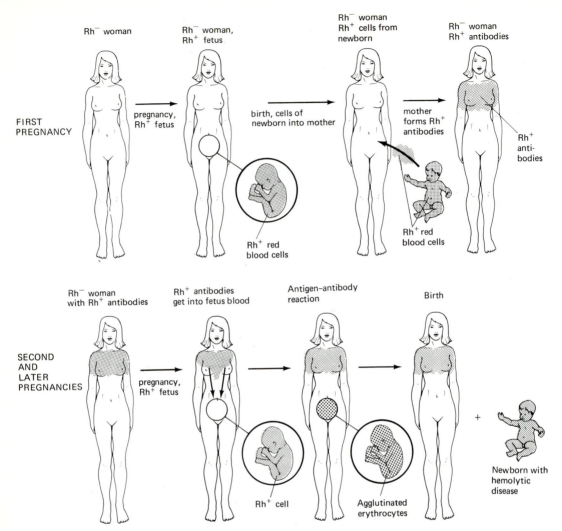

Figure 11-6 The series of events leading to hemolytic disease of the newborn (HDN). An Rh⁻ woman is exposed to the Rh⁺ antigen. If she develops Rh⁺ antibodies and produces an Rh⁺ fetus, the antibodies may pass from her circulation into the fetus, destroying erythrocytes in the fetal circulation and causing the premature loss of the fetus or the birth of a child with HDN.

recipient Rh⁻ but lacking antibodies at that time. If an Rh⁻ woman is carrying an Rh⁺ fetus, some fetal cells may get into the mother's bloodstream during pregnancy, but there are usually too few to induce a new (primary) immune response. The most common time for primary immunization to occur is at the time of labor and the delivery of an Rh⁺ child, when a relatively large number of fetal cells may enter the maternal circulation. About 15 percent of Rh⁻ women with their first Rh⁺ fetuses are immunized at this time. The first Rh⁺ child, responsible for that primary immunization at birth, is not affected because the mother has not had the antibodies during pregnancy. However, during the next Rh⁺ pregnancy, the few fetal cells that cross the placenta during pregnancy may be sufficient to act as a "booster," greatly increasing the antibody titer in the mother. These maternal antibodies belong to a molecular species (IgG) that can readily cross the placenta and destroy the red cells of the fetus, leading to the clinical symptoms of HDN.

DETECTION AND TREATMENT OF HDN. The realization that HDN resulted from an antigen-antibody reaction made it possible to take measures to circumvent or at least decrease its ill effects. First of all, the pregnant woman is tested to determine whether she is Rh⁻ or Rh⁺. If Rh⁺, there is no problem. If she proves to be Rh⁻, then her husband should be tested. If he is also Rh⁻, again there should be no problem because two Rh⁻ parents cannot produce an Rh⁺ child.

The woman's serum may be tested periodically during pregnancy for any rise in the antibody titer that would cause a miscarriage or a stillbirth, with the possibility of the induction of a premature birth in order to save the child. Nevertheless in the 1950s about 20 percent of all erythroblastotic babies died before birth and another 10 percent shortly afterward. *Amniocentesis*, the process of puncturing one of the membranes surrounding the fetus (the *amnion*) for the removal of fluid from the womb, was initially developed to diagnose the status of a potentially erythroblastotic child by measuring the amount of the toxic pigment *bilirubin* being released by the destruction of the fetal red blood cells. In severe cases premature delivery might be the best hope for saving the child.

In a still more heroic technique, the blood of the fetus may be replaced *in utero* with Rh⁻ blood, which will not react to the mother's antibodies. Such a transfusion is hazardous because it involves an abdominal incision through which the umbilical cord is drawn out and exposed.

Immediately after birth, when the oxygen demands of the infant suddenly increase and any deficiency of red blood cells becomes more acute, a whole-body transfusion (*exchange transfusion*) may be made, in which all of the normal positive blood of the newborn is replaced with nonreacting Rh⁻ blood, so that the antibody is gradually removed as the child slowly manufactures more Rh⁺ erythrocytes. It has been estimated that in one Boston hospital, exchange transfusion has reduced the death rate of erythroblastotic children from brain damage from 25 percent to 5 percent. Exposure of the newborn jaundiced infant to bright fluorescent light (*phototherapy*) destroys the toxic yellow pigment bilirubin or converts it to an unstable intermediate, and such exposure can be used in addition to exchange transfusion. In one controlled study, 45 infants suffering from HDN required, among them, 69 exchange transfusions when treated with phototherapy for four days, whereas another 78 infants not given phototherapy required 224 exchange transfusions.

PREVENTION OF RH ANTIBODY FORMATION. An effective method of circumventing the Rh problem was developed both in the United States and in England in the early 1960s. Two teams, working for the most part independently but with some intercommunication, devised a successful procedure for reducing the Rh hazard based on a simple injection that would inhibit

Table 11-7 Results of clinical tests to prevent Rh immunization by treatment of mothers with Rh⁺ gamma globulin. (J. C. Woodrow, Some Aspects of Immunogenetics, in *Selected Topics in Medical Genetics*, ed. C. A. Clarke, Oxford University Press, London, 1969.)

	Controls		Treated	
Place	Total	Number immunized	Total	Number immunized
U.S.A. and Canada	814	73	984	1
West German group	756	29	487	2
Liverpool group	320	35	315	1
Edinburgh	101	9	87	0
Sweden	45	3	43	0

the formation of Rh antibodies in the mother in the first place. Because the immunization of the Rh⁻ mother ordinarily occurs during the birth of an Rh⁺ child, when some Rh⁺ cells get into her bloodstream, it is possible to prevent the mother from manufacturing antibodies at that time. That fraction of the blood that carries antibodies, the *gamma globulin*, is extracted from a person who is producing Rh⁺ antibodies, is concentrated, and is injected into Rh⁻ mothers of Rh⁺ children within seventy-two hours after birth. Antibody formation is suppressed because the injected antibodies coat the Rh⁺ fetal cells in the mother's circulation and prevent them from inducing more antibodies. Less than one half of 1 percent of such treated mothers are immunized (Table 11-7), as opposed to 7 percent of those Rh⁻ mothers of Rh⁺ children who are left untreated.

The time limit of seventy-two hours for the treatment is based on the period of time arrived at during initial tests of the effectiveness of the procedure. However, the seventy-two-hour period was chosen for a curious nonmedical reason. The original clinical tests of antibody suppression in the United States were made on male inmates of Sing Sing prison in New York, and it was decided by the investigators, who wished to inject the external antibody some period of time after the injection of erythrocytes (to parallel the procedure that might follow birth), that two visits to the prison separated by seventy-two hours would minimize the possibility that the investigators' regular schedule could be incorporated by inmates into an escape plan or be used as an excuse to riot!

DIFFICULTIES IN THE ERADICATION OF HDN. This treatment is based on the prevention of antibody formation in the mother in the first place. If an Rh⁻ mother has been immunized either by a previous birth, by a miscarriage, or by an earlier transfusion of Rh⁺ blood, the treatment is not effective. Rh-immune globulin fails to prevent immunization in from 10 to 15 percent of women at risk; the most common reason is that in these cases, Rh⁺ fetal cells do cross the placenta in sufficiently large numbers to induce antibodies in the maternal circulation prior to birth and therefore the Rh-immune glob-ulin administered after birth cannot be effective. Administration of Rh⁺ globulin also has the disadvantage of requiring the injection of Rh⁺ gamma globulin after the birth of every Rh⁺ child. In order to prevent the immuni-zation of Rh⁻ mothers following abortion or amniocentesis, injections of Rh⁺ gamma globulin are recommended. For these reasons, some 6,000 in-fants affected to some extent by HDN were born during the year 1975 in the United States.

Medicolegal Applications

The simple, unambiguous transmission of the alleles of the ABO, MN, and Rh (as well as other) blood groups makes it possible to apply genetic principles to legal problems. In the most common application, blood group-ing may answer the question of accused paternity. If an accused male has a genetic constitution inconsistent with those of the child and the mother, then he is excluded from possible paternity. Thus in one case a woman of group O accused a man of fathering her child. The child proved to be of group A and the man of group B. Clearly he had to be exonerated in this instance, because an A child of an O mother must receive the *A* allele from the father, who must then be either A or AB. On the other hand, if the man had actually been A or AB, this evidence would hardly serve to convict him because a sizable fraction of the population of males are either A or AB.

With the combined knowledge of the many cell-surface antigenic differ-ences of the ABO, MN, Rh, and other blood groups, as well as other antigenic differences such as those important in transplantation, it is now possible to ask a different type of question. What percentage of males have a genotype such that they might have been the father of the child? Suppose that the set of alleles that the father must have contributed is found in only one male in a hundred. The fact that the putative father's alleles fit when so few men's would can be used to argue, as positive evidence, that he stands a very good chance of being the father. Of course, with many additional loci added to the analysis (and there are now more than seventy available), the possibility of either exclusion or "positive" identification becomes more and more feasible.

In other applications, blood grouping has been used to determine mater-nity, when an infant is wrongfully claimed by a woman to be her own, and to untangle cases of mix-ups of infants in maternity hospitals. Children who were thought to be ordinary twins have, in several cases, been shown to have different fathers. Identification by blood grouping has helped to iden-tify impostors in inheritance cases. Because the properties of the antigen do not disappear after drying, grouping can be of service in criminal cases where the perpetrator has left a few drops of blood behind. In some cases, saliva or perspiration will contain the antigens. Dried blood has even been used by anthropologists interested in the constitution of ancient person-ages; thus more than 300 Egyptian mummies have been classified according to their ABO groups.

Biochemistry of ABO or ABH System

The basic antigens in the ABO complex consist of large molecules (lipids or proteins) to which are attached modified-sugar side chains, to form glycolipids (on the erythrocyte surface) or glycoproteins (in solution). These modified sugars determine the antigenic specificity of the molecule. The precursor to the A and B antigens has the sugars galactose and glucose attached at intervals along the backbone (Figure 11-7).

The gene *H* is responsible for an enzyme that adds another sugar, fucose, to the basic precursor (Figure 11-8). In discussions of this sort, it is very often convenient to omit mention of the enzyme whose role in the intermediary step in the process is implied. Thus, in Figures 11-7 and 11-8 and in the description of the activity of different alleles, the role of the enzyme is omitted.

The *H* locus is independent of the ABO locus. The *H substance*, produced by the *H* gene, is then acted upon by the *A* or *B* allele, or both, to produce further derivatives. The *B* allele adds a galactose to the chain; the *A* allele adds instead a modified galactose, abbreviated to Gn (Figure 11-8). The *O* allele produces an inactive enzyme that adds nothing to the H substance. This inactive enzyme can, however, act as an antigen and produce an antibody that, when tested against erythrocytes from persons of group A or B, agglutinates only those red cells that are *AO* or *BO* and not those that are *AA* or *AB*. Thus it is possible to differentiate between the homozygotes and the heterozygotes in the A and B groups.

THE H SUBSTANCE. The H substance is antigenic; that is, it can stimulate production of the anti-*H* antibody. However, the H substance is found in most people, because almost everyone is either *HH* or *Hh*, and it is recognized by the antigen-antibody system as a normal component of the body. Therefore no antibody is ordinarily produced.

THE BOMBAY PHENOTYPE. Persons homozygous for the recessive allele *h* do not manufacture the H substance; therefore they do not produce the AB antigens, which depend on the H substance as a precursor. Persons who are homozygous *hh* appear superficially to be of group O, irrespective of their allelic composition at the AB locus. Such homozygotes were first found in Bombay, India, in 1952, as two individuals who needed transfusions after blood loss; the *h* allele is often referred to as the *Bombay allele*. Figure 11-9

Figure 11-7 The basic structure of the molecule responsible for the antigenic specificity in the ABO blood group system. A galactose molecule (Ga) and a glucose molecule (Gl) together are attached at intervals along the protein backbone to produce a combination known as *glycoprotein*.

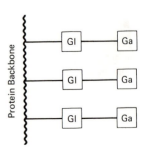

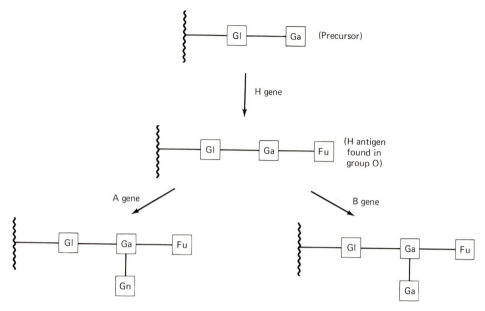

Figure 11-8 The series of steps that convert the precursor into the A and B antigens. The *H* gene, present in almost all people, adds a fucose molecule (Fu) to the galactose, whereupon the *A* gene, when present, adds a modified galactose (Gn) and the *B* gene another galactose (Ga) to the galactose already present.

shows the pedigree of a family with this allele. The propositus appears to be O, although her AB origin and offspring both indicate that she must have carried the *B* allele and must have been of the genotype *BO, hh*.

Persons with the Bombay phenotype have cells that are not agglutinated by either anti-A or anti-B antisera, suggesting that they are of group O, but neither are their cells agglutinated by anti-H, which would be expected of group O cells. Their serum contains anti-H as well as anti-A and anti-B antibodies. The wisdom of performing crossmatches of the donor and recipient blood prior to a transfusion is well illustrated by the problem that arises when the recipient has the Bombay phenotype. A simple agglutination test of Bombay erythrocytes with A and B antisera would prove negative, indicating that the recipient is of group O. If, however, O blood were transfused, agglutination would occur because the anti-H antibody ordinarily present in

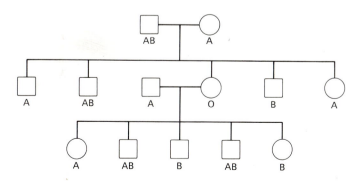

Figure 11-9 Pedigree of a family in which the mother appeared to be group O (although her mother was AB), and who, with her A husband, produced two AB children. It seems clear that she must have carried the *B* allele, which was not expressed; she was homozygous *hh*.

the *hh* individual would react with the H antigen normally found on the erythrocytes of group O donors. Crossmatching prior to transfusion would reveal this potential danger.

ABO Incompatibility

If immunization of the mother to the Rh antigen can give rise to HDN, then why does the ABO group not react similarly? In fact, one might expect that in some mother–child combinations the results would be disastrous because the maternal antibody would already be present and not have to be induced, as in the case of Rh incompatibility.

Consider the very common case, for instance, in which an O woman has an A or B child. Her anti-A or anti-B antibodies, some of which are capable of crossing the placenta, should get into the fetal circulation and agglutinate the fetus's erythrocytes. Although some cases of HDN in late pregnancy or at delivery have been traced to this cause, it is relatively rare.

Consider the very common case in which a woman of group A has a child of group B. Because she already has anti-B antibodies, why do not these attack the erythrocytes of the fetus? The reason appears to be that in women of groups A and B the natural antibodies (presumably present because of early exposure to microorganisms with antigens like A and B) are of large molecular weight, in a class designated as IgM. These do not cross the placenta. In group O women, however, these natural antibodies can be of a smaller molecular weight, IgG, and can cross the placenta to attack the fetal erythrocytes. In those cases where they do, they may destroy normal development so early than an abortion results. Thus there are fewer medical problems *at birth* with HDN caused by the ABO groups than by the Rh factors.

We can find evidence for an ABO maternal fetal incompatibility by comparing the number of stillbirths and fetal deaths from O mothers, most of whose children will be incompatible, being A, B, or AB, with the frequency of similar loss from AB mothers, all of whose children will be compatible. Such studies suggest that ABO incompatibility may increase the number of early fetal deaths by as much as 15–40 percent, although the number of late fetal deaths is not affected appreciably. One in twenty-five pregnancies may be affected by ABO incompatibility at an early stage, but only one in a thousand pregnancies results in HDN produced by ABO incompatibility.

Another interesting aspect of ABO incompatibility is its protection against HDN caused by the Rh factors. Some studies have shown that Rh-caused erythroblastosis is found more frequently when the mother–child combinations are ABO-compatible than when they are incompatible. One explanation is that when the child's erythrocytes enter the mother's bloodstream, they are destroyed by her natural A and B antibodies before her immune system has a chance to manufacture anti-Rh antibodies.

The Xg Blood Groups

An unexpected antibody was found in a patient in Grand Rapids, Michigan, who suffered from severe nosebleeds and had therefore received many transfusions. This antibody was evoked by a previously unidentified antigen present in some of the erythrocytes the patient received during his transfusions. Persons with this antigen have an allele Xg^a on their X chromosome, if male, or on at least one X, if female. The alternative allele, for which there is no known antigenic property yet, is Xg. A simple count of males in England and the United States with (65 percent) and without (35 percent) this antigenic property gives us an immediate estimate of the allele frequencies in the population. The estimate of the proportion of homozygous Xg females is .35 × .35, or about .12. This is very close to the observed frequency of females who lack the Xg^a antigen in these populations.

FAILURE OF Xg LOCUS TO INACTIVATE.   Heterozygous females might be expected to have two different erythrocyte populations as a result of X chromosome inactivation. Those cells with the Xg^a allele, should have the corresponding antigen and should be agglutinated by the anti-Xg^a antibody; the cells with the Xg allele, which is antigenically inactive, would not be agglutinated. In fact, all cells of the heterozygote are agglutinated, indicating that they all carry the antigen. One explanation might be that the antigen is manufactured somewhere else in the body and subsequently coats all erythrocytes. However, this is not the case. Some individuals have been found who have two distinct erythrocyte populations arising from genetically different sources (i.e., they are *chimeras*). In one case, one fraction of the erythrocytes were group AB, the rest group O. Those that were group O could be agglutinated by the anti-Xg^a antibody, and so were Xg^a. Those cells in group AB could not be so agglutinated and so were Xg. This proved that each erythrocyte was expressing its own genotype and that the Xg^a antigen was not synthesized elsewhere and was coating all red cells. The best explanation is that the Xg locus is located on the X chromosome in a region that is not inactivated.

Other Blood Group Systems

A dozen more loci responsible for erythrocyte surface antigens are known. Their presence becomes known only when an antibody is produced that cannot be attributed to any previously known group. These can show up when an unexpected case of HDN appears, or when an unexpected cross-reaction occurs in the preliminary test for a transfusion or during one. When such a new group appears, it can be described as *public* if the antigen is carried by the vast majority of people or *private* if it is carried by relatively few people. If it is public, its discovery usually depends on the formation of antibodies in one of the rare individuals not carrying it, followed by some clinical indication of the existence of that antibody. If private, then individuals carrying it are relatively infrequent and its discovery would depend on detection of the antibody in someone who had been immunized by the rare antigen.

A list of other less important blood groups, along with the designations for their major alleles, is found in Table 11-8.

LECTINS AND PHYTOHEMAGGLUTININS.   Lectins, substances with antibody-like properties may be found in unexpected places. Beans and peas (i.e., the

Blood System	Designations of Alleles
P (= Q of Furuhata)	P^1, P^2, p
Kell	K, k, k^P
Lutheran	Lu^a, Lu^b
Duffy	Fy^a, Fy^b
Kidd	Jk^a, Jk^b
Lewis	Le, le
Diego	Di^a, Di^b
Yt	Yt^a, Yt^b
Dombrock	Do^a, Do
Auberger	Au^a, Au
Stoltzfus	Sf^a, Sf

Table 11-8 Some other blood group systems and their alleles. (From *Principles of Human Genetics*, Third Edition, by Curt Stern. W. H. Freeman and Company. Copyright © 1973.)

legumes) in particular provide a diverse source of unusual extracts that react with specific antigens. These are not antibodies in the usual sense, and, when they are found in plants, they are called *phytohemagglutinins*. One such lectin agglutinates cells of group O; another is specific for A. Another lectin from the bean *Vicia* reacts with M but very weakly with N. These lectins have the great advantages of being easily prepared, inexpensive, and very potent.

Substances with agglutinin-like properties are found in other organisms. An anti-B agglutinin occurs in mushrooms, an anti-H in eels, and an anti-M in the horse and the cow. The existence of substances with these properties from unexpected sources should not be interpreted as an infallible indication of any basic biological identity with the antibodies induced in mammals after exposure to a specific antigen. Rather, they should probably be viewed as compounds with some specific function in those organisms in which they occur that, by coincidence, have a molecular configuration allowing them to combine with human antigens.

Persistence of Allelic Diversity

After all has been said about the characteristics of these erythrocyte cell-surface antigens, some questions still remain unanswered. (1) Why does there appear to be so much allelic variation compared with other loci? (2) What purpose do these loci serve in the first place?

Loci that show allelic forms occurring with moderately high frequencies are said to be *polymorphic*. It is not known either why there are so many loci producing surface antigens or why so many of them are polymorphic. Possibly the loci have some important function that is not affected by the polymorphic system of alleles. The antigenic properties that we observe unambiguously in transfusion mishaps, or in erythroblastosis, and can readily test for experimentally may be incidental to some other important but unknown function. Certainly the viability of homozygous types such as *O/O*, is strong evidence against the overwhelming importance in any specific individual of the antigenically active alleles at these loci.

Table 11-9 The association of duodenal ulcers with blood group O. (R. B. McConnell, in *Selected Topics in Medical Genetics*, ed. C. A. Clarke, Oxford University Press, London, 1969.)

Blood Group	Control Population	Nonbleeding Duodenal Ulcer	Bleeding Duodenal Ulcer
O	3,146	351	329
A	2,648	244	157
Total	5,794	595	486
Percentage O	54.0	59.0	67.7

One of the easiest checks is to compare the distribution of the blood group in the overall population with that in a subgroup selected for some medical reason. Large numbers of such comparisons have been made, and it is now fairly well established that there is an excess of 0 persons among those suffering from duodenal ulcers (Table 11-9) and an excess of A among those with stomach cancer. Although these correlations are interesting and unquestionably valid, one is left with the subjective impression that these may be incidental to the basic functions of the polymorphic loci, functions that remain to be discovered.

We know that the capability of the body to produce antibodies against the erythrocyte cell-surface antigens not only varies from one individual to the next but also shows great variation from one antibody to the next. Possibly the loci (but not necessarily the specific alleles) for the cell-surface antigens are involved in a complex developmental set of interactions of great importance during embryonic or fetal life. They could be part of a system that provides for an extensive immune system capable of identifying and perhaps destroying an unusual tissue development, such as a cancer, during later life.

Additional Reading

CLARKE, C. A. 1968. The prevention of "rhesus" babies. *Scientific American,* **219**:46.

EDELMAN, G. M. 1970. The structure and function of antibodies. *Scientific American,* **223**:34–42.

MANN, J. D. 1962. A sex-linked blood group. *Lancet,* **1,** pt 1:8–10.

McDONAGH, F., L. A. PALMA, and D. A. LIGHTNER. 1980. Blue light and bilirubin excretion. *Science,* **208**:145–151.

NUSBACHER, J., and J. R. BOVE. 1980. Rh immune prophylaxis: Is antepartum therapy desirable? *New England Journal of Medicine,* **303**:935–937.

WATKINS, W. M. 1966. Blood-group substances. *Science,* **152**:172–181.

ZIMMERMAN, D. 1973. *Rh—The Intimate History of a Disease and Its Conquest.* New York: Macmillan.

Questions

1. Can you recall any instance in which your own immune system was clearly incited to action?
2. Do you know of any cases in which the immune system of a person has been responsible for an exaggerated reaction?
3. What are antigens? Antibodies? Where are they found? What is meant by *agglutination*?
4. How would you demonstrate that there are four major blood groups in humans? When the blood of a person of one blood group is transfused into the body of a person belonging to a different group, there may or may not be a severe agglutination reaction. How does the antigen–antibody constitution of each of the two individuals determine whether agglutination occurs?
5. How do we know that the ABO blood groups are inherited? Why are the A and B alleles called *codominant*? What is meant by *multiple alleles*?
6. List all possible kinds of marriages that can occur between persons of the four blood groups, and show what kinds of progeny they might have.
7. Why did it take such a long period of time after the ABO groups were described to discover the MN groups?
8. If two parents are both MN, what kinds of children, and in what relative proportions, might they have?
9. Why is the Rh blood group usually less important in transfusion than the ABO group?
10. How are the Rh factors inherited?
11. Describe the course of events that leads to HDN. Why did the discovery of the Rh blood groups not lead to the complete elimination of HDN as a cause of infant mortality?
12. Why is amniocentesis sometimes used in cases of Rh maternal–fetal incompatibility?
13. List the various measures that might be taken to reduce the incidence of Rh-caused HDN, whether before or during pregnancy, or after birth.
14. How can the blood groups be used in legal cases?
15. What are the chemical steps resulting from the action of the *A* and *B* alleles? How are these related to the H substance?
16. What is meant by the *Bombay phenotype*? How might a person of this constitution run into serious difficulties after a routine transfusion?
17. How does incompatibility with respect to the ABO groups seem to confer some protection against the incompatibility of the Rh groups?
18. On which chromosome is the locus of the *Xg* alleles? How have chimeras been used to prove that this locus is not inactivated in X chromatin as most other X-linked loci are?
19. What are phytohemagglutinins? They are useful to immunologists and cytologists for slightly different reasons. Can you recall these reasons?
20. What is known about the reason for the relatively high variability in the alleles of the blood group loci?

12

The Immune System: Transplantation and Related Problems

It has been only since the mid-1950s that transplantation of organs from one human to another has been shown to be a safe and practicable operation—sometimes. Since the unusual (for that time) success in Boston of transplanting kidneys from one identical twin to another, kidney transplants have exceeded the 30,000 mark. This operation is now accomplished daily with little fanfare in countries all over the world and is generally considered routine. Transplantation of other organs, although generally not so successful as kidney transplants, is becoming increasingly frequent. In this chapter we look into the problems surrounding these procedures and consider the biological basis for the reaction of one body against an organ from another—the phenomenon known as *rejection*. Some of the organs commonly involved in transplantation operations, as well as those concerned with the function of the immune system, are shown in Figure 12-1.

Transplantation

The widespread practice of transplantation has already given rise to a host of new problems. Usually the supply of organs of all types is limited, and there are more persons near death awaiting transplants than there are transplantable organs available. In many cases some kind of decision must be made as to who will receive the transplant (and live) and who will not (and

Figure 12-1 The location of a few of the transplantable organs and of those involved in the rejection reaction.

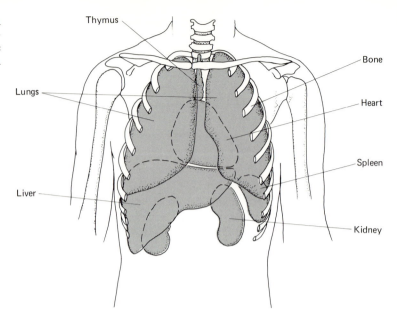

Thymus

Bone

Lungs

Heart

Spleen

Liver

Kidney

die). This decision is necessarily influenced by subjective judgments regarding the worth of the person to the particular society in which the person happens to be living, his or her potential contribution to society, the expectations of future behavior, and so on. When the personal characteristics of patients are taken into account in deciding who, among the large waiting list, will be selected as a recipient, is there any implicit discrimination based on social class, education, or economic resources? Does a hospital have the right to withhold a transplant from a well-qualified patient who lacks the financial resources, or the medical insurance, to pay for the operation?

Every day, scores of such life-and-death decisions are being made by doctors and by committees who are only too conscious of the serious consequences of their judgments. In many cases it is not possible to select the "best" recipient, because, on the availability of an organ, the decision about which person will be the recipient may depend on compatibility tests. In such cases, those with the responsibility for choosing the recipient may feel relieved that other, impersonal factors enter into the final decision.

HEART. The first human heart transplant was made in 1967; it was followed the next year by more than a hundred similar operations in the United States (54), Canada (14), France (10), South Africa (2), and in twenty-one other countries (Table 12-1). By 1978 a total of 406 heart transplants had been carried out (in 395 patients) and 100 were alive at that time. Those patients who survived the critical three-month period after the operation had a 70 percent chance of surviving the first year, and a 95 percent chance of surviving each subsequent year. This is a distinct improvement over the earlier experience with heart transplant patients, who had an average life span of only 300 days.

Table 12-1 The distribution of human heart transplants by year and country. The numbers in parentheses indicate the numbers of patients alive on January 1, 1976. (By permission of J. Bergan, Director of the Organ Transplant Registry, and *Journal of the American Medical Association;* unpublished data.)

Year	World Totals	U.S.A.	Canada	France	South Africa	Other Countries*
1967	2	1	0	0	1	0
1968	101 (2)	54 (1)	14	10 (1)	2	21
1969	47 (2)	34 (1)	1	0	4 (1)	8
1970	17 (3)	16 (3)	1	0	0	0
1971	18 (3)	13 (2)	1	0	3 (1)	1
1972	18 (5)	15 (5)	0	0	2	1
1973	33 (6)	21 (5)	1	8 (1)	1	2
1974	29 (12)	17 (9)	0	9 (3)	1	2
1975	31 (19)	23 (16)	0	5 (2)	2 (1)	1
Total	296 (52)	194 (42)	18	32 (7)	16 (3)	36

*Argentina, Australia, Belgium, Brazil, Chile, Czechoslovakia, England, Germany, India, Israel, Japan, Peru, Poland, Spain, Switzerland, Turkey, U.S.S.R., Venezuela.

Recipients between forty and fifty years old tolerate the necessary immunosuppression poorly and are only half as likely to survive a heart transplant as younger patients. Those aged from fifty to sixty are only half as likely to survive as the forty to fifty age group. Surgeons would obviously prefer to operate on younger patients with a greater probability of survival (to say nothing of the longer time a successful operation would be enjoyed!). Such a decision has, however, been described as "age discrimination" and has been advanced as one of several reasons that the federal government should cease funding these operations.

Because it is necessary to remove the donor's heart while it is still capable of functioning, the donor must be declared dead for reasons other than heart impairment, such as permanent and irreversible brain damage in an accident. This necessity has given rise to new legal and ethical problems with respect to the definition of death, because cessation of heartbeat has historically been used as the primary indicator of death. Many states have responded by redefining legal death to exclude heart stoppage by itself but to include permanent, irreversible brain damage along with a half dozen other primary indications of death, such as absence of breathing and reflexes.

Acute rejection of transplanted hearts usually occurs in the first several months after the operation and is followed by a long-range chronic rejection that may not occur until two or three years later. More of the recipients die of rejection than from any other cause, with infection ranking second. The third most frequent cause of death is the immediate failure of the transplanted heart itself.

Liver, Lung, and Bone Marrow Transplants. The first human liver transplant was carried out in Denver in 1963, and by 1978 360 such operations had been performed throughout the world. The one-year survival rate for liver transplants was 40 percent in 1978. The longest-term survivor was an eleven-year-old girl who had received her transplant at the age of three. The liver continues to be one of the most difficult organs to transplant successfully for a long-range survival.

As of 1978 some 50 lung transplantations had been performed (Table 12-2), but only one patient survived as long as ten months.

Bone marrow is made up of the primitive cells that give rise to blood cells and to the mature differentiated cells found inside the bones. When these primitive cells are defective, as when the patient has a disease of the immune system, antibodies may be produced in insufficient quantities or not at all. In such cases bone marrow cells of normal persons may be transplanted, usually after an attempt is made to destroy the original defective cells. It is usually difficult to determine whether the donor cells become established in the recipient, unless the donor and recipient cells differ in their chromosome constitution, their antigenic properties, or some other identifi-

Table 12-2   Totals of five different organ transplants, throughout the world, up to 1976. (By permission of J. Bergan, Director of the Organ Transplant Registry, and *Journal of the American Medical Association;* unpublished data.)

Year	Heart	Liver	Lung	Pancreas	Kidney
1953–1961					123
1962					67
1963		6	2		157
1964		4	0		359
1965		7	3		453
1966		3	1	2	561
1967	2	8	6	1	832
1968	101	39	6	6	1,245
1969	47	46	7	7	1,538
1970	17	31	2	9	1,990
1971	18	15	4	1	2,904
1972	18	23	3	5	3,486
1973	33	24	2	5	3,828
1974	29	30	0	7	3,620
1975	31	18	1	4	2,756
Total	296	254	37	47	23,919

able cell characteristics. In a group of 50 patients in whom the graft took, 55 percent survived longer than ninety days, whereas of a parallel group of 65 patients in whom it did not take (or at least in whom it could not be shown to have taken), only 20 percent survived that long.

KIDNEY. Of all the organ transplants, those that achieve the best success involve the kidney. There is precise information about the details of the operation and its outcome in more than 10,000 cases. Of the recipients, more than half are still alive with the graft functioning, and another sixth are still alive with a nonfunctioning graft. Almost two thirds of all grafts are made from cadaver donors; this figure, however, varies from one country to the next. Slightly more than half of the grafts in the United States are from cadavers, whereas in Australia almost all are from this source.

Recent technical advances have led to a sharp decline in mortality: only 5–8 percent of all patients transplanted with a kidney from a relative do not survive as long as a year, and only 14 percent who receive a kidney from a cadaver die within that time.

It should be kept in mind that the failure of a kidney graft does not necessarily mean that the patient is doomed, as is usually the case for heart transplant patients. In many cases the recipient may have one kidney that is still functioning to some extent or may be kept alive by a kidney dialysis machine, which removes the blood and purifies it as kidneys do, once every few days. In fact, patients undergoing dialysis at home have survival rates equal to those of patients who received transplants from living related donors, and they have better survival rates than recipients of kidneys from cadavers.

One of the reasons that kidney transplants are more successful than other organ grafts is that because each individual has two kidneys, it is very often possible to obtain a more compatible one for transplantation from a close relative who may be willing to part with one of two kidneys. Another important reason is that this particular organ does not elicit the same massive rejection reaction that other organs, such as liver or lung, do. On the other hand, one of the problems in kidney transplantation is that the original disease the transplant is intended to alleviate sometimes appears in the grafted kidney, at an average of two years after surgery and as late as six years after. (Similarly, in heart transplants, the primary disease requiring the transplant can affect the grafted heart, although this may not always be obvious as the transplanted heart has had nerves severed; thus it is impossible for the patient to feel heart lesions, though they can be detected by electrocardiograms and other coronary observations.) Table 12-3 shows the survival rate of transplant recipients for various organs.

Cellular Basis of the Immune Reaction

Surgical techniques have advanced beyond the point where the success of a transplant depends primarily on the skill of the surgeon. Instead, the fundamental problem is that of rejection. In order to understand rejection, we must consider in some detail the cellular basis of antibody formation.

Table 12-3   Total cases reported to the American College of Surgeons/National Institutes of Health, showing the survival of recipients of grafted organs. (By permission of J. Bergan, Director of the Organ Transplant Registry, and *Journal of the American Medical Association;* unpublished data available only to 1978.)

Organ Transplantation in the World	Heart	Liver	Lung	Pancreas	Kidney
Recipients of transplants	346	318	37	57	25,108
Alive with functioning grafts	77	47	0	0	approx. 13,384
Longest survival with functioning graft	8.7 yr	7.5 yr	10 mo	4.2 yr	20 yr*

*Identical twin.

IMMUNE RESPONSE.   When very large molecules such as proteins or complex sugars are introduced into the body and are recognized as being foreign, the immune system reacts in two distinctly different ways. Certain white blood cells, called *lymphocytes,* manufacture antibodies, molecules that react specifically with the foreign molecules, or antigens. Other white cells become primed against the antigens and bind to them. If the antigen is found on the coating of a bacterium or other microorganism, those lymphocytes attack the foreign invader.

Antigenic molecules are found in every living cell, from bacteria to human tissue, and there are many that are identical in all living cells; against these there is no antibody production. On the other hand, from one species to the next, and from one individual to the next, there are chemical differences that can be distinguished by the immune system, a good example being the cell-surface antigenic differences of the blood groups. It is this system for recognizing foreign substances that lies at the basis of the antigen-antibody reaction in its many different manifestations, such as rejection and allergy.

We only partly understand the way in which a lymphocyte manufactures an antibody specific against a foreign antigen. There are two basic questions: (1) how can a cell that has never been in contact with a particular antigen be induced to manufacture antibodies specific to that antigen, in great quantities? And (2) how is it possible for the immune system to carry the potential for producing many hundreds, perhaps thousands, of different antibodies?

The earliest theory was the "lock and key" hypothesis: that the antigen came to rest on the surface of a lymphocyte, which somehow sensed its molecular configuration and then produced many molecules with a complementary configuration. These, then, could react specifically with that antigen because of the complementarity of the surfaces. It was later hypothesized that within each individual there exist a very large number of cells, each with a predetermined specificity, covering the range of just about all of the important foreign antigens that might be introduced, so that when a specific antigen enters the bloodstream the corresponding type of white cell is "awakened" and stimulated to undergo rapid mitosis to produce large numbers of cells with large quantities of antibody specific against the antigen.

More recent studies show that the loci responsible for the production of antibodies have the capacity to produce many variant forms, which they do constantly. When a foreign antigen reacts with a lymphocyte, a change occurs on the surface of the cell and causes that cell to undergo rapid proliferation and at the same time to produce much more of that particular antibody. Thus the immune system is characterized by two outstanding features: the ability to react to an antigen by producing specifically reactive antibodies, and a sensitization of that system to any subsequent exposure to the same antigen.

STRUCTURE OF THE ANTIBODY MOLECULE. It is estimated that the immune system can produce hundreds of thousands of different antibodies, corresponding to that many kinds of antigenic structures. If one gene produced one and only one antibody, a disproportionate share of the loci in the human genome (around twenty thousand) would have to be dedicated to this vast array. Therefore we can surmise that each gene responsible for antibody production generates at least several different antibodies.

Study of antibody structure has been impeded because, unlike hemoglobin, antibodies are difficult to obtain in quantity and in pure form. Each person's gamma globulin fraction consists of many thousands of immunoglobulins (or antibodies), which are chemically so similar as to be inseparable by ordinary chemical means.

This problem has been solved in several ways. In a type of cancer called *myelomatosis,* a single antibody-forming plasma cell starts uncontrolled growth. If the cell had differentiated to produce one particular antibody prior to its malignant growth, that antibody would be found in great abundance in the circulation. Some victims of myelomatosis therefore provide us with blood samples in which one specific antibody may be studied with little disturbance from the comparatively low concentrations of other structurally different antibodies.

In another approach, the loci responsible for antibody production are cloned in *Escherichia coli,* in ways described in Chapter 10, to produce large quantities of the actual DNA sequences responsible for specific antibodies.

Still another method involves, first, the injection of an antigen into an animal to stimulate antibody production; second, the isolation of a single lymphocyte induced to antibody production; and, third, the fusion of this cell with a cancer cell that is capable of rapid unrestrained growth in culture. These hybrid cells, called *hybridomas,* generate one specific type of antibody in great quantities. Because the antibodies are produced by clones derived from a single cell, they are referred to as *monoclonal antibodies.*

By these means we have been able to study antibody molecules closely and now have a fairly good picture of what they are like. The overall structure of an antibody molecule is that of a Y, with two long (heavy) chains of amino acids, or polypeptides, associated with two shorter (light) chains (Fig-

Figure 12-2 A schematic for the gross structure of the immunoglobulin molecule showing the two heavy chains, the two light chains, and the portion of the molecule with variable structure responsible for the antibody specificity of the molecule.

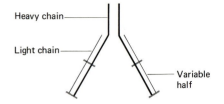

ure 12-2). Most of the amino acids making up the chains are constant from one molecule to the next, except those at the two tips of the Y. The variations in the amino acid sequence in these so-called variable portions give rise to different three-dimensional configurations when the polypeptide naturally folds on itself. It is this conformation that endows the molecule with the specificity that enables it to react with antigens that have a complementary three-dimensional structure.

The immunoglobulins have been classified into five groups: IgG, IgM, IgA, IgD, and IgE. The last two occur at a very low frequency in the blood plasma, and not much is known of their function. Of the other three, IgG is the smallest and is the only type that regularly crosses the placenta. Hemolytic disease of the newborn is the responsibility of the IgG antibodies, whereas hypersensitivity (allergy) involves the IgE antibodies.

In all immunoglobulins, two heavy and two light polypeptide chains form the basic structural units. There are five different classes of heavy chains, and the light chains come in two varieties, which are both associated with all five heavy classes. IgM has a high molecular weight because five of these units are held together by common sulfur bonds; IgG consists of two structural units hooked together. IgA, the major antibody found in secretions, may consist of a single unit, or of two, three, or four units joined with sulfur bonds.

How the numerous immunoglobulin-chain sequences arise is still not understood. Those theories that have gained wide acceptance at the present time involve the assumption either (1) that all of the chain sequences are inherited, the diversity of specificity being derived from the manner in which the peptide sequence is determined (constant and variable regions might be spliced together in some fashion) and from the particular combination of light and heavy chains in each antibody; or (2) that only a few chain sequences are inherited and that diversity arises from somatic mutation or from recombination of similar sequences.

CELL RESPONSE. The blood cells that respond to antigenic stimuli are of two types, both derived from a common ancestor. One type of cell develops in the lymphoid tissue in the thymus gland, the lymph nodes, the tonsils, and the spleen and is referred to as the *T cell* (Figure 12-3); the other type develops primarily in the bone marrow and is referred to as the *B cell*. When a foreign antigen enters the body, the molecules adhere to the surface of a large white cell, the *macrophage* (Figure 12-3A). Both the B and the T cells come into contact on the macrophage. The T cells produce several cell types. Among them are killer cells, which migrate to the source of the antigenic stimulus, giving the white discoloration characteristic of infections, and an additional type, primitive cells that "remember" the specific anti-

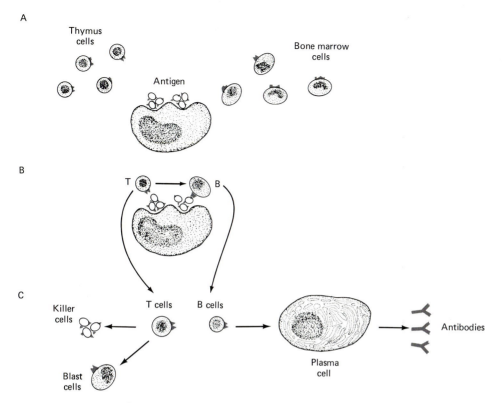

Figure 12-3 The interaction of antigens and the lymphocyte system in producing humoral and cell-mediated immunity. *(A)* The antigen molecules come to rest on the surface of the macrophage. *(B)* The T and B cells interact with the antigens (and with each other), resulting *(C)* in the stimulation of the B cells to differentiate into plasma cells that manufacture antibodies, while the T cells form killer cells that attack the foreign antigen and blast cells that remain in a primitive state and form the long-term memory of the immune system. (Adapted from Immunological Unresponsiveness, by W. O. Weigle, *Hospital Practice* **6**:121, New York, 1971, and *Immunobiology*, R. A. Good and D. W. Fisher, eds., pp. 123–34, Sinauer Associates, Inc., Stamford, Conn., 1971.)

genic stimulus and are primed to react quickly in the event of a second stimulus much later. It is this latter class of "remembering" cells that goes into action when we are given a booster shot. They can be quiescent for a long period of time, as is seen when we become immune for life after having had a childhood disease. The B cells differentiate to produce a large white blood cell called the *plasma cell*, and these plasma cells produce the specific antibody against the antigen (Figure 12-3 *B, C*). The antibody is released into the circulatory system, and for this reason this release is referred to as the *humoral* (or fluid) *response*, as opposed to the whole-cell or *cell-mediated response* of the T cells.

SURVIVAL VALUE OF IMMUNE RESPONSE. It is obvious that the immune system has a very high survival value; that is, any change in our genetic constitution that makes us better able to ward off a disease would be found with a high frequency in the survivors of an epidemic of that disease and would eventually become established as a permanent feature of our biological makeup. On the other hand, an infectious organism can "mutate" to produce essentially brand-new antigenic types every few years, creating a new epidemic each time, as the influenza virus does. Those new types must still be capable of evoking an immune response, or the human race would have been in danger of extinction throughout its history. A highly virulent strain of a common infectious organism would suffer the disadvantage of jeopardizing its own existence as it kills off its hosts.

These considerations might not apply, however, to organisms to which humans have never been previously exposed and against which they may have no immune protection. When the astronauts returned from the first visit to the moon, they were kept in isolation for a long period of time so that the possibility could be eliminated that they had, by chance, become infected with some unknown organism to which the human race has no immune defense and which could therefore cause an enormous loss of life. When H. G. Wells wrote *War of the Worlds,* he ended his story with the extinction of the alien invaders by Earth-born microorganisms against which they had no resistance. He could equally well, however, have ended it in the reverse fashion, with the human race made extinct by microorganisms brought here by Martians. In any case, if one is ever approached by a creature from outer space, it would be well to keep one's distance, if for no other reason than to minimize the possibility of an uncontrolled infection. Furthermore there may be a strong argument for holding in isolation for a month or two those people who claim to have had any physical contact with such beings . . .

Overcoming the Immune Reaction

Except for cases of identical twins, every individual is almost certain to have some antigenic differences from any other individual. Not all of these are important in transplantation rejection, as we shall see later, but, briefly, when a transplant is made from one person to another, some of the antigens of the transplanted tissue will evoke an antibody response, with the production of specific killer lymphocytes against that foreign tissue and the gradual destruction of the transplanted tissue. The solution to this problem is to avoid or suppress the immune reaction. This procedure can be approached in a number of ways.

Certain kinds of tissue are relatively immune to rejection. Corneal transplants, for instance, are quite often highly successful operations, perhaps because the cornea does not release antigens to the recipient's system, or because the absence of blood vessels in the cornea makes it impossible for those lymphocytes primarily responsible for tissue rejection to reach the cornea. The testes are also immune, protected apparently by the associated circulation, which is isolated from the main blood supply. Two other kinds of tissue that are relatively inert are bone (from which all live cells are removed prior to transplant) and blood vessels. In both cases, it is the ability of the graft to function without the presence of live cells that makes it acceptable to the recipient.

RADIATION. One of the first techniques developed for suppressing the immune reaction was to apply a dose of radiation of a fairly high intensity, sufficient to kill or to suppress the mitotic activity of the primitive stem cells that give rise to lymphocytes. Although this procedure is effective up to a point, it can also be very dangerous. In fact, it is a common cause of death in persons accidentally exposed to high doses of radiation, because the destruction of the immune system leaves the patient vulnerable to simple infectious disease. Of course, in the case of bone marrow transplants, where an attempt is made to kill off the patient's normal lymph stem cells and replace them with those from another person, this may be an effective procedure.

DRUGS. There are a number of drugs available that suppress the immune reaction: actinomycin, azathioprine, methotrexate, prednisone, Imuran, mitomycin C, and cyclosporin A. It is the effectiveness of the last-named that has led to an increase in successful transplants, particularly heart transplants, in recent years. A balance must be struck in the application of these drugs, so that a sufficient concentration is given to suppress the immune reaction, but not enough to suppress it completely and to allow the patient to be defenseless against disease. As an overall total, five sixths of all patients who die a short time after a transplant operation die of a subsequent infection, particularly of the lungs. Not only is there danger from infection when the immune system is suppressed, but the normal control of malignant processes is reduced, so that a transplant recipient is more susceptible to cancers and other malignant growths.

Another advance in the suppression of the immune system for transplantation involves the use of *antilymphocyte serum (ALS)*. This serum carries antibodies *(antilymphocyte globulin*, or *ALG)* that react against lymphocytes and suppress the proliferation of those that would attack the transplanted organ.

OTHER TECHNIQUES. Because of the importance of the problem of the immune reaction to transplantation, a large amount of research effort is going into its solution, and a number of new approaches are currently being tested.

First, let us consider one obvious approach, which, unfortunately, will not work. A preliminary test might be made, whereby the recipient of an organ would receive a patch of unessential skin transplanted from the potential donor. Whether that skin transplant is rejected or whether it takes, and, if rejected, how quickly, might then serve as an indication of the probable fate of the organ transplant.

This kind of test, however, would itself sensitize the recipient, unless by some remote chance the two individuals were antigenically identical.

MIXED-LYMPHOCYTE CULTURE TEST. Although it is obviously not possible to test for incompatibility in the actual recipient, tests can be made in culture to determine the similarity of the loci of the two individuals. A basic feature of the immune reaction involves the proliferation of lymphocytes in the presence of foreign antigens. In the *mixed-lymphocyte culture test,* lymphocytes of the two persons involved are mixed together and cultured. Evidence of growth can be taken as a clear indication that there are dissimilar antigens differentiating the two lymphocyte samples and that a graft between the two individuals would be in jeopardy. An absence of mitosis, on the other hand, suggests that the two sets of lymphocytes do not recognize any foreign antigens from each other and therefore predicts a more viable transplant. Furthermore the degree of mitotic activity may serve as an indication of the number of antigenic differences.

LYMPHOCYTOTOXICITY TEST.   Another test for compatibility between donor and recipient is the *lymphocytotoxicity test.* If antibodies to a specific antigen are added to a culture of lymphocytes carrying that antigen, the antibody may succeed in killing the lymphocytes. If a large number of sera carrying different antibodies are added to both donor and recipient lymphocyte cultures, the cultures will respond in similar ways to all of the different antibody-containing sera if both sets of cells are antigenically similar. On the other hand, if the two lymphocyte samples are quite different, they will react differently to the various antibody preparations. These reactions are more easily observed if special staining techniques that differentiate between living and dead cells are used. Because this test, unlike the mixed-lymphocyte culture test, depends on the specific antigens carried on the lymphocytes reacting with the corresponding antibodies in the sera, it can be used for tissue typing and for determining the specific antigens carried by different individuals.

Genetics of Compatibility Systems

Compatibility of two sets of tissue is referred to as *histocompatibility* and incompatibility as *histoincompatibility.* That there is a genetic basis for these reactions has already been implied in the actual figures for rejection, which show that success depends on the closeness of relationship between the donor and the recipient, being highest between identical twins.

To understand its genetic basis, we may take advantage of the mouse, which has a system not very different from that of humans and with which we can make transplantations at ease, without the same degree of concern for the consequences as we would have for humans. This problem is best approached by means of mouse strains that have been highly inbred for many generations, usually by brother-sister matings, until they are homozygous at virtually all loci. Inbred strains of different origins are homozygous for different alleles; thus one inbred strain might be of composition $A_1A_1 \ B_3B_3 \ C_8C_8 \ldots \ X_2X_2$, and a second may be of composition $A_2A_2 \ B_4B_4 \ C_1C_1 \ldots \ X_8X_8$, and so on. Then, if we consider two such inbred strains, we first notice that any transplant made within a strain (an *isograft*) takes readily and permanently without any difficulty but that a transplant between two individuals of different strains (an *allograft)* does not take. We can now breed an F_1 hybrid and make grafts in all possible combinations of parents and hybrid progeny to complete this simple analysis. When this is done, it is found that a graft from either parent or from the hybrid will take when the recipient is a hybrid but that a graft from the F_1 hybrid will not take in either of the parents, although it will in sibs.

This experiment tells us that rejection has a genetic basis; furthermore we can surmise that there are codominant alleles involved because the progeny exhibit antigenic properties of both parents. These interactions are easily understood if we simply imagine that one strain is of composition A_1A_1, the other A_2A_2, and the hybrid A_1A_2, and that a graft will take only if the recipient has the antigens present in the donor tissue.

How many loci are involved? We can make an estimate of this number by crossing two F_1 individuals and obtaining a large number of F_2. What proportion of the F_2 segregants have tissue that will take when grafted into the original parental strains? If there were only one locus involved, with each inbred strain having its own specific alleles, then in the F_2 we should have one fourth of the segregants like one of the two parents, and one-fourth like the other; therefore grafts from 25 percent of the F_2 in each original parental strain should take. On the other hand, if a very large number of loci were different in the two strains, then the chance that we would get a genotype in the F_2 precisely like one of the parents would be negligibly small, and we would expect virtually zero success. Clearly we can work out the precise expectations for various numbers of loci, based on an observed frequency of takes between the F_2 segregants and the parental stock. From such analyses, involving many inbred strains of the mouse, it can be shown that there is one locus that is particularly potent and that there are a dozen or more additional loci of lesser importance, each of these loci having a large number of alleles. There is also in all mammals an antigen produced by a locus on the Y chromosome, but this must be of minor significance in antibody production in humans because male organs can be grafted to females just about as readily as can female organs.

Obviously similar tests cannot be performed on humans, but extensive studies in families, combined with the two-cell culture tests described earlier, show that there are several loci important to transplantation. The first of these is the *ABO* locus: the potential donor should not carry antigens produced by this locus for which the recipient has the corresponding antibody. It is important to keep this fact in mind because it limits the availability of certain kinds of potential donors. However, ABO compatibility considerations are so taken for granted in transplantation studies that they are often not explicitly mentioned, but merely implied.

Major Histocompatibility Complex (MHC)

On the short arm of chromosome 6 are located a group of loci, the *Major Histocompatibility Complex*, (MHC), involved in the production of antigens important in transplantation. Of greatest interest are the histocompatibility loci, with the symbol *HLA*, with five subloci distinguished by the letters *A* to *D*: *HLA-A*, *HLA-B*, *HLA-C*, *HLA-D*, and *HLA-DR*. It has been determined that the order of the five is not the sequence above; they are

Figure 12-4 The order of the *HLA* loci on the short arm of chromosome 6, along with an estimate of the genetic distances between them, based on recombination studies.

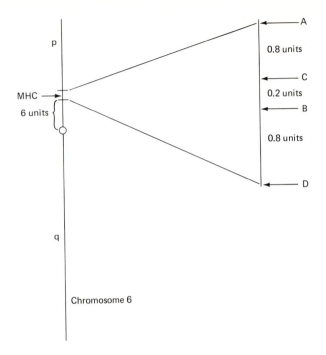

found on the chromosome in the order shown in Figure 12-4. A list of the alleles found at these loci is given in Table 12-4.

Of the five, two, *HLA-A* and *HLA-B*, are most often checked in determinations of compatibility of prospective recipient and donor tissues, but some tests show that compatibility at the *HLA-D* and *HLA-DR* loci may also be very important. Compatibility at the *HLA-C* locus does not appear to affect transplantation success.

Each person has an *A*, *B*, *C*, *D*, and *DR* locus on one homologue of chro-

Table 12-4 The alleles at the *HLA* loci in the European white population with frequencies greater than 5 percent. At each locus there are alleles as yet undetectable; totals may therefore add up to less than 100 percent.

	HLA-A	HLA-B	HLA-C	HLA-D	HLA-DR
	A1 .158	B5 .059	Cw2 .054	Dw1 .079	DRw1 .062
	A2 .270	B7 .104	Cw3 .094	Dw2 .095	DRw2 .112
	A3 .126	B8 .092	Cw4 .126	Dw3 .095	DRw3 .089
	A9 .112	B12 .166	Cw5 .084	Dw4 .051	DRw4 .078
	A10 .059	Bw16 .055	Cw6 .126	Dw5 .090	DRw5 .151
	A11 .051	B17 .057		Dw6 .115	DRw6 .086
	A29 .058	B18 .062		Dw7 .058	DRw7 .156
		Bw35 .099			
		B40 .081			
	Plus 8 others	Plus 14 others	Plus 1 other	Plus 4 others	
Totals	15	23	6	11	7

mosome 6 and a similar set of five on the other homologue. Thus one chromosome might have the alleles *A2, B12, Cw4, Dw6* and *DRw5* on one homologue and *A1, B7, Cw6, Dw2* and *DRw7* on the other. Each of these sets of five is referred to as a *haplotype*. Every individual has two homologues, and therefore two haplotypes. If a person's lymphocytes were shown to carry the antigens produced by the ten alleles given above, it would not be obvious what the haplotypes were because the same result would come from a person with other combinations, such as *A2, B7, Cw4, Dw2* and *DRw7* on one homologue and *A1, B12, Cw6, Dw6* and *DRw5* on the other.

We can determine the precise constitution of each of the two haplotypes only by finding out which antigens are found in the children (or the parents). A simplified pedigree, involving only the *HLA-A* and *HLA-B* loci, is shown in Figure 12-5. From this pedigree it can be seen that if the mother carries the antigens *A1, A9, B5,* and *B12* and the father *A2, A3, B8* and *B13*, the haplotypes must be *A1, B5* and *A9, B12* in the mother, and *A2, B8* and *A3, B13* in the father. Any other combination would give different types of offspring from the four types observed. Note also that because there can be four combinations in the offspring, if there are as many as five children, at least two will be alike.

The amount of crossing over between these loci is very low, 1 percent or less (Figure 12-4). Therefore, if a person inherits a specific allele from one parent, he or she is almost certain (with a probability of about 99 percent) to inherit the others that happen to be on that same chromosome in that parent. This probability provides an explanation of why organs of sibs who are *HLA-A* and *HLA-B* identical with a patient are rejected less often than organs from cadavers. Those sibs who are *HLA-A* and *HLA-B* identical, as shown by tests, are usually *HLA-D* and *HLA-DR* identical because they have received the same chromosomes from the parents. An unrelated individual, however, might be identical at the *A* and *B* loci but quite different at the *D* and *DR*.

If we take the number of allelic differences at the *A* locus to be fifteen, we can calculate that there are one hundred and twenty possible different combinations of two (15 homozygotes, plus $(15 \times 14)/2$ different heterozygotes). A similar calculation gives us two hundred and seventy-six different combinations for the twenty-three alleles at the *B* locus, sixty-six at the *D* locus and twenty-eight at the *DR* locus. (The C locus is omitted because of its relative unimportance in transplantation.) The total number of different

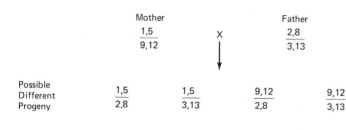

Mother 1,5 / 9,12   X   Father 2,8 / 3,13

Possible Different Progeny: 1,5/2,8   1,5/3,13   9,12/2,8   9,12/3,13

Figure 12-5 Hypothetical scheme showing possible constitutions of two parents, with the four different kinds of progeny. The first number of each pair represents a specific allele at the *HLA-A* sublocus and the second number, the allele at the *HLA-B* sublocus. One haplotype found on one chromosome is given above the line, and the other haplotype, on the homologue, is placed below the line.

combinations for all four loci is the product, $120 \times 276 \times 66 \times 28$, or 61,205,760 possible combinations of alleles at the four loci.

Of course, not all of these genotypes occur with equal frequency, because the alleles found at each of the two subloci themselves do not occur with equal frequencies. This multiplication gives an indication of the magnitude of the task of finding two compatible individuals taken at random from the population—as would be the case, for instance, between a patient awaiting a heart transplant and a potential donor suffering irreversible brain damage in an automobile accident.

When the transplantation antigens were first being investigated, it was hoped that the identification of all of the antigens at these loci would make it possible to approach complete success in matching donors with recipients, just as identification of the erythrocyte cell-surface antigens and antibodies of donor and recipient makes it possible to ensure success in virtually all blood transfusions. This has, unfortunately, turned out not to be the case. Although compatibility of the *HLA* antigens, along with *ABO* compatibility, plays a predominant role in the success or failure of transplants, the different antigens appear to have different strengths for reasons that are not yet completely clear, and there appear to be other uncontrollable factors that modify the rejection reaction.

In addition, there is evidence that other factors affect success. Patients who have had several blood transfusions prove to accept subsequent transplants more readily than those who have not. The reason is unclear. The probability of a take cannot be predicted with certainty on the basis of the similarity of the *ABO* blood groups and the known *HLA* alleles. There is reason to believe that the *D* locus of the Major Histocompatibility Complex system may also be very important; compatibility of donor and recipient at this locus improves the chance of success appreciably. As a matter of practical procedure, it is necessary to make one of the direct tests in cultures, such as the mixed-lymphocyte test, to assess the probability of a take.

Other Antigen–Antibody Reactions

GRAFT-VERSUS-HOST (GVH) REACTION. Ordinarily our concern is with the formation in the recipient of antibodies and killer lymphocytes specific against the antigens present in the grafted tissue, causing its rejection. However, in those cases where the transplant involves lymphocyte-forming tissue (as in bone marrow transplants), the reverse can occur. If a transplant of such tissue is made after all of the recipient's own has been destroyed, or if the recipient was born lacking the capacity to form such tissue, it would seem a simple matter to transplant comparable tissue into the defective individual to restore missing immune capacity. Rejection should, in principle, not be a problem because the recipient lacks just that tissue responsible for rejection. However, what happens in such transplants is that the transplanted tissue itself forms antibodies against the recipient's antigens, with the result that the graft attacks and destroys the host tissue. When a newborn mouse is injected with a sizable number of spleen cells from an incompatible animal, a poorly developed individual, or runt, is the result; it is said to have *graft-versus-host* or *GVH disease*.

ACQUIRED TOLERANCE. Under ordinary circumstances, the body does not form antibodies against its own antigens. If a mouse is injected with cells of another strain very early in development, that mouse will not develop antibodies against the foreign antigens, hence the term *acquired tolerance*. It has been suggested that individuals could be made compatible with a wider array of antigens—and would therefore have less difficulty in finding a suitable donor if a transplant should become necessary—by the injection of newborn infants with incompatible tissues from several different origins, so that they will acquire a tolerance for the antigens on those tissues. In theory they would then be more likely to accept an otherwise incompatible graft later because of their acquired tolerance.

ANAPHYLAXIS AND ALLERGY. One dramatic manifestation of the responsiveness of the immune system is found in the phenomenon of *anaphylaxis,* the sudden and violent, sometimes fatal, reaction of an animal to a second or subsequent exposure to an antigen after the initial sensitization. This reaction results from the release of a powerful muscle contractant, leukotriene C, which causes constriction of the small airways of the lungs, giving rise to the wheezing and labored breathing characteristic of such a reaction.

A more moderate form of this severe reaction is found in *allergy,* in such common disorders as hay fever, asthma, and eczema, which can be responsible for fever, hives, rashes, headaches, cramps, and other symptoms. Different *allergens* may produce different effects. Ragweed pollen ordinarily affects the nose and the eyes, whereas allergic reactions to foods are more commonly seen as hives or skin rashes.

Allergy may be treated successfully if the specific allergen is identified through exposure of the allergic person to a wide variety of potential offenders, and then if the offending factor (sometimes a household pet or an overstuffed article of furniture) is removed, or when this is not possible, if the person is desensitized by a series of injections of small doses of the specific allergen. In addition, because one of the characteristics of the allergic reaction is the liberation of *histamine*—a powerful substance that causes contraction of the airway muscles, relaxation of the capillary walls, and a fall in blood pressure—*antihistamines* are very effective in relieving the symptoms of allergy.

AUTOIMMUNITY. It is a truism that under ordinary circumstances a normal individual does not form antibodies against antigens naturally present in his or her own body. There are, however, some cases in which this rule breaks down; in these instances *autoimmunity* is said to exist. Some cells may be normally isolated from the rest of the body, and if an accident, disease, or other misfortune should allow the antigens of such cells to be released and to stimulate the immune system, then antibodies may be produced against those antigens (and those cells), leading to an autoimmune reaction.

It has been suggested, as another cause of autoimmunity, that some change may appear in those genes that produce specific antibodies, so that the antibodies are directed no longer against the original antigen but against some antigen found normally in the body. Still another possibility is that an antigen introduced from the outside may induce the production of an antibody that also reacts with antigens normally present in the body.

Some diseases that have been shown to originate in an autoimmune reaction include pernicious anemia, rheumatoid arthritis, disease of the thyroid, and, in some cases, multiple sclerosis, chronic gastritis, and infantile eczema. This list continues to grow as our knowledge of autoimmunity increases.

A striking example is found in injuries to the eye, where damage to one eye may cause the formation of antibodies specific to antigens ordinarily held captive within it, with the result that these antibodies sometime later may attack the other eye and damage it as well. It is a not uncommon medical observation that a person suffering serious injury in one eye may, months later, lose the sight of the second, uninjured eye.

Males who have undergone sterilization by vasectomy (which prevents the release of mature sperm, so that they must instead be absorbed by the body) may develop antibodies to those sperm. These antibodies are of two types: those that agglutinate the sperm and those that immobilize them. Whereas only about 1 percent of all males have antibodies in their plasma against sperm prior to vasectomization, more than half of all vasectomized males show sperm-agglutinating antibodies after six months, and more than 61 percent show them after a year. Furthermore roughly 40 percent of all vasectomized males have in their plasma, one year after the operation, antibodies that immobilize sperm.

In a small sample of infertile women, it has been shown that more than half have antibodies against the gelatin-like covering of their eggs. It has been hypothesized that of the very large number of eggs (over several million) produced in the female at about the time of birth, most break down, releasing an antigen that evokes an antibody response. These antibodies might then attach to eggs as they mature, preventing sperm entry and causing infertility. From 5 to 10 percent of all men and women who are infertile appear to have developed antibodies against sperm.

A widely held hypothesis is that of *autosurveillance,* according to which an autoimmune reaction may limit the growth of cancers. According to this idea, when malignant tissues produce new antigens, the antibodies and killer lymphocytes formed in response may be effective in destroying the growth. This would suggest that every individual who has survived beyond middle age has successfully fought off a cancerous growth without being aware of its presence. It is also proposed that the weakening of the immune system in old age allows tumor cells to proliferate with diminished challenge or allows an infection, which would be successfully fought off in youth, to terminate life.

Immunological Deficiency Diseases

In 1952 an account appeared in a pediatric journal of a boy who was originally diagnosed as having acute rheumatic fever because of a painful left knee joint; after treatment with antibiotics he recovered completely but subsequently relapsed. Over the next four years he suffered from severe infections nineteen times, each time being only temporarily cured with antibiotics. Because of his high rate of infection, it was thought that perhaps he might have produced an unusually high level of antibodies in his serum.

Figure 12-6 The separation of a heterogeneous mixture of blood serum proteins by electrophoresis, showing the main categories identifiable in this way.

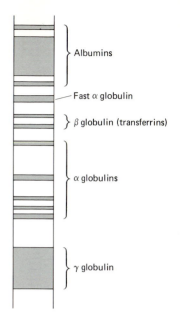

Albumins

Fast α globulin

β globulin (transferrins)

α globulins

γ globulin

Antibodies may be detected and identified as proteins of the gamma globulin class by *electrophoresis,* a technique in which a heterogeneous mixture of proteins is placed in an electrical field and the molecules are allowed to migrate to the poles at their own characteristic speeds (Figure 12-6). (The name *gamma globulin* is derived from the fact that it is the third major group of globulins in such a separation , alpha and beta globulins being the first and second, with the word *globulin* referring simply to the large "globule" structure of the molecule.) Electrophoresis of the proteins in the boy's serum showed a blank spot at the level where the gamma globulin fraction was expected. In fact, at the first clinical test it was thought that the electrophoresis apparatus had broken down. When later tests confirmed the original results, this deficiency of gamma globulin was named *agammaglobulinemia.*

In the classic case of agammaglobulinemia, now known to be sex-linked, the lymphocytes are normal, and cellular immunity is present, but free antibodies are absent and, therefore, so is humoral immunity. Although individuals so afflicted are unusually susceptible to many diseases (e.g., diphtheria and typhoid), they are easily able to withstand some viral infections, such as measles and chickenpox, suggesting that the mechanism of defense may be humoral in the former set and cell-mediated in the latter. Furthermore it has been found that the gamma globulin fraction containing the antibodies may not be completely absent from the serum, as a crude preparation by electrophoresis would suggest, but may simply be greatly decreased in quantity.

Alleles at more than a dozen different loci, mostly autosomal, are known to affect the immune system. However, there are five X-linked recessive syndromes; these appear as diseases affecting the male only. This existence suggests that the X chromosome carries a number of loci involved in the immune response. It is known that women have higher antibody concentrations and better ability to form antibodies against infectious diseases than men. It is possible that the more efficient immune system of the female, either because of the presence of two X chromosomes or because of the action of female hormones in promoting antibody production, plays a substantial role in the greater longevity of the female than the male.

If a child with agammaglobulinemia is not diagnosed at an early age, before the time when the protection provided by the maternal antibodies that crossed the placenta into the fetus is disappearing and the child's own immune system should be taking over, the child will very likely die. Unusual susceptibility of a young child, particularly a male, to common infections should give rise to a question about the possibility of this defect's being present.

These individuals may be protected by injections of gamma globulin, as such serum fractions obtained from normal people contain protective antibodies against the common diseases. Although gamma globulin therapy may reduce the frequency of acute infections, it affects chronic infections and progressive pulmonary disease hardly at all. Therefore it is necessary to continue regular treatment with antibiotics, and the most effective treatment appears to be a combination of antibiotics and gamma globulin. One unfortunate aspect of this treatment is that whatever immune defenses the patient has may produce antibodies against the gamma globulin being introduced, because gamma globulin, being protein, can also act as an antigen.

The patient may also be protected by the transplantation of bone marrow from another individual. When transplantation is attempted, however, it becomes obvious that the immune reaction is not completely absent, because transplantations induce a typical, although somewhat delayed, immune response. In some cases bone marrow transplants have been successful; more often they have not. Not only is the delayed rejection a hazard, but the transplanted lymphoid tissue may develop antibodies against the host (graft-versus-host or GVH disease). The GVH reaction may be bypassed if the transplanted lymphoid tissue is not *immune-competent* and therefore cannot react against the host. Fetal liver contains many primitive *immune-incompetent* white cells; such cells taken from abortuses and injected into children with an immune-deficiency disease may eventually provide a source of white cells for a functional immune system. It is worth noting that among patients receiving bone marrow transplants, those with an immune-deficiency disease experience a higher frequency of successes than those without such a disease.

Another approach involves identifying the potentially affected child prior to birth (usually after an older sib has proved to be affected). The delivery of the child is then made under strictly germfree conditions. If the child is kept germfree by the most stringent precautions and is isolated in a special chamber, bacteria and viruses, the causes of the most common infections, will not reach the child. When the disease is of the type in which the immune system is slow to develop, the child may eventually be released into the outside world. If the disease is permanent, however, release from this germfree isolation may be postponed indefinitely until an effective cure is found.

Allele	Disease	Affected Organ	Increased Risk
B27	Ankylosing spondylitis	Joints	88x
Bw35	Thyroiditis	Thyroid	22x
B8	Diabetes	Pancreas	2x
B27	Anterior uveitis	Eye	15x
Bw37	Psoriasis	Skin	5x

Table 12-5 The association of alleles at the *HLA* loci with disease. Only a few out of twenty-five reported correlations are listed here.

HLA Alleles and Disease

ASSOCIATION OF *HLA* ALLELES WITH DISEASE. Ankylosing spondylitis is characterized by low back pain and stiffness, particularly in the early morning. It is an inflammatory arthritis that affects young men in particular. More than a million persons in the United States are affected with this disease. It has been found that about 96 percent of all affected persons carry the *B27* allele of the *HLA* system, whereas only about 8 percent of the general population carry it. This association of *HLA* alleles with certain specific diseases has been found in a number of other cases. Some of these are given in Table 12-5.

MATERNAL–FETAL INCOMPATIBILITY. Almost every fetus inherits from its father *HLA* alleles not present in the mother. These could be responsible for mother–fetus incompatibility; if an antibody were induced in the mother, it could attack the tissues of the fetus and kill it. In this sense the fetus would behave like a graft.

We know, however, that this reaction does not happen as a rule; if it did, the human race might have difficulty surviving. It has been found that women normally possess an antibody that blocks the production of immune cells against the paternal antigens of the fetus. Some women who chronically miscarry do not have this blocking antibody, so that they reject the fetus just as they would an incompatible graft.

Medicolegal Applications of HLA

The great diversity of alleles at the *HLA* loci make it possible to identify individuals with more precision than is possible with the erythrocyte cell-surface antigens. Suppose, for instance, that a woman has the antigens A9, A28, B12, and B17 and her child carries A1, A9, B5, and B12. The child could not have received the alleles for A1 and B5 from the mother (because she does not have those alleles), so it must have received them from the father. But it is known from population studies that the haplotype A1B5, which the child must have received from the father, is found in only 1 percent of the population. Thus there is an excellent chance, in a paternity suit, that a falsely accused man will not have this haplotype. Such exclusion would be decisive. On the other hand, if the accused should have them, this is sometimes used as positive evidence that he stands a good chance of being the father. It is estimated that in only one case in twenty will there be consistency between the HLA antigens of the child and an accused man when that man is not in fact the father of the child. Of course, in combination with information from the ABO, MN, Rh, and other blood groups, the possibility of exclusion, on the one hand, or indication of paternity, on the other, becomes much greater.

Additional Reading

AREHART-TREICHEL, J. 1975. Restoring immunity to deficient patients. *Science News,* **107**:42–43.

AREHART-TREICHEL, J. 1980. Marrow transplants. *Science News,* **119**:104.

EBRINGER, A. 1978. The link between genes and disease. *New Scientist,* **79**:865–867.

HOPSON, J. L. 1980. Battle at the isle of self. *Science 81,* **210**:77–82.

KOFFLER, D. 1980. Systemic lupus erythematosus. *Scientific American,* **243**:52–61.

MARX, J. L. 1981. Antibodies: Getting their genes together. *Science 80,* **212**:1015–1017.

MAUGH, T. H., II. 1980. New techniques for selective immune suppression increase transplant odds. *Science,* **210**:44–46.

MAUGH, T. H., II. 1980. Transplants (II): Altering the donor organ. *Science,* **210**:177–179.

PERKINS, H. A. 1979. Concise review: Current status of the HLA system. *American Journal Hematology,* **6**:285–292.

PERKINS, H. A. 1980. The human major histocompatibility complex (MHC). In *Basic Clinical Immunology,* 3rd ed. (Fudenberg, et al., eds.) Los Altos, Ca.: Lange Medical Publications.

POPAY, J., et al. 1980. Transplanting priorities. *New Scientist,* **86**:136–138.

ROGERS, J. 1980. The genes that make antibodies. *New Scientist,* **86**: 155–157.

SEIDMAN, J. G., et al. 1978. Antibody diversity. *Science,* **202**:11–17.

TALMAGE, D. W. 1979. Recognition and memory in the cells of the immune system. *American Scientist,* **67**:173–177.

YANCHINSKI, S. 1981. Where next for monoclonal antibodies? *New Scientist,* **90**:102–104.

YELTON, D. E. and M. D. SCHARFF. 1980. Monoclonal antibodies. *American Scientist,* **68**:510–516.

Questions

1. In what ways has organ transplantation raised new ethical problems?
2. How has the legal definition of death become an issue since the advent of organ transplantation?
3. Which organ transplants are most successful and which are least successful?

4. How is the immune system able to produce many—thousands—of different antibodies?

5. The study of chemistry of the antibody molecule has been impeded by the low concentrations of specific antibodies circulating in the body. How has this problem been bypassed by (a) the use of cells from a type of cancer called *myelomatosis,* (b) cloning, and (c) the production of monoclonal antibodies in hybridomas?

6. What is the basic structure of the antibody molecule? Into how many major groups have antibody molecules been classified? Which is involved in HDN? In allergy?

7. What are the two primary mechanisms by which the immune reaction responds to the presence of a foreign antigen?

8. How may the immune response have any survival value to the human species?

9. Is there any way that the immune reaction may be overcome or bypassed? Describe two different tests that can be applied to increase the likelihood of an organ transplant's "taking."

10. What is the difference between the mixed-lymphocyte culture test and the lymphocytotoxicity test?

11. How can it be shown that the compatibility of a graft depends on the genetic relationship of the donor and the recipient?

12. Describe the major histocompatibility complex (MHC) in humans. Draw a segment of the short arm of chromosome 6 showing the order of the *HLA* loci.

13. Why can it be said with certainty that if a couple has five children, at least two will have identical HLA compositions? What is a haplotype?

14. What is the graft-versus-host (GVH) reaction?

15. How might acquired tolerance theoretically prove beneficial to a person in later life?

16. What is the difference between anaphylaxis and allergy?

17. Rheumatoid arthritis is one of a dozen or so diseases that are believed to be caused by autoimmunity. Explain what is meant by this.

18. If one reads in the newspaper of a child who is protected from the outside world because of an immune-deficiency disease, that child is likely to be a male rather than a female. Can you suggest a reason?

19. A person with an immune deficiency has a reduced capability of rejecting transplants. Why, then, is a bone marrow transplant not a simple solution for repairing the innate defect in such a case?

20. Why is the injection of gamma globulin not a simple universal cure for agammaglobulinemia?

21. Do the *HLA* alleles have any relation to disease?

22. A fetus carries some *HLA* alleles from the father; these alleles are not present in the mother. Why, then, does the mother not reject the fetus as she would an organ transplant?

23. How can the HLA antigens be used in forensic medicine?

24. Is cessation of brain function a better criterion of death than heart stoppage? If so, could this criterion lead to a prolongation of the existence of otherwise dead persons for the purpose of making organs available for transplantation?

25. Mr. R. desperately needs and wants a kidney to replace his dialysis treatments. His wife suggests the following solution: she will become pregnant and after five or six months have an abortion so that the kidney of the fetus may be transplanted to her husband. The transplant surgeon knows that Mr. R. has threatened to commit suicide if he has to remain on dialysis. Should the surgeon agree to this suggestion?

13

Mutation

The variability that is found between one individual and another and is passed on from one generation to the next must originate in some way. Strictly speaking, the process by which these genetic changes occur is known as *mutation;* the changes themselves are called *mutants.* It is common practice, however, to use the word *mutation* for both the process and the end result of that process. Mutational changes may be classified as either *somatic* (body) or *germinal* (germ line). Somatic mutations affecting the body cells at some point after the cleavage divisions have begun may be of great importance to the individual involved but are of little importance to the species, because only germinal mutations affecting the gametes can be transmitted to the next generation. It is this latter category that is invariably meant when the term *mutation* is used without further qualification.

Somatic Mutation

Cell division is a remarkably exact process, and mistakes such as nondisjunction occur only rarely. The replication of the genetic material of a cell at interphase is even more precise; errors of replication that change the phenotype of the individual are almost never seen.

There is plenty of opportunity for such errors to occur and be observed. For example, consider albinism. More than one person in a hundred is heterozygous for albinism, of genotype *Aa*. During the hundreds of mitoses that occur from fertilization of the egg to the completion of the embryo, one might expect that in a few cases the normal *A* gene would fail to replicate itself properly and instead produce an inactive allele, similar to *a*, which would then give rise to a clone of cells effectively *aa*, or albinotic tissue. Such an event would be obvious because there are millions of cells exposed to the surface (in the skin, hair, and eyes) where such a clone would be easily noticed as a light patch. And this is not the only recessive gene found with a high frequency in heterozygotes in the human population. Nevertheless such *mosaic spotting*, suggesting the ocurrence of body or somatic mutations, is very rare.

Occasionally a person is found with eyes of two different colors, usually one brown and the other blue, green, or hazel. The frequency of such somatic mosaics is very low compared with the known frequency of heterozygotes *Bb* (brown), in which the cell or cells in the early embryo giving rise to one eye might have become genotypically *bb* (blue) by the mutation of the *B* allele to *b*. However, because there are other possible explanations for mosaic tissue in heterozygotes, such as the loss of the dominant allele in a clone of cells giving rise to one eye, thereby unmasking the recessive, the frequency of mutation must be even lower than the low incidence of somatic mosaics would suggest.

Germinal Mutation

When Gregor Mendel got his garden pea strains from nurseries, he could not have understood why they differed from each other, although judging from his own words in the introduction to his classic paper, he was clearly aware of the evolutionary importance of such characteristics.

Their significance was emphasized by Hugo de Vries, a professor of botany and curator of the botanical gardens at the University of Amsterdam, and one of several workers who "rediscovered" Mendel's paper in 1900. As de Vries bicycled from his home to the university, he observed that the evening primrose, *Oenothera lamarckiana*, was characterized by some unusual variations from one individual to the next. This plant was American in origin and had been brought to Europe for cultivation in gardens, but a number of plants escaped and rapidly proliferated in some areas. De Vries brought several of the plants into his garden to study and found that although these variant types usually bred true, they occasionally produced yet other types that were true-breeding generation after generation. De Vries referred to these changes as *mutations* and in 1901 published a book on mutation theory in which he postulated that these mutational changes were the ones necessary for evolution.

In its broadest sense we can regard as a mutation any sudden alteration in the genetic composition of a species. Included in this broad spectrum are changes in chromosome sets, or polyploidy; changes in chromosome number, or aneuploidy; and changes that affect the arrangement of chromosome parts either between or within chromosomes. The fourth category of mutation refers to changes in the gene itself, gene mutations. Generally the word *mutation* refers to the category of gene mutation. As the result of the elucidation of the structure of the DNA molecule in the early 1950s by Watson and Crick, and the genetic analysis of fast-growing viruses and bacteria by sophisticated physical and chemical methods by a vast number of others, the chemical basis for gene mutation is now understood.

Alterations in DNA Structure

We have seen (in Chapter 5) that DNA is made of a sequence of base pairs and that these bases, taken three at a time, specify amino acids, which are then assembled in long polypeptide chains. These make up proteins with very diverse structures and functions. If the sequence of base pairs is altered in some way, then the final product, the protein, may be changed so that its new activity is detectable as a mutation. There are two main ways in which the DNA is altered: by substitution of one base pair by a different pair, and by the deletion (or addition) of one or more base pairs.

BASE SUBSTITUTION. Consider a length of DNA of which one strand is transcribed to yield mRNA, which in turn is translated into an amino acid chain. If there is a triplet CTT in the transcribed strand of DNA, then the mRNA contains the codon GAA, which will be translated as glutamic acid. During DNA replication, if a mistake occurs to change the DNA triplet to CAT, the RNA triplet will become GUA, which codes for the amino acid valine (see Figure 13-1). Valine will replace glutamic acid at that position in the protein.

Furthermore this mistake in DNA replication will be perpetuated, because when the strand containing the A subunit reproduces, the complementary strand will carry T instead of A, and all subsequent double strands will perpetuate the mutational error.

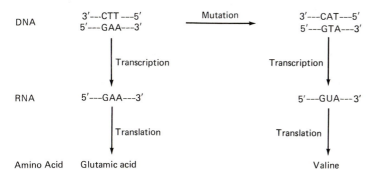

Figure 13-1 In the original gene, one of the triplets is CTT (*upper left*). This is transcribed as GAA, which codes for glutamic acid in the polypeptide that is being synthesized. If a mutational error occurs such that the triplet becomes CAT (*upper right*), the mRNA will carry, instead of GAA, the triplet GUA, which is translated as valine. Thus the single base substitution in the DNA is responsible for the incorporation of a different amino acid into the polypeptide.

The effect that such a substitution will have on the overall molecular structure of the protein is somewhat unpredictable. Some such changes may destroy the function of the protein; others will have almost no effect.

BASE CHANGES IN HEMOGLOBIN. Base changes in the hemoglobin molecule are the best known; the total number of genetically different variants that have been analyzed comes to several hundred.

Some hemoglobin variants are easy to detect. They may produce the symptoms of anemia, and so are a matter of medical concern. These base changes may have some phenotypic characteristics that are dominant in their effect, so that they can often be spotted in a single person. In some instances, the lower oxygen content of the blood imparts a bluish color to the patient's skin, simplifying the identification of the presence of those hemoglobin variants. A single individual can provide an ample supply of material for biochemical analysis (limited only by the patient's patience in being struck frequently with a hypodermic needle). Finally, as this is a disease of medical importance, found with rather high frequencies in different ethnic groups, funds are available for the study of the disease to a greater extent than they are for the study of other genetic problems.

Sickle-cell anemia is caused by a hemoglobin variant resulting from base substitution. Hemoglobin is made up of two pairs of different polypeptide chains, named alpha (α) and beta (β). The sixth triplet in the beta chain gene which codes for normal hemoglobin is transcribed as GAA, which is then translated into the amino acid, glutamic acid. The DNA strand that codes for this protein in sickle-cell patients has a single base change in the triplet. This is transcribed as GUA, which is then translated into valine. In total, the beta chain has 146 amino acids; this single change in the sixth in the sequence changes the overall structure sufficiently to cause the hemoglobin to behave abnormally. Thus the entire genetic difference between normal persons and those with sickle-cell anemia is the result of this one base substitution.

More than 250 abnormal hemoglobins have been described with an alteration in which the usual amino acid at one or another of the 287 positions in the alpha and beta chains is replaced by another. It is worth noting that these changes are consistent with the genetic code. In various hemoglobins, aspartic acid has been found to be replaced by asparagine, histidine, tyrosine, valine, alanine and glycine, and never by certain others, such as leucine or proline. The reason is not hard to see (Figure 13-2). Aspartic acid is coded for by GAU (or GAC). The first six amino acids mentioned above can be coded for by AAU, CAU, UAU, GUU, GCU and GGU. Note that the first group of three of the six involves a change in the first base of the codon and the second group of three involves a change in the second base of the codon. Thus it is possible to get from the codon for aspartic acid (GAU) to each of the other six by a single base substitution. On the other hand, leucine has the code CU plus any of the other four bases, and proline has CC plus any of the other four. It is not possible to go from aspartic acid (GAU or GAC) to these two by one simple substitution.

There are so many variant hemoglobins now known that fall into this pattern of single base substitution that if the genetic code had not been worked out first by experiments with synthetic DNA sequences and later

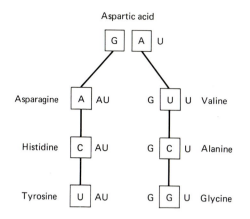

Aspartic acid

But GAU⇻ UUC, CUU, AUU, UCU, CCU, ACU, UGU or AGU because <u>two</u> base changes are required. Therefore aspartic acid is not replaced by phenylalanine, leucine, isoleucine, serine, and so on.

Figure 13-2 Consistency of the amino acid variants found in hemoglobin with the genetic code. Aspartic acid, which occurs normally in more than a dozen different positions in the hemoglobin molecule, may be substituted by the three amino acids on the left, when G in the first position is replaced by A, C, or U. Similarly it may be replaced by the three amino acids on the right, when the A in the second position is replaced by U, C, or G. Other amino acid substitutions, like those listed underneath, are not found because they would require the unlikely occurence of *two* changes within the same triplet.

with lower organisms, an equivalent relationship could have been deduced from the changes in amino acids resulting from single base substitutions in the hemoglobins.

FRAME-SHIFT MUTATIONS. If a DNA base is deleted, then the process of, translating sets of three subunits in sequence, gets out of phase after the deletion and the rest of the amino acid sequence is garbled (Figure 13-3). The addition of an extra DNA base would also put the triplet translation out of phase. Because the reading of the DNA bases three at a time is analogous to viewing them through a frame three units long, which is moved in jumps of three units along the DNA molecule, a change that throws this precise reading out of phase is called a *frame-shift mutation*. Deletions or insertions of larger numbers of DNA bases could similarly cause gross changes in the polypeptide produced (unless a deletion or addition involved three—or a multiple of three—subunits, in which case one or more amino acids would

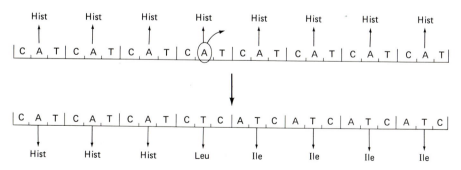

Figure 13-3 The effect of a base deletion on protein synthesis. In this simplified model, when the A is removed, the sequences, when counted off by threes from left to right, become ATC, and the new coding has completely different consequences.

be missing from or added to the polypeptide chain, but the translation would continue accurately after the deletion or insertion). When the reading is thrown out of register in this way, the new triplets specify a new set of amino acids.

Hemoglobin Wayne was the first frame-shift mutation reported in humans. Codon 139 of the alpha chain is AAA, followed by UAC, CGU, and UAA. Note that UAA terminates the reading. The third nucleotide of AAA was lost so that a new sequence was set up: AAU, ACC, GUU, and AAG, followed by others. Thus, not only did the frame-shift change the coding so that different amino acids were added to the polypeptide, but the stop code UAA was split up so that additional amino acids were added on until the ribosome reached a new stop code, accidentally manufactured by the frame shift.

EVOLUTIONARY IMPLICATION OF FRAME-SHIFT MUTATIONS. After a frame-shift mutation has ocurred, the polypeptide chain may show no similarity whatsoever to the original, at least with respect to the section occurring after the change. Such a change, particularly if it occurs near the beginning rather than near the end of the gene, may produce an inactive product. No doubt many simple recessives originate in this way. On the other hand, the newly mutated gene might, in rare instances, prove to have some unexpected usefulness to the cell or to the organism. In this case it would act as a beneficial dominant, if only one such new allele would be necessary to produce the new product. Whether such a new allele could replace the old in the population might very well depend on whether the old was dispensable (and most alleles are not) or whether there were other loci in the genome with similar functions. The speed of incorporation of beneficial dominants would be very high, and we would not be likely to catch one in the process of being incorporated unless there were some additional reason that it could not reach a frequency of 100 percent. This is so for the gene for sickle-cell anemia, which happens to be quite deleterious in the homozygous state.

From this point of view, organisms with a large amount of genetic material present in duplicate (i.e., with *redundancy*) would be at an advantage in evolution. This redundancy could take the form of duplications of genes inevitably found in polyploids or in the duplicated segments within the chromosome set. In this connection, it is worth pointing out that a calculation of the amount of genetic material in a cell of humans, compared with an estimate of the number of genes, leads to the conclusion that less than 5 percent or so of the genetic material consists of active genes, and the rest is made up of superfluous DNA, so that there is plenty of DNA available for redundancy.

REVERSE MUTATION. Another point of interest comes out of the consideration of base changes, particularly deletions, leading to mutational changes. It is clear that a single base loss can lead to the production of a protein (e.g., an enzyme) with partial or complete inactivity. This loss can occur at any one of a large number of positions along the DNA molecule, and the frequency of mutation from this event would depend on the total of such losses leading to a demonstrably different allele. However, a reverse mutation that

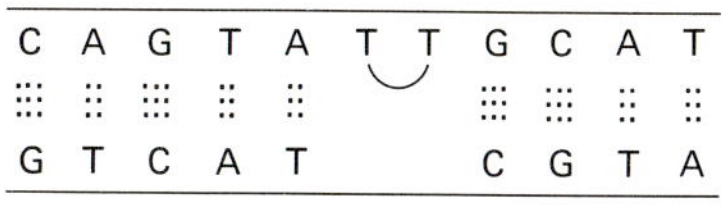

```
C A G T A T T G C A T
::: :: ::: :: ::    ⌣    ::: ::: :: ::
G T C A T        C G T A
```

Figure 13-4 The linkage of two T bases to form a dimer. The strand produced by replication will be defective at this point.

would restore the initial condition of the DNA precisely would require a specific change that would essentially restore the original series of bases. For this reason we might expect the frequency of true reverse mutations to be very much lower than that of forward mutations—perhaps so low that their effect in modifying gene frequencies in the human population would be insignificant.

FORMATION OF DIMERS. One defect that commonly occurs in DNA is the chemical change in two Ts adjacent on a single strand, forming what is called a *dimer.* Such a dimer is produced when ultraviolet light hits the DNA. Clearly, when two adjacent T units are linked to each other (Figure 13-4), reproduction of the complementary strand is impossible at this point and a defective strand is produced. There is an enzyme, normally found in all cells, that recognizes such dimers in the DNA molecule and excises them from the affected strand of the double-stranded DNA. Then other enzymes resynthesize the correct sequence from the normal complementary sequence on the other strand.

Some individuals lack one or more of the enzymes needed to repair the dimer. After exposure to ultraviolet light these people develop severe lesions, which eventually become cancerous and lead to death. This condition is called *xeroderma pigmentosum* (Figure 13-5). It is possible that a good part of the known carcinogenic effect of an excess of ultraviolet light in normal persons may originate in the production of so many dimers that the repair enzymes cannot take care of them; these then produce defective DNA strands in the progeny cells.

Problems in Identifying New Mutations

In experimental plants or animals it is not too difficult to detect new mutational changes as they occur. A laboratory line of animals that has produced no unusual offspring over a large number of generations may be assumed to have suffered a new mutation if a new transmissible type suddenly appears. Indeed, it is common practice to arrange experimental matings so that any new mutations that may occur will be obvious. With humans, however, the unambiguous identification of a new mutational event is more difficult.

DOMINANT VERSUS RECESSIVE ALLELES FROM MUTATION. When a new genetic type suddenly appears as a single affected individual from unaffected parents, the immediate question is whether the affected person is the result of a new dominant mutation, or of the fortuitous homozygosity for two similar recessive alleles for which the parents happened to be heterozygous. This question can be resolved if the new phenotype can be unambiguously

Figure 13-5 Xeroderma pigmentosum, caused in most cases by a demonstrable defect in a DNA repair system. In the early life of homozygotes, it is manifest as unusually heavy freckling in parts of the skin exposed to light. Heterozygotes are also heavily freckled, although not so severely as the homozygote, and have a normal life expectancy. (Courtesy of P. E. Polani, Guy's Hospital, London.)

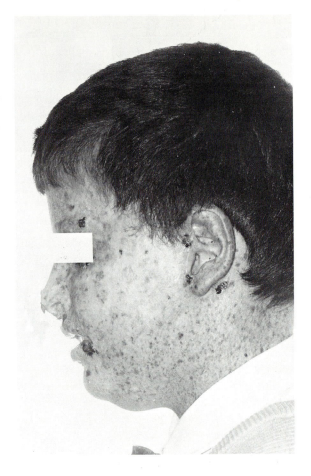

identified as a previously known one with a specific type of inheritance, or if sufficient pedigree information becomes available to discriminate among the alternatives. In rare cases, such as the sudden appearance of hemophilia in the descendants of Queen Victoria, we can be sure that the allele was not present in the earlier history of the royal line, and that a new mutational event occurred either early in her germ line (because several of her offspring inherited the defective allele) or that she received it as a new mutation from one of her parents. In those cases where the phenotype is caused by homozygosity for a recessive, it is most likely that the two alleles were transmitted from ancestors of both parents. It is possible, but less likely, that one of the two alleles is newly mutated and much less likely that both represent new mutational events.

DEVELOPMENTAL VARIABLES. The course of human embryonic and fetal development is subject to numerous influences, which lead to a wide range of variability in the newborn. These variations may appear as congenital defects, or phenocopies, or, if there are specific genetic factors involved, as characters with variable expressivity or incomplete penetrance. There are many cases known in which apparently similar gene defects are caused by alleles at quite different loci. (These aspects of development have been covered in Chapter 5.) In addition there is always the possibility to be considered that when a child carries an allele not present in either of the two legal parents, the child might, in fact, be the offspring of some other woman or, more often, of some other man.

Determination of Mutation Frequencies

If we limit ourselves to simple dominant mutations with complete penetrance and invariable expressivity, we find that there are still other problems. In the first place, attention is quite naturally focused on changes that occur with some frequency, because we cannot make a count of nonexistent cases! However, we may rightfully feel uneasy about the possibility that we are selecting for loci that are more mutable than average. Clearly, if there are many loci that do not mutate to dominant alleles (and most loci do not), they will be excluded from our investigation. It goes without saying that our estimate of the average rate at which genes mutate to dominant alleles will then be spuriously high.

THE MUTATION RATE. The frequency with which normal alleles mutate in the course of one generation is called the *mutation rate*, and it is indicated by the Greek letter μ *(mu)*. For dominant alleles, by far the simplest approach is to make an actual count of the new mutations that have appeared in the course of a generation. For instance, we might check all of the children born at maternity hospitals in the United States in the last ten years for a specific defect, to see whether or not their parents had the same defect. After an effort is made to eliminate the spurious cases, a certain number of individuals can be classified as probably carrying new dominant mutations. A correction must be made at this point to account for the fact that each person has two alleles, and so our population of alleles under observation is twice as large as the number of people.

When these steps are taken, the rates come out in the vicinity of several such new mutations per 100,000 alleles in each generation; that is, the mutation rate is roughly between 10^{-4} and 10^{-5} (between 1 in 10,000 and 1 in 100,000) per gene per generation. However, because of the problems mentioned previously, including phenocopies, misclassification, incomplete penetrance, and unconscious selection of the data for those cases with abnormally high mutation rates, it seems safe to say that on the average a human gene has a probability of spontaneous mutation of about one per million per generation. Inaccurate as this estimate must be, it is still worth making because the numerical value can be used in evaluating mutation risks in populations, particularly where those populations are exposed to agents, such as radiation, that markedly increase the mutation rate.

INDIRECT CALCULATIONS. For loci that produce recessive mutations, the procedure outlined above cannot be used because it is usually not possible to know whether a homozygote is homozygous because of the occurrence of a new mutation in one of the parents, the other being heterozygous, or because of the heterozygosity of both parents. In virtually all cases the cause is the latter. A different system for estimating the mutation rate has been devised, based on the following argument: We can assume that the human population is at *equilibrium* with respect to its genetic composition, that is, that over the last several thousand years, as new alleles have been added to the human gene pool by mutation, other similar alleles have been lost from the pool by inviability, infertility, or other aspects of "unfitness" of the affected individuals. At the present time a balance exists, more or less, between these two opposing factors. We can reasonably assume that the recent advances in medical science have not yet had sufficient impact to alter the overall balance.

A nice analogy for this state of equilibrium is given in Figure 13-6, which shows water being poured into a container with holes in the side. As the water level in the container gets higher, the water pressure at the opening in the bottom becomes greater, and the rate of loss is increased until the amount being lost is equal to the amount coming in. When the level is constant (i.e., is at equilibrium), we need to know neither the amount of fluid in the container nor the size of the holes in the side, but simply that the amount leaving the container must be equal to the amount coming in.

Thus, if we assume that the human population is in equilibrium at the present time with respect to allele frequencies, then the number of mutant alleles being lost in each generation is a measure of the number entering the gene pool by mutation. Genes are ordinarily lost by the decreased productivity of the individuals carrying them.

For estimates of mutation rates of recessives, we examined homozygotes to determine what, on the average, their reproductivity has been, compared with normality. For the normal comparison, sibs are preferred, because they have been subjected to roughly the same external influences from family and society with respect to reproduction as the affected homozygotes. The difference between the reproductivity of the affected persons and those normal persons chosen for comparison is the loss of reproductivity *(L)*.

The following is a typical illustration. The number of mutant alleles in-

Figure 13-6 Analogy showing the state of equilibrium reached when the outflow equals the input in a simple water system. (From *Principles of Human Genetics*, Third Edition, by Curt Stern. W. H. Freeman and Company. Copyright © 1973.)

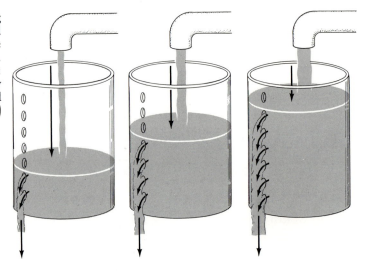

troduced into the population in each generation equals the mutation rate (μ) times the total number of individuals (N_T) times 2 (because each individual is diploid). The number of alleles lost equals the number of affected individuals in the population (N_{Aff}) times the loss of reproductivity (L) times the number of alleles lost with each reproductive loss. For a simple dominant gene, these relations can be stated algebraically as

$$\mu \times N_T \times 2 = N_{Aff} \times L \times 1$$

or

$$\mu = L \times N_{Aff}/2N_T.$$

Thus, if 40 individuals are found with a specific identifiable dominant allele (achondroplasia, for instance) in a population of 600,000 persons, and they produce only one quarter as many progeny $(L = .75)$ as their unaffected sibs, μ equals $(.75 \times 40)/ (2 \times 6 \times 10^5)$ or $2.5 \times 10^{-5.}$

In the case of recessive genes, we shall assume for simplicity that the reproductivity of the heterozygote is the same as that of the homozygous normal. Here the reduced fitness of the affected homozygote causes the loss of two alleles instead of just one, the appropriate relations are

$$\mu = L \times N_{Aff} / N_T.$$

For rare recessive X-linked genes, the equation becomes

$$\mu = L \times N_{Aff}/3N_T.$$

All that is needed is some knowledge about the frequency of the defect in the population (N_{Aff}), about the loss of reproductivity caused by the defect (L), and about how the defect is inherited, in order to apply the proper equation. In cases as severe as muscular dystrophy, for instance, the value of L must be very close to 100 percent, or 1. Thus, if we know that the frequency of Duchenne muscular dystrophy is about 15 per 100,000 in the population, we can immediately estimate the mutation rate of the normal allele to that recessive to be one third as great, or 5×10^{-5}.

Tables 13-1, 13-2, and 13-3 list the mutation rates calculated for a number of dominant, recessive, and sex-linked alleles. For the reasons given previously, these rates are undoubtedly very much higher than the rates for the average, run-of-the-mill gene. One estimate suggests that three quarters of all normal alleles have rates less than 1 per 1 million and that as many as half may have rates of 1 per 10 million, or less.

MUTATION RATES PER GAMETE. What is the chance that, by mutation, a child will have an allele that neither of the parents carried? This chance is very simply calculated. If we accept the spontaneous mutation rate as 10^{-6} per gene per generation, and the number of genes in the haploid set as 2×10^4, then we have a product of 2×10^{-2}, or roughly 2 percent. For the diploid, we multiply this result by 2, coming out with a figure of 4 percent. That is, a newborn child stands a chance of 1 in 25 of carrying a new mutation that was not carried by either of the two parents. Because each of the values entering into such calculations is subject to considerable error, the calculation might be in error by a considerable factor but probably not by as much as a factor of 10 too much, or too little.

Table 13-1 Estimates of spontaneous mutation rates at some human autosomal loci: dominant diseases. (From L. S. Penrose, in *Recent Advances in Human Genetics,* ed. L. S. Penrose, Churchill Livingstone, Edinburgh, 1961.)

Trait	Mutation Rate per Gamete per Generation $(\times 10^{-6})$	Region	Source	Date
Epiloia	8	England	Gunther and Penrose	1935
Chondrodystrophy	70	Sweden	Böök	1952
Aniridia	5*	Denmark	Møllenbach	1947
Microphthalmos without mental defect	5	Sweden	Sjögren and Larsson	1949
Retinoblastoma	4	Germany	Vogel	1954
Partial albinism and deafness	4	Holland	Waardenburg	1951
Multiple polyposis of the colon	13	U.S.A.	Reed and Neel	1955
Neurofibromatosis	100	U.S.A.	Crowe, Schull, and Neel	1956
Arachnodactyly	6	Northern Ireland	Lynas	1958
Huntington chorea	5	U.S.A.	Reed and Neel	1959

*This estimate differs by a factor of 2 from that given by the author, but it is based on his material.

Table 13-2 Indirect estimates of spontaneous mutation rates on the assumption of recessive inheritance. (From L. S. Penrose, in *Recent Advances in Human Genetics,* ed. L. S. Penrose, Churchill Livingston, Edinburth, 1961.)

Trait	Mutation Rate per Gamete per Generation $(\times 10^{-6})$	Region	Source	Date
Juvenile amaurotic idiocy	38	Sweden	Haldane	1939
Albinism	28	Japan	Neel et al.	1949
Ichthyosis	11	Japan	Neel et al.	1949
Total color blindness	28	Japan	Neel et al.	1949
Infantile amaurotic idiocy	11	Japan	Neel et al.	1949
Microcephaly	49	Japan	Komai et al.	1955
Phenylketonuria	25	England	Penrose	1956

Table 13-3 Estimates of spontaneous mutation rates at some human sex-linked loci. (From L. S. Penrose, in *Recent Advances in Human Genetics*, ed. L. S. Penrose, Churchill Livingstone, Edinburgh, 1961.)

Trait	Mutation Rate per Gamete per Generation ($\times 10^{-6}$)	Region	Source	Date
Hemophilia	20	England	Haldane	1935
Hemophilia	32	Denmark	Andreassen	1943
Hemophilia	27	Switzerland and Denmark	Vogel	1955
Pseudohypertrophic m.s. (muscular dystrophy)	60	Northern Ireland	Stevenson	1958
Pseudohypertrophic m.s. (muscular dystrophy)	47	England	{ Blyth and { Pugh	1959

Our Genetic Load

Among the mutations present in the population, a large number are recessives with lethal effects on the developing embryo and fetus, or if they do not cause death at such an early age, they may be responsible for death or debilitation early in postnatal life. It would be worth knowing how many such deleterious genes the average person carries. Is modern humanity so protected by medical science and the advantages of our society that we have accumulated large numbers of such deleterious genes? Or are we relatively free of "bad" genes? Further, do different groups of people differ in the number of such genes they carry? Clearly any answers to these questions, however rough, are of considerable importance.

Obviously we cannot make direct breeding tests for recessive lethal or semilethal genes, but we can take advantage of one commonly occurring type of inbreeding, that of first-cousin marriages. First cousins have a pair of grandparents in common. If either of the two grandparents carries a defective recessive, it stands a good chance of becoming homozygous in any child who is a product of the consanguineous marriage (Figure 4-10). To make this statement meaningful, however, it is necessary to state it quantitatively.

Figure 13-7 diagrams a first-cousin marriage with one of the common

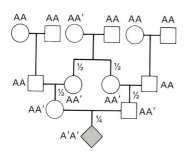

Figure 13-7 Pedigree showing the basic relationships in a first-cousin marriage along with the probabilities that a specific allele present in one of the common grandparents will appear in successive generations.

grandparents (*I-3*) heterozygous for a specific allele. The chance that one of his progeny (*II-2*) will receive it is one half, and the chance that the grandchild (*III-1*) will receive it from his parent is also one half. The chance that the grandchild will receive the allele designated in the grandparent is therefore $1/2 \times 1/2$, or 1/4. A similar argument applies to the other grandchild, *III-2*. Therefore the chance that the two grandchildren, who are first cousins, will both be heterozygous is $1/4 \times 1/4$, or 1/16. When this is the case, the chance, then, that an offspring of a first-cousin marriage will be homozygous for a specific allele present in one of the grandparents is $1/16 \times 1/4$, or 1/64. However, there are four alleles present in the two grandparents, each of which has this same probability of homozygosity. The total chance of homozygosity is therefore $4 \times 1/64$, or 1/16.

Disregarding the much smaller probability that homozygosity can also result from the fortuitous combination of two alleles of independent origin, the offspring of a first-cousin marriage should be homozygous for a particular allele one sixteenth as often as the allele is found in the grandparents or, more generally, one sixteenth as often as the allele is found in the population at large. From the frequency of homozygotes from first-cousin marriages, it is possible to make a simple estimate of allele frequencies in the population: if PKUs were to be found among the offspring of first-cousin marriages with a frequency of 1 in 800, one would conclude that the frequency of the recessive allele in the population is $16 \times 1/800$, or 1/50.

By modifying the logic slightly, it is simple to estimate the number of recessive lethal genes that the average person carries. If one of the common grandparents carries such an allele, the chance (and therefore the frequency) of a lethal homozygote in the offspring of first cousin marriages is one sixty-fourth. As the other common grandparent might equally well carry the allele, we must multiply 1/64 by 2 to get 1/32, or about 3 percent. Thus, for each 3 percent of mortality found in offspring of first cousins in excess of that found in unrelated marriages, we may conclude that the grandparents, and therefore the population at large, carry one recessive lethal effect. From Table 13-4, it can be seen that the mortality rate in first-cousin marriages exceeds that of the general population by about 6 percent prenatally and another 6 percent postnatally. If one lethal allele corresponds to 3 percent mortality, then the total of about 12 percent suggests that the average person carries about four such alleles.

It is not necessary to postulate completely lethal genes to account for the increased mortality observed, which might equally well be the result of eight different genes, each of which when homozygous would cause mortality in half the offspring, as well as other combinations having comparable effects. To take such possibilities into account, the term *lethal equivalent* has been coined, indicating any gene or group of genes that, acting separately or together, cause one death when homozygous.

The data available do not give information on very early losses of the embryo, a likely time for homozygous lethals to exert their effect, or, for that matter, on premature deaths later in life. When these are taken into account, several more lethal equivalents appear to be part of the average person's makeup. An educated guess is that each person carries five to seven lethal equivalents.

Table 13-4 For mortality at different periods, difference in mortality of progeny, when parents are unrelated or are first cousins. The numerical excess can be attributed to homozygosity for defective alleles. (By permission of N. Morton, *Progress in Medical Genetics* **1**:261–291, 1961. Grune & Stratton Inc., New York.)

| | | Relationship of Parents | | Difference in |
Type of Mortality	*Authority*	*Unrelated*	*First Cousins*	*Mortality*
Miscarriages	Slatis et al. (1958)	.129	.145	.016
Stillbirths and neonatal deaths	Sutter and Tabah (1958)	.044	.111	.067
Postnatal deaths	Slatis et al. (1958)	.024	.081	.057
Infant and juvenile deaths	Sutter and Tabah (1958)	.089	.156	.067
Juvenile deaths	Bemiss (1958)	.160	.229	.069
Early deaths (Hiroshima)	Schull and Neel (1958)	.031	.050	.019

Not only is viability affected by consanguinity, but the frequency of other defects, both physical and mental, also increases (Figure 13-8). One study in Chicago shows that whereas the frequency of physical and mental defects runs at about 10 percent in the normal population, it goes as high as 16 percent in first-cousin marriages. Once again, such studies must be limited to differences that are clear and unmistakable. Although the above difference of 6 percent would correspond to an average of only two detrimental genes per person, some workers in the field suggest that the total number of lethal equivalents plus detrimental genes carried by the average person is in the vicinity of ten. This number probably does not differ significantly from one population group to another. It is the totality of these detrimental genetic conditions that constitutes our *genetic load.*

Additional Reading

CAVALLI-SFORZA, L. L., and W. F. BODMER. 1971. *The Genetics of Human Populations.* San Francisco: Freeman.

CROW, J. F. 1961. Mutation in man. *Progressive Medical Genetics,* **1**:1–26.

JUKES, T. H. 1980. Silent nucleotide substitutions and the molecular evolutionary clock. *Science,* **210**:973–977.

NEEL, J. V. 1974. Developments in monitoring human populations for mutation rates. *Mutation Research,* **26**:319–328.

SCHULL, W. J., ed. 1968. *Mutations.* Ann Arbor: University of Michigan Press.

VOGEL, F. and G. RÖHRBORN, eds. 1970. *Chemical Mutagenesis in Mammals and Man.* New York: Springer.

WATSON, J. 1976. *Molecular Biology of the Gene;* 3rd ed. New York: Benjamin.

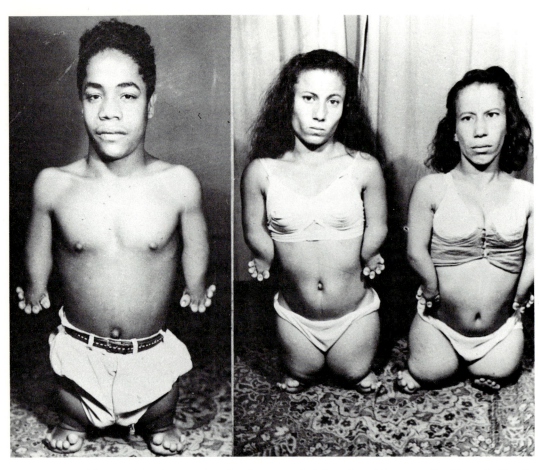

Figure 13-8 A recessive developmental defect found in a small village in Brazil; more than 60 percent of the marriages, which produced 68 such affected individuals, were consanguineous. (From A. Quelce-Salgado, A New Type of Dwarfism. *Acta Genet.*, S. Karger AG, Basel, **14:**63–65, 1964.)

Questions

1. Which is most important to future generations, somatic or germinal mutations?
2. What were the observations that led de Vries to coin the word *mutation?* Do we now use the word in the same sense that he did?
3. What are the two main ways that DNA can be altered to produce an effect that we might recognize as a new mutation?
4. Why are changes in hemoglobin structure useful in the study of mutation in humans? What is the difference between normal adult hemoglobin and the hemoglobin S found in sickle-cell anemia?
5. What do we mean when we say that the ways in which the 230 or so hemoglobin variants differ from normal are consistent with the genetic code?

6. What kind of mutation is responsible for hemoglobin Wayne? Why is the length of the alpha chain changed in this hemoglobin?

7. Can you show why frame-shift mutations might, in some cases, have a profound effect on the final gene product?

8. Can you argue that reverse mutations, which restore mutated DNA to its original state, must occur with a frequency far lower than that of mutations in general?

9. What is a dimer? Do we ever see evidence of this type of defect in the DNA in humans?

10. What are some of the difficulties in identifying an unusual human characteristic as a new mutation?

11. If you wished to make an estimate of the mutation rate in humans by observing the frequency of its appearance in the newborn, would you choose a dominant or a recessive trait?

12. What is the basic argument behind the calculation of mutation rates by the indirect method?

13. If the frequency of cystic fibrosis is about 1 in 2,000, what would you estimate the mutation rate to be? (Assume that L, the loss of reproductivity of the homozygote, is 100 percent, or 1).

14. If we assume the average mutation rate of all genes to be about 1 per 1 million per generation (10^{-6}/generation) and the number of genes in the haploid set to be about 20,000, how often, on the average, would we expect a newborn to carry a new mutation? Would this figure also be an estimate of a newborn with a congenital defect?

15. What is meant by our *genetic load?* Can you reconstruct the argument that leads to the conclusion that the average person may carry about four recessive lethals that have a detectable prenatal or postnatal effect? Why does the calculation of our load have considerable social significance?

16. What is meant by *lethal equivalent?* How many deleterious genes, including those contributing to mental defects, make up our genetic load?

14

Chemical Compounds Causing Genetic Damage

Carcinogenesis, Teratogenesis, and Mutagenesis

The adverse effects of agents on cells and their constituents, particularly on the genetic material, have been designated as *carcinogenic* (producing cancer), *teratogenic* (producing developmental abnormalities *in utero* and *mutagenic* (producing mutations).

The first evidence of a carcinogenic effect goes back to 1775, when Percivall Pott identified occupational cancer of the scrotum in chimney sweeps in England. The high incidence of this carcinoma can be attributed to the unusual concentration of carcinogenic compounds in coal tars. At the present time more than 500 chemical carcinogens are known, most of them relatively rare synthetic organic compounds.

DISCOVERY OF CHEMICAL MUTAGENS. Following the reports of the increased frequency of mutation after experimental X-ray treatments by H. J. Muller in 1927 (in *Drosophila*) and by L. J. Stadler in 1928 (in barley), a large number of investigators made serious attempts to discover mutagenic chemicals on the assumption that information about the composition of the gene might be obtained if it were known what kinds of chemicals could specifically alter gene structure. Most of the chemicals tested were those previously known to be carcinogenic in mammals. In general, the tests gave negative or rather uninteresting slightly positive indications of an effect.

It was not until 1941 that a powerful chemical mutagen was found. This happened during the course of work in England with mustard gas, a chemical highly poisonous to humans that was being manufactured secretly in the event that circumstances forced its use as a weapon. It was noted that individuals who were accidentally exposed to mustard gas during its manufacture reacted with inflammation, blistering, and burning, consequences somewhat similar to those produced by X rays. The similarities interested a group of pharmacologists, who, with the aid of a *Drosophila* geneticist, Dr. Charlotte Auerbach, showed that mustard gas, like X rays, was indeed a powerful mutagen. This information was not reported to the scientific world until after the conclusion of World War II.

In retrospect it should be noted that, prior to that war, observations had been made that colchicine could produce changes in chromosome number in plants and that nitrous acid could produce mutations in the mold *Aspergillus*, commonly found on bread, cheese, and other foods. We also learned that during the war the Germans and the Russians had discovered, in animal and plant experiments, that other compounds such as urethane and formaldehyde increased the mutation rate somewhat.

It is known that there exist a vast number of compounds with a wide variety of chemical configurations that have adverse biological effects. The differences in the chemical structure of a few of the better-known ones are given in Figure 14-1.

CHROMOSOME BREAKAGE AS AN INDEX. It is impossible to determine the activity of any potentially dangerous agent—a newly synthesized organic compound, for instance—by testing it on humans, because of the under-

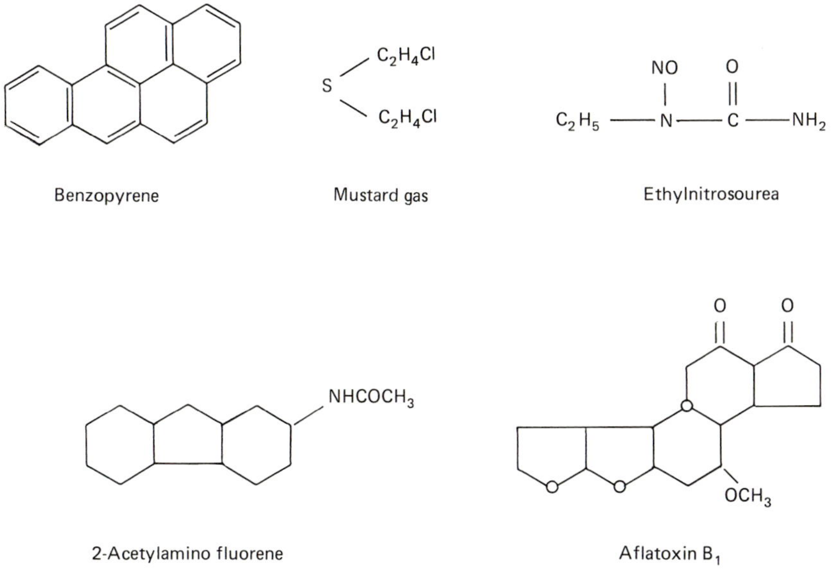

Benzopyrene

Mustard gas

Ethylnitrosourea

2-Acetylamino fluorene

Aflatoxin B$_1$

Figure 14-1 The structural formulas of a few of the better-known carcinogens and mutagens. The great diversity in structure makes it difficult to predict that any new compound will be completely safe.

standable reluctance to expose humans deliberately to large quantities of a new chemical. Of the three major categories of deleterious change in humans, potential mutagenicity of a new agent is the most difficult to determine. In the first place, most mutational changes are recessive and cannot be detected until homozygotes are produced by chance many generations later. Second, even for rare dominants that could be detected, in theory, by the immediate appearance of affected newborn infants, the newly induced mutants would be so rare that they would not make an appreciable increment to those occurring spontaneously.

Breaking of chromosomes by exposure to some test substance can, however, be observed very quickly both in the body (*in vivo*) and in the test tube (*in vitro*). Because agents that cause chromosome damage may also be responsible for one or more of the three major categories of biological effects, its occurrence is a valuable, but not an infallible, index of biological potency. However, the reverse is not the case. Many substances now known to be mutagenic and carcinogenic do not break chromosomes. When an agent can produce chromosome breaks, it is said to be *clastogenic.*

SPECIES DIFFERENCES.   When tests are made *in vivo*, it becomes necessary to resort to organisms other than humans for these biological tests. One of the hazards in making deductions concerning the biological potency of a chemical comes from the inconsistency that is shown from one organism to the next. A chemical that may be strongly mutagenic in bacteria may show no effect whatsoever in the fruit fly, and one that is mutagenic in both of those may not show any effects in experimental mammals. Some agents, such as radiation, are known to be mutagenic, carcinogenic, teratogenic, and clastogenic. Thalidomide is teratogenic and is not known to have any of the other properties. However, all chemical carcinogens seem to be mutagenic but not necessarily teratogenic. Conversely, most mutagens are potential carcinogens; much of the preliminary testing for possible carcinogens in humans is actually performed as testing for mutagenicity in lower organisms.

Not only may mammals differ in their sensitivity to these agents—as, for instance, from mice to humans to hamsters (Table 14-1)—but in some instances compounds shown to be biologically active in one strain of a given species also prove inactive in another strain of the same species, implying genetic variability in sensitivity. Nevertheless these organisms must provide the experimental material for specific information about possible dangers to humans, particularly when prior knowledge is required before a compound is released for general use or consumption. It is no wonder that the field is highly controversial, with quite contradictory claims being made by different investigators whose objectivity and impartiality are not open to question.

LINEAR-DOSE RESPONSE.   One of the most vexing problems in assessing chemical damage involves the nature of the dose–response curve. Ordinarily one can obtain information about the effects of a chemical on an experimental animal when it is given in moderately high doses (Figure 14-2). As the amount of chemical being tested decreases, the effect does also, so that many more animals must be used to find any effect at all. In fact, in most

Species	Smallest Dose Producing Defects	Largest Dose Producing No Defects
Man	0.5–1.0	?
Baboon	5	—
Monkey, cynomolgus	10	—
Rabbit	30	50
Mouse	31	4,000
Rat	50	4,000
Armadillo	100	—
Dog	100	200
Hamster	100	8,000
Cat	—	500

Table 14-1 The effectiveness with which thalidomide produces birth defects. Note that not only is its teratogenicity quite variable from one species to the next but also that a rather small dose may be effective in some cases (first column), whereas a much larger one may be ineffective in others (second column). (From H. Kalter, *Teratology of the Central Nervous System,* University of Chicago Press, Chicago, 1964.)

cases it is prohibitively expensive to test very low concentrations of a chemical on laboratory animals, a problem that is compounded because there will always exist a small but variable frequency of spontaneous changes of the sort being tested for. This experimental "noise" makes it difficult, sometimes impossible, to carry out low-dosage experiments even if the question of financial practicality were not important.

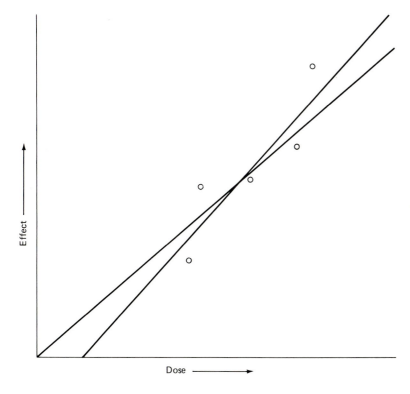

Figure 14-2 Illustration of the difficulty in determining dose–effect relationships from data using high doses. The observations fit both a line with a threshold and a line with no threshold equally well. These two possibilities have completely different implications in human populations, the first being "safe" at low doses and the second not.

If there exists a threshold below which a compound has no effect, and humans consume less than that amount, there is no cause for concern. On the other hand, if the effect is linearly related to dose, and there is no threshold (see Figure 14-2), then every amount consumed, no matter how small, will produce its proportional effect, and the contribution to biological damage can be readily calculated simply if one knows the total amount of the offending material ingested. Thus, if in an actual test a total of twenty cancers are produced after 100 animals each eat 1 gram of a test compound, if linearity is assumed without a threshold, twenty cases of cancer will be induced for every 100 grams ingested, irrespective of whether that amount is divided among 100, 1,000, or 1 million animals. Those who argue for the existence of a threshold would maintain that 100 grams total of a carcinogen or a mutagen, divided into 1 million doses and taken by that many animals, would not produce twenty cases of cancer, but many fewer and probably none at all. Arguments for and against the banning of such compounds as cyclamates, saccharin, and other commonly used additives center on this question about the nature of the dose-response curve at very low doses.

INFLUENCE OF THE METABOLISM. Another important reason for conflicting results is that the metabolism of the organism may change the compound into closely related compounds, which may or may not be biologically active. In one class of mutagens the compound itself is inactive but its metabolic breakdown products are potent. These are called *promutagens*. Phenobarbital is a suspected promutagen.

In another class, the compounds are active without the intervention of any metabolic change. These are called *direct-acting mutagens*. A direct-acting mutagen may be degraded into nonmutagenic products by the metabolism of the particular organism in which it is found. Therefore in each case the mutagenic activity would depend on the exact metabolic pathways present in that species, and it would not be surprising to find that the same compound is mutagenic in one species and not in another. It has been shown in the guinea pig, for instance, that one compound (acetoaminofluorene or AAF) is mutagenic in one strain and not in another, and that the second strain lacks an enzyme present in the first that normally converts nonmutagenic AAF into a highly mutagenic product. As another example, 2-naphthylamine is a potent carcinogen in humans but is not a carcinogen in rats.

These effects may be further complicated by the fact that two compounds together may produce a result not found from each individually. Thus simultaneous treatment of lymphocytes *in vitro* with a clastogenic agent and an antioxidant (vitamins C and E, for instance) gave a 30–60 percent reduction in the number of induced chromosome breaks, compared with the results of a control experiment that used the clastogen only. It has been postulated that the declining incidence of stomach cancer in the United States may be an accidental result of the inclusion of antioxidants in foods as nutritional additives (vitamins) or as preservatives. Such a beneficial effect would be a refreshing departure from the almost universally valid generalization that preservatives and other food additives may be good for the food but bad for the human body.

TUMOR PROMOTERS. Compounds may interact in such a way as to increase the likelihood of a carcinoma. Some substances, not carcinogenic in themselves, enhance the effect of a small amount of a carcinogen that would ordinarily be ineffective (or have an effect too small to be observed). The list of suspected promoters includes the drugs phenobarbital, saccharin, sodium cyclamate, and even the bile acids that are normally present in the intestine.

THE AMES TEST. A common test for mutagenicity is one developed by B. Ames of the University of California. It makes use of a mutant strain of a bacterium *Salmonella typhimurium,* that is unable to synthesize the essential amino acid histidine. The mutant strain must therefore be provided with histidine to survive. When this strain is exposed to a mutagen (Figure 14-3), the DNA may be altered so that the defect is corrected and the bacteria can now grow in a histidine-free medium because they manufacture their own histidine. The number of colonies growing after treatment, which indicates the number of mutations that occurred, is therefore an indicator of the relative potency of the test chemical as a mutagen in this bacterium.

Bacterial metabolism lacks many enzymes found in the mammalian liver and other tissues. The Ames test in its simplest form would not detect promutagens that require mammalian enzymes for their activation. This problem is easily taken care of by exposure of the test chemical to ground-up liver cells that contain the enzymes necessary to activate promutagens.

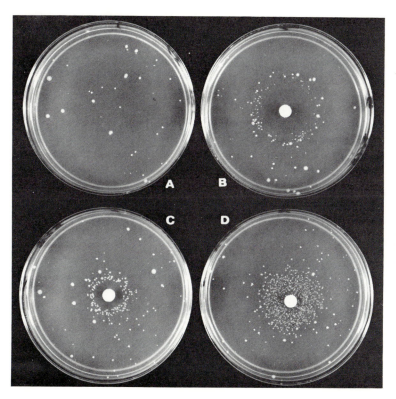

Figure 14-3 The Ames test. A control plate, showing the usual number of spontaneous mutants *(A)* and the greatly increased number of mutants after the addition of the Japanese food additive furylfuramide (AF-2) *(B),* the mold carcinogen aflatoxin B *(C),* and 2-aminofluorene *(D).* (Reprinted with permission from B. N. Ames, I. McCann, and E. Yamasaki, *Mutation Research* **31**:347, 1975. Copyright © 1975 Elsevier Scientific Publishing Company, Amsterdam.)

This test for mutagenicity agrees with previous knowledge of the carcinogenicity in mammalian studies in about 60–70 percent of all cases. It is so simple and inexpensive, however, that it serves a useful function as a first step in a more extended and expensive series of tests, involving *Drosophila*, mammalian cells in culture, and, finally, tests on laboratory mammals.

Substances Found in Food

NITROSAMINES.   Sodium nitrite has been very widely used as both a preservative and a color enhancer in meat and fish products, such as frankfurters, bacon, smoked meat and fish, and Vienna sausage. It is considered a hazard because it is converted into nitrosamines in the human stomach, and that class of compounds has been shown to be highly mutagenic in bacteria, viruses, molds, and other organisms. However, the great benefit of sodium nitrite in retarding bacterial growth, particularly the growth of spores of *Clostridium*, responsible for botulism, is viewed as outweighing its potential mutagenic effects, and tentative plans to outlaw the use of sodium nitrite as a preservative have been shelved.

Scotch whiskey and beer, with one or two exceptions, have been shown to contain small quantities of nitrosamine produced when barley is roasted into malt by a direct-heat method. Brewers have modified their procedures to eliminate this potential problem. Some commonly used drugs (Librium and Antabuse) have chemical compositions that promote the formation of nitrosamines in the body.

ARTIFICIAL SWEETENERS.   For more than twenty years, but particularly in the last ten, sodium cyclamate has found widespread popularity as a sugar substitute in diet foods and soft drinks. Because of their universal use, these have been subjected to extensive tests by both manufacturers and governmental agencies.

Sodium cyclamate has been shown to have a carcinogenic effect in the bladders of mice in which pellets containing it were implanted. In one experiment the proportions of mice developing carcinomas were 61 percent of those exposed; carcinomas developed in only 12 percent of control mice exposed only to the cholesterol used as a carrier for the cyclamate. The bladder of the male mouse appears to be very sensitive to cyclamate; this organ was examined in individuals given cyclamate orally, and it was found that in such cases carcinomas were also produced. On the other hand, rats fed diets containing as much as from 1 to 5 percent of sodium cyclamate (and also saccharine) for two years appeared to suffer no effect at all at the lower dose and no toxic effect at the higher dose.

It seems likely that the technique of implanting pellets in the bladders of mammals may not be a valid procedure for testing the carcinogenicity of compounds that are ordinarily taken by mouth. There is no evidence that the use of cyclamate has been responsible for any congenital abnormalities in children, any carcinomas, or any other biological effect except possibly for a rare skin hypersensitivity. Nevertheless, in accordance with the rule that any suspicion of a deleterious effect should be sufficient reason for the removal of the product from general consumption, cyclamate was banned from general use. It was banned even though the mice that developed bladder tumors after feeding had consumed a daily amount per body weight equivalent to the daily human consumption of 250 bottles (of 16 ounces each) of a typical diet drink. Another report shows that in certain rat cells the metabolic breakdown product of cyclamate, cyclohexylamine, is clastogenic. A person would have to consume about a pound of cyclamate a day for life, however, to bring the exposure up to the level given the rats in this experiment. Possibly the move to ban cyclamates was aided because other artificial sweetners (like saccharin) exist that can replace cyclamates.

Saccharin has also come under suspicion as a carcinogen. When it was fed to 100 rats as 7 percent of their diet (the equivalent of a human's consuming 800 cans of diet drink daily), 17 developed tumors and 14 percent of their offspring were similarly affected. However, the incidence of bladder cancer in diabetic humans, who consume large doses of saccharin, is no higher than in the general population. Because of the obvious benefits to diabetics and to persons with weight problems (along with the usual arguments against accepting results from animal tests), this substance has not been banned.

CAFFEINE. One of the most widely ingested drugs is caffeine, found not only in coffee and tea but also in chocolate and in some soft drinks. It is a common component of a very large number of headache remedies and pain relievers that have aspirin as their principal ingredient. In 1948 it was reported that caffeine produces mutations in bacteria as well as in fungi and, a year later, that it produces chromosomal breaks in the root tips of onions. At extremely high concentrations it has been shown to be weakly mutagenic and clastogenic in *Drosophila* and clastogenic in cultured human cells. However, there is evidence that the action of caffeine is not to initiate the primary break but rather to inhibit the normal repair of breaks that have already occurred for some other reason. For this reason caffeine may be most damaging when applied in combination with some other agent that produces chromosome breaks.

Large-scale studies on mice fed amounts of caffeine in excess of those ordinarily taken by humans did not reveal any increase in the mutation rate. When heavy doses are given to male mice, litter size is not reduced, indicating that extensive chromosome breakage in the male mouse's germ cells is not a result of high caffeine treatment. In another experiment, involving female mice fed caffeine, the birth rate dropped from the customary five to seven per litter to one or two. The reason for the difference between the male and female sensitivities is not known. It is a reasonable deduction, however, for other chemical agents, that the decrease in litter size after treatment of the male parent can be attributed to chromosome breakage in the sperm exposed prior to the mating, but a decrease after treatment of the female can have several causes, including chromosome breakage as well as the loss of embryos in an environment rendered unpleasant by the drug.

Because of its universal consumption and its known ability to penetrate to the human germinal tissue and, through the placental barrier, to the fetus, where it is found in the same concentrations as in the maternal blood-stream, caffeine is thought by some to be one of the most potentially danger-ous compounds now widely used by humans. One estimate of its effective-ness is that the average coffee drinker, during her or his lifetime, is exposed to a total mutagenic action equivalent to several X-ray examinations.

There is evidence for a carcinogenic agent in coffee itself. Persons who drink three cups of coffee per day have three times the risk of developing pancreatic cancer as noncoffee drinkers. That this is not an effect of caffeine is likely because, on the one hand, tea, which also contains caffeine does not appear to have any carcinogenic effects but, on the other hand, decaffeinated coffee does. In the latter case, it is possible that the chemical process of removing the caffeine inadvertently introduces or synthesizes a carcinogen.

FOOD DYES. Amaranth, the most common red food coloring in the United States and Great Britian, used in jam, ice cream, lipstick, candy, sausages, fish, and even dog food, has been suspected of being both carcinogenic and teratogenic. Its use has been banned in Russia and West Germany, and in 1972 the World Health Organization recommended strict limits on its use. How difficult the interpretation of experimental data can be is shown by the fact that whereas one group concluded that there was no evidence of danger to the fetus, another group, on the basis of the same evidence, concluded that a pregnant woman might exceed the safe level if she consumed as much as a third of a can of cherry soda a day.

IRRADIATED FOODS. Radiation causes the formation in water of a large number of highly reactive compounds such as peroxides. Studies on the feeding of heavily irradiated foods to rats and dogs show that there may be an effect on the number of offspring they produce, as well as on the body weight and life span of the experimental animals. The reasons are unknown, but it is likely that a family of active compounds (such as the peroxides) produced by radiation are responsible.

AFLATOXIN. In 1960 an outbreak of disease killed more than 100,000 tur-keys and large numbers of other fowl in England, and at the same time fish hatcheries in the United States reported unusual losses of fish from cancer of the liver. These problems were traced to peanut meal from Brazil, which was contaminated by the mold *Aspergillus flavus*. On investigation it devel-oped that certain strains of this mold, found on beans, nuts, corn, grains, and other agricultural crops, produce very potent mutagens, which were named *aflatoxins*. Aflatoxin B1 has been shown to be highly effective in producing mutations in fungi and wasps and is clastogenic in human cell cultures. In pregnant mice it has been shown to increase the number of fetal deaths tenfold, probably by the gross breakage of chromosomes. It has been impli-cated in the high incidence of liver cancer in Thailand, where it is responsi-ble for a high degree of contamination of food. Buyers and processors of susceptible foods, particularly peanuts, have established rigid controls to prevent the dissemination of products contaminated by this mold.

CARCINOGENS IN FOODS. It would be tempting to attribute the mutagenicity of foods only to the additives placed in them during processing. However, mutagens occur naturally as well. Quercetin and kaempferol, both highly mutagenic, are found in many edible plants, vegetables, and fruits. The residents of certain areas of China have an extraordinarily high (up to a hundredfold increase) incidence of cancer, much higher than in Western populations; both of these have been attributed to the way in which food is prepared in those countries. In the United States, on the other hand, Seventh Day Adventists and Mormons, who live on a more restricted diet than the average American, have a significantly reduced cancer incidence.

Cooked foods may have their share of carcinogens that are synthesized by the combination of several organic compounds present naturally. Smoked fish and charcoal-broiled steak are a few of the foods that contain carcinogens produced by the cooking process. Tests of hamburgers fried on a grill, in a frying pan, or in an electric hamburger cooker showed that they had much greater mutagenicity than did hamburgers cooked under a broiler or in a microwave oven. The critical factor is the cooking temperature, which is higher when heat goes to the meat through direct contact with hot metal.

Drinking water that has been chlorinated shows a low level of mutagens and carcinogens; these may be reduced by treatment with sulfate. Tests have even shown that there are mutagens in beet and carrot juice!

Activity of Drugs

LSD. Few compounds have received such widespread attention in the past ten to fifteen years as the hallucinogenic drug *lysergic acid diethylamide (LSD)*. Over the past decade, more than a hundred studies have been published related to the adverse biological effects of this compound. Some of the early studies involved subjecting leukocytes *in vitro* to varying concentrations of LSD, and a number of workers have reported increased chromosome breakage, although there appears to be no simple relationship between the dosage and the extent of the clastogenic effect.

In evaluating results, it should be kept in mind that such tests involve cells that have been artificially stimulated to divide rapidly, a condition that does not exist in the body, and that therefore some chromosome breaks may show up in cultured cells that might not be present in a comparable *in vivo* test. In addition, the concentration or exposure used is usually greater than the human dose, and the amount of breakage is not outside of the range induced by many common agents such as antibiotics, caffeine, and aspirin. In addition, in the *in vitro* tests the effects may be exaggerated over those found in the body, where excretory and detoxification mechanisms might serve to reduce the response of the cells to the chemical.

When studies are made of the chromosome breaks of individuals who have taken LSD, conflicting results are found. If the subjects under study are not in a medically controlled situation, the purity of the LSD cannot be assured (because it may have been "cut" with such chemicals as belladonna, arsenic, and cyanide), and the substance taken may not even have been LSD.

Of 126 subjects given pure LSD, 18 showed a frequency of chromosome breaks higher than that of the controls, whereas 90 of 184 persons taking illicit LSD showed an increased frequency of aberrations. This latter frequency may be high for several additional reasons. Users of LSD may also regularly take other drugs. Furthermore there is among drug abusers a high frequency of viral disease resulting from infection; viruses are known to produce chromosome damage.

Several reports have suggested that LSD may be carcinogenic. These reports are based on the observed breakage of chromosomes in the early studies reported above and the apparent similarity of chromosomes known to be broken in certain kinds of leukemia and other carcinomas. Although a few cases have been found in which leukemia has developed in individuals treated with pure LSD, at present there does not appear to be any strong evidence that LSD is a carcinogen. It is generally true that cancerous tissue shows a high level of chromosomal abnormality, but in most cases this is likely to be a consequence of abnormal mitoses and differentiation of those cells, rather than the cause.

Experiments with *Drosophila* and fungi indicate that treatment with relatively massive doses of LSD does increase the mutation rate slightly. It is not known whether LSD has any mutagenic effect in humans.

A large number of studies have been made of rats, mice, and hamsters, on the effects of LSD on their progeny, and it has been shown that there appears to be a wide variation, depending on the species used and particularly on the time at which the LSD was applied during pregnancy. No effect has been reported when exposure occurs late in the pregnancy in those animals.

In one instance in which LSD was injected into pregnant rhesus monkeys, one female delivered a normal infant; two were stillborn, with abnormalities of the face; and the fourth died after one month of life. Two control animals produced normal progeny. The lowest dose used in these experiments was 100 times greater than the usual experimental dose in humans. In some studies in humans it was found that the ingestion of illicit LSD by pregnant women resulted in larger numbers of chromosomal breaks in their offspring but that the children were in apparent good health and without any ill effects at the time of birth and immediately thereafter. On the other hand, a number of cases have been reported of malformed infants born to women who used illicit LSD before or during pregnancy, the defects generally involving the arms and legs. However, in most of the cases it is known that these women had also ingested other kinds of drugs, such as marijuana, barbiturates, and amphetamines. There is no evidence that pure LSD is teratogenic at moderate doses in humans.

Several investigations have questioned the medical significance of the chromosome breaks reported in some of the studies with LSD. First of all, untreated people usually show some small frequency of lymphocytes with chromosome breaks, of the order of 2–5 percent; second, after certain diseases—especially those involving viruses, such as measles, smallpox, chickenpox, mumps and hepatitis, and possibly including the common cold—the observed frequency of chromosome breakage may go up several-fold.

SMOKING. There have been reports that nicotine is clastogenic in the dividing cells of some plants. There are low concentrations of radioactive polonium and lead concentrated in insoluble smoke particles, and these can become imbedded in lung tissue or, after being ingested by macrophages, may be transported to the liver, the spleen, and the bone marrow. One of the many carcinogenic products of smoking is benzopyrene, which is formed after the burning of tobacco or other organic materials. Extremely low concentrations of cigarette smoke—the amount that would be produced by 1/400 of a cigarette—can produce detectable numbers of sister strand exchanges in DNA. The number of fetal deaths found in the litters of female mice mated to males treated with this substance is more than ten times greater than that of females mated to untreated males. The statistical evidence that cigarette smoking causes cancer of the lung has been sufficient to convince the U.S. government to require a warning to be placed on all packages of cigarettes. In recent years a marked increase in the number of women dying of lung cancer has been noted; this increase has been attributed to the increase in the number of women who took up smoking after World War II, followed by a twenty- to thirty-year lag period before the effects appeared. Deaths from lung cancer have increased fivefold since 1958 and are now the leading cause of cancer death in women, overtaking breast malignancies.

Because benzopyrene is a common product from burning plant materials, particularly at lower burning temperatures, it is introduced into the atmosphere at a high rate, 1,300 tons per year in the United States alone. As energy costs mount, the introduction of wood stoves and closed fireplaces to decrease heating loses from burning wood will increase the production of this carcinogen.

MARIJUANA. Marijuana has been implicated in chromosome breakage. In one study forty-nine marijuana users showed an average of 3.4 percent cells with broken chromosomes, whereas there were only 1.2 percent in the twenty controls. It has been suggested that the chromosome breaks found in individuals who have been exposed to LSD may in fact be the result of the marijuana very often taken by the same groups of persons who take LSD. In addition, recent observations indicate that cultures of leukocytes taken from marijuana smokers have an increased frequency of fragmented nuclei that do not have a full complement of chromosomes and result from abnormal mitoses. The incidence of hypoploid cells, with fewer than thirty chromosomes per nucleus, was significantly higher in blood samples of heavy marijuana smokers (averaging thirteen to fifteen joints per day) after a heavy smoking period than it was before, or than it was in a control population. Studies show the rate of replication of DNA in lymphoblasts of marijuana smokers to be considerably depressed compared with that of normal controls, and at a level corresponding to that found in patients suffering from cancer or with a medically suppressed immune reaction after a transplant. Other, more recent studies deny these effects.

Following a lead suggested after it was reported that several young male marijuana users showed unusual breast development, one team of workers found that the male sex hormone testosterone was on the average 44 percent lower in marijuana smokers than in controls and that in several cases the sperm count was so low that sterility resulted. In addition to the possibility that the drug may alter the course of normal masculine development in young males, it has been questioned whether it might not have a similar effect on male prenatal development when the mother uses marijuana heavily during the early months of gestation. Female monkeys fed daily doses of tetrahydrocannabinol (THC, the effective agent in pot), in an amount equivalent to four or five joints a day, lost 42 percent of their fetuses or newborn offspring, compared with 11 percent in the controls.

It should be noted that in many cultures the use of marijuana or closely related drugs in concentrations many times higher than those found in an ordinary marijuana cigarette has been a socially acceptable activity for centuries, with no reported ill effects to those populations. On the other hand, because marijuana smoke contains benzopyrene as well as the dozens of other carcinogenic compounds produced when vegetable matter is burned, it is possible that future years will see an increment in cases of lung cancer from this source.

So the controversy will continue. The daily press will continue to feature articles describing biological damage caused by marijuana alternatively with articles denying any such effects. It is clear that the final judgment will depend on more information than is now available.

ALCOHOL. As in the case of tobacco, the almost universal use of alcohol over many centuries makes it unlikely that any pronounced mutagenic or carcinogenic effect would have gone unnoticed. However, it has been widely reported as a cause of teratogenesis in pregnant women, and it has been reported that men who consume excessive amounts (four or more drinks per day) may also have abnormal offspring more frequently than expected (see Figure 6 in Chapter 16). The presence of nitrosamines in alcoholic beverages was noted earlier.

ANTIMALARIAL DRUGS. The antimalarial drug quinacrine (also called *atabrine*) was used on a very extensive scale during World War II in the South Pacific, where as many as 7 million persons may have been exposed to large quantities of the drug. This compound, which has subsequently been found to be very useful for staining chromosomes, is mutagenic in bacteria and viruses as well as in fruit flies. No reports have yet been made of any similar adverse effect on humans.

ANTICANCER DRUGS. Highly suspect are anticancer drugs, because of the ways in which they destroy living cells. In some cases, their effectiveness depends on chromosome breakage in mitotically active cells, producing anaphase bridges with subsequent loss of such cells. In other cases, they may be effective in nondividing cells, producing their effect in some other way, as in enzyme inactivation. In any case, any anticancer drug stands a reasonable chance of being either mutagenic, teratogenic, carcinogenic, or clastogenic, or any combination of these. However, because most individuals being treated with such drugs are critically ill, the question of the potential hazard becomes secondary to the possibility of destroying the cancer and saving the life of the patient. In addition, it can be argued that such individuals are usually beyond the childbearing years and so will not pass mutations on to subsequent generations even if such mutations should be induced. On the other hand, youngsters should not be subjected to such drugs unless it is a matter of life and death. In particular, the drugs should not be used on young people for diseases that are not serious or that can be treated otherwise. For instance, the anticancer agent methotrexate, which has been shown to be clastogenic, is used on occasion to treat extreme cases of the nonmalignant skin disease psoriasis. Obviously such chemical treatments, like those involving radiation, should be avoided.

SEDATIVES. Methylpyrilene is used in sleeping aids, sedatives, nasal sprays, and decongestants. It induces liver tumors in mice when given in concentrations equivalent to sixteen to twenty pills per month for twenty-five years in humans. Anticonvulsants (such as phenobarbital) taken during pregnancy have been shown to be clastogenic in both mother and child.

Hazards of Technology

PESTICIDES. Compounds used for combating insect pests include one group that sterilizes (TEPA and TMA) and anther group with direct lethal action (captans, DDT, and DDVP, or vapona). Because the former compounds sterilize insects by causing an increased frequency of chromosome breakage in the germ cells, which then causes the zygote to die, it is clear why there would be concern about exposing humans to such compounds. Both of them increase the frequency of fetal death in the mouse by a factor of 30 when the male parent has been exposed.

Much controversy surrounds the widespread use of DDT as a pesticide. Claims have been made that this compound is both carcinogenic and mutagenic. Part of the apprehension concerning DDT stems from its relative indestructibility and its tendency to accumulate in the food chain, where its concentration may reach values hundreds of times greater than initially. Nevertheless its widespread, even indiscriminate, use in worldwide programs to eliminate malaria, where it may have been ingested by tens of millions of people, sometimes in considerable quantities, does not support any substantial carcinogenic effect in humans.

Although some studies made on mice and rats have suggested such a carcinogenic effect of DDT, other studies have not. These contradictions may be caused by metabolic differences of the sort described earlier; it is known the DDT may be metabolized by at least three different pathways. In one legal examination of the situation, the conclusion of the hearing examiner was that on the basis of the evidence available, there appeared to be no evidence that DDT was either mutagenic, carcinogenic, or teratogenic in the amounts to which humans may ordinarily be exposed. This may be a fair statement; on the other hand, lack of evidence on these points at the present time does not preclude the possibility that future work may show such effects. Because as long a period as twenty to thirty years can separate the exposure and the appearance of a cancerous growth, it may take some time before any ill effects appear.

DIOXIN.   An extemely potent carcinogen, dioxin, is found as a contaminant in the herbicide 2,4,5-T. It also has a teratogenic effect, which is discussed in detail in Chapter 16.

FORMALDEHYDE.   Used in the manufacture of house insulation, plywood, particle board, carpeting, and some plastics, formaldehyde has been shown to be carcinogenic. Formalin, an aqueous solution containing 37 percent formaldehyde, is used in disinfectants and as a preservative in embalming fluid.

ASBESTOS.   The industrial asbestos worker may develop some disease (often lung cancer) related to this substance, on the average, within thirty years. Those who smoke in addition increase their risk tenfold. Often used in construction material because of its fire-resistant properties, asbestos is now found in the walls and ceilings of many school classrooms.

TRIS.   Once used as a flame retardant in children's sleeping pajamas, Tris-BP (tris(2,3-dibromoprophyl)phosphate) has proved to be carcinogenic and has been prohibited from such use.

HAIR DYES.   2,4-Diaminoanisole sulfate, which was once used in black, dark brown, and ash brown hair dyes, was found to be mutagenic by the Ames test and has proved to be carcinogenic in rats. Before the cosmetic industry was prohibited from its use, this dye was removed from preparations made for men because it penetrated the scalp sufficiently to turn the urine black or brown. This chemical is not found in temporary tints, rinses, or semipermanent hair dyes; these may have their own characteristic carcinogens. Some manufacturers have substituted compounds with a slightly altered composition; these compounds, which were introduced without testing, may eventually prove to be carcinogenic as well.

SOLVENTS.   Some solvents, such as those used in dissolving glue, have been found to be clastogenic. These substances are found in a very wide variety of mixtures in commercial products so that the specific active agent may be difficult to identify, although benzene has been named as a contributor to chromosome breakage. A study of twenty-five persons in Italy who had been exposed to benzene poisoning revealed that they suffered from a greatly increased frequency of chromosome breaks.

Vinyl chloride, which has appeared in a wide variety of plastic articles and aerosol sprays, has been correlated with a high frequency of cancer of the liver in persons exposed to high concentrations. It has been shown to be mutagenic in *Drosophila* at concentrations as low as 50 parts per million of air, and workers in some plants have been exposed to 20,000 parts per million. Its use has been banned in applications where people may receive high exposures, or low exposures over a prolonged period of time. It is no longer used in aerosol sprays.

TOXIC DUMP AT LOVE CANAL. From 1947 to 1952 a chemical company near Niagara Falls dumped toxic chemical wastes at Love Canal, covered them over, and sold the site to the local school board. After a school was built on the land, it was resold to developers, who built houses on it. Since that time there has been a high number of miscarriages and evidences of ill health in the residents of that area. A chromosome study of thirty-six persons living in that area has revealed that eleven had an unusually high incidence of acentric fragments, rings, and other abnormal chromosomes (Figure 14-4). Although this study is open to criticism because the government agency sponsoring it did not make adequate provision for a well-planned control for comparison, the existence of some unexpected chromosome abnormalities in the preparations used in the study has been confirmed by other workers as well. Whether the frequency of chromosome breakage is significantly higher than that of a carefully selected control group from that same general area awaits further study. However, an exhaustive study of the incidence of cancer among the residents has shown no greater frequency among the residents than among a control group.

It has been estimated that well over a million Americans could be similarly exposed to toxic wastes from dumps, which may number well over 30,000 and in many cases may be located in obscure and unmarked places.

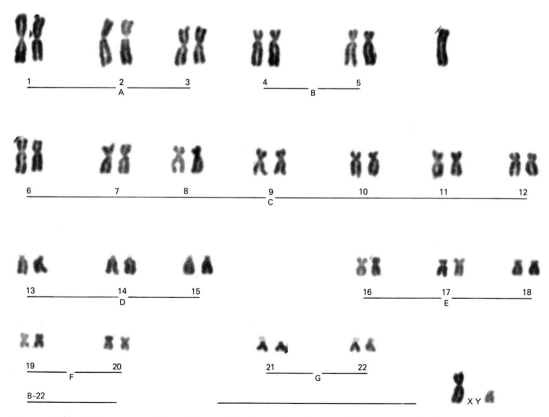

Figure 14-4 A karyotype of a resident of the Love Canal area. Of interest is the large unpaired acentric chromosome at the right in the first row. (Courtesy of D. Picciano, Biogenics Corporation.)

PCBs. Polychlorinated biphenyls are characterized by a high electrical resistance and excellent cooling characteristics and so have been used in electrical equipment, such as transformers, for many years. In a number of instances this liquid has contaminated food for poultry and livestock, which then present a hazard to humans. PCBs have been linked to congenital defects and to nervous system disorders. A study of 132 male Florida State University students showed that their mean sperm count was only 60 million per milliliter of semen (compared with 100 million/ml, found in a 1929 study) and that a quarter had only 20 million/ml, an amount that is considered by many fertility experts to indicate functional sterility. A search for foreign chemicals in the sperm showed excessively high levels of PCBs (as well as other toxic and unidentified compounds).

BURNING ORGANIC MATERIAL. Biologically active compounds are produced when organic material is heated. Carcinogens ingested when tobacco is smoked or charcoal-broiled meat is eaten are generated in this way. Home heaters which use oil or coal emit carcinogens; diesel engines are particularly polluting. Use of the wood stove, which is of necessity becoming increasingly widespread, produces carcinogens at a high rate, particularly when the oxygen supply to the fire is limited so that the wood burns at a lower temperature.

COST-BENEFIT RATIOS. One of the important considerations in reducing occupational exposures to such materials involves a comparison of the amount of money that the reduction will cost with the good that can be expected. Thus it has been calculated that to bring workers' exposure to benzene down to safer levels that might prevent one death from leukemia every two or three years would cost $500 million annually. Equating human lives with dollars and cents in this way may seem morally reprehensible, but it is the kind of decision that has to be faced when a large number of socially desirable needs compete for a fixed amount of money.

Overall Impact of Mutagenesis

The frequency with which chemical agents produce transmissible genetic changes in humans is an open question at the present time, nor is there any simple means of arriving at a reasonable estimate, unlike the case for radiation effects about which we can make some rough guesses. The problems encountered in analyzing the effects of new chemical compounds seem almost insurmountable; we do not even know whether the dose-effect relationship is linear at very low doses for any chemical in humans. The analogy with physiological poisons, which invariably have a threshold, is difficult to dismiss, even though some compounds appear to behave linearly at low doses. Clearly this is the most important consideration when hundreds of millions of humans may be ingesting small amounts of a mutagenic compound such as caffeine.

When other methods are developed permitting a more precise assay of mutagenicity, we shall undoubtedly learn that some of our apprehensions have been unfounded and, unfortunately, that some compounds now considered innocuous are having more influence than now realized. If anything seems clear, it is that no new compound should be put into widespread use without a thorough testing on a variety of animals in the most scientifically controlled fashion possible. With the synthesis of tens of thousands of new compounds each year and with the commercial production of about a thousand, new hazards are constantly being added to the environment with no means available at present for testing each one prior to its introduction to the population. It can be stated as a general rule that although the evidence that specific chemicals are biologically deleterious can be confused, contradictory, and highly inconclusive, it would be well for humans, given freedom of choice, to ingest as few new synthetic or suspected chemicals as possible. This is particularly important for women in the early stages of pregnancy.

The exposure of humans to chemical carcinogens is largely self-administered (Figure 14-5). Although industrial chemicals make their own contribution to the incidence of cancer, they are not the major cause: smoking, ultraviolet light from sunshine, and natural chemicals in food and water

Figure 14-5 Our indulgence in countless self-administered biologically active chemicals undoubtedly accounts for the greatest exposure to the hazards of pollution. (Courtesy of A. H. Wiebenga, University of Amsterdam Medical School.)

make a greater contribution to cancer incidence in humans. Smoking causes about 30 percent of all deaths from cancer; another 30 percent are related to nutritional factors.

To summarize the information on the wide variety of chemical compounds discussed in this chapter, it should be noted there are four different ways by which biological damage may be manifest (carcinogenic, teratogenic, mutagenic, and clastogenic), and species and strain differences make it difficult to extrapolate experimental data from laboratory experiments to humans. Compounds that are suspect include sodium nitrite, sodium cyclamate, caffeine, food dyes, aflatoxin, pesticides, and some solvents. The data on certain drugs (e.g., LSD and marijuana) are contradictory, whereas the evidence that other more socially acceptable drugs (alcohol, tobacco) cause biological damage is less controversial.

Additional Reading

AMES, B. N. 1979. Identifying environmental chemicals causing mutations and cancer. *Science, 204:*587–593.

AUERBACH, C. 1974. Smoking and cancer. *New Scientist, 61:*711.

BROSS, I. D. J. 1980. Muddying the water at Niagara. *New Scientist, 89:*728–729.

COMMONER, B., et al. 1978. Formation of mutagens in beef and beef extract during cooking. *Science, 201:*913–916.

GEHRING, P. 1977. The risk equations: The threshold controversy. *New Scientist, 75:*426–428.

LIJINSKY, W. 1977. How nitrosamines cause cancer. *New Scientist, 73:*216–217.

LOVE CANAL: How culpable is the city? 1981. *Science News, 119:*42.

MAUGH, T. H., II. 1974. Marihuana (II): Does it damage the brain? *Science, 185;*775–776.

MAUGH, T. H., II. 1978. Chemical carcinogens: The scientific basis for regulation. *Science, 201:*1200–1205.

MAUGH, T. H., II. 1978. Chemical carcinogens: How dangerous are low doses? *Science, 202:*37–41.

MAUGH, T. H., II. 1981. New study links chlorination and cancer. *Science, 211:*694.

McGINTY, L. 1979. Stemming the tide of toxic chemicals. *New Scientist, 84:*432–435.

NAHAS, G. G., et al. 1974. Inhibition of cellular mediated immunity in marihuana smokers. *Science, 183:*419–420.

PAIGEN, B. 1981. Public health issues: Bendectin, Love Canal. *Science, 211:*7–8.

SMITH, R. J. 1980. Latest saccharin tests kill FDA proposal. *Science, 208:*154–156.

TULLIS, R. H. 1981. Carcinogens and regulation. *Science, 211:*332–334.

VOGEL, F., and G. RÖHRBORN, eds. 1970. *Chemical Mutagenesis in Mammals and Man.* New York: Springer.

WHITE, S. C., S. C. BRIN, and B. W. JANICKI. 1975. Mitogen-induced blasto-genic responses of lymphocytes from marihuana smokers. *Science,* **188:**71–72.

WINGERSON, L. 1980. What's the matter with marihuana? *New Scientist,* **85:**458–459.

Questions

1. What are the three major categories of chemical compounds that act on biological systems?
2. Briefly describe the history of the events leading to the discovery of chemical mutagenesis.
3. Why is the clastogenic effect of compounds *in vitro* a profitable means of testing for biological activity?
4. Why is the shape of the dose-response curve important in an assessment of the hazards of a potential carcinogen or teratogen? Would not experiments with large numbers of animals treated with low doses settle this question?
5. What is a *promutagen?* Can the metabolism of an organism affect the potency of a promutagen?
6. How is the Ames test carried out? What additional step may be added to increase the likelihood of detecting promutagens?
7. Discuss the occurrence and the action of the following: nitrosamines, artificial sweeteners, caffeine, food dyes, aflatoxin.
8. Compare the possible effects of LSD, tobacco, marijuana, and alcohol. Do you think that all these substances should be prohibited from general use? If not, might your attitude reflect a personal bias rather than a scientific judgment?
9. Why are drugs and pesticides always suspect as possible mutagens, teratogens, or carcinogens?
10. List some other substances that are now known to have severe adverse effects on humans.
11. What did the study of the chromosomes of residents of the Love Canal district at Niagara Falls show? Why has the government agency that sponsored this study been criticized?
12. What is meant by *cost-benefit ratio analysis?* Is it unethical to equate a human life with a dollar value? Can you think of cases in which society regularly does this?
13. One government agency has estimated the value of an average human life to be $287,175, the amount lost by a death in a motor vehicle accident. Can you think of cases where the figure might be higher? Lower? Are there situations in which such a valuation appears to be completely meaningless?

15

The Basis For Induced Genetic Change

Early Attempts to Induce Mutation

Just as the alchemist, in the early beginnings of chemistry, dreamed of the transmutation of one element, lead, into another, gold, so did the first geneticists recognize the importance of inducing new heritable changes in living things. The efforts of those early geneticists to modify the genetic systems of experimental animals and plants were doomed to failure, not because of the incorrectness of the idea itself, but because the knowledge essential to carrying out the experiments and detecting the resulting changes was lacking. This knowledge had to be developed systematically over a long period of time. Curiously some of the agents tested unsuccessfully during the early days of experimental genetics, in the 1910s, such as high-energy radiation (from radium and X rays) and certain chemicals, proved later to be among the most effective mutation producers, or *mutagens*. We now know that other physical agents (heat and ultraviolet light) and chemical compounds, running into the many hundreds, are mutagenic.

X RAYS. The first announcement of success in attempts to induce gene mutation came in 1927, when H. J. Muller reported that he had succeeded in producing mutations in *Drosophila* with a high frequency after X-ray treatment. This research was followed by a similar report in 1928 by L. J. Stadler, who, having treated barley with X rays and radium, had found large numbers of new mutants present in the treated plants, and not in unirradiated control plants. In each case the mutations appearing were random with respect to which locus was affected and the type of allele produced; that is, these first beginnings were far from approaching the goal (still unattained) of causing a specified kind of change in a particular gene.

ULTRAVIOLET LIGHT.   It was simply a matter of time before other mutagenic agents were uncovered. Ultraviolet light, radiant energy like X radiation but of lesser potency, also produces mutation in those cases where it can reach the genetic material. Because *transmissible* mutations can be produced only if the agent reaches the nuclei of the germ cells in the gonads, ultraviolet light can be ruled out as a potential mutagenic agent of any genetic importance in humans. Ultraviolet light cannot penetrate very deeply into living tissue; its energy is efficiently converted into heat when it hits the epidermis of the skin, as every person who has suffered from severe sunburn can unhappily testify, and it is responsible for an increased frequency of skin cancer in persons who receive high exposures. Such doses of ultraviolet, which do not reach the gonads, would not, of course, have any mutagenic importance to future generations.

The Nature and Action of X Rays

Of considerable interest and importance to humans at the present time is the high-energy radiation originating from X-ray machines, from cosmic rays, and from radioactive isotopes resulting from bomb blasts and the production of nuclear power. Before considering the effects on human tissue, we must first understand the nature of this energy.

THE ELECTROMAGNETIC SPECTRUM.   The most obvious section of the radiation spectrum (Figure 15-1) is, of course, that which we see as visible light. A convenient way of classifying radiation is in terms of its wavelength, which for visible light is about a millionth of a meter, or 1 μm. Radiant energy comes in indivisible units called *photons*, which have very specific amounts of energy, called *quanta* (singular, *quantum*). For a photon of a specific wavelength there exists a corresponding quantum energy value. Because the product of the wavelength of a given part of the spectrum and its frequency is a universal constant (the speed of light), wavelength and frequency are inversely related. Radio waves, which have a long wavelength (measured in many feet), have a low energy per quantum, whereas X rays, gamma rays, and cosmic rays have high values.

Because electromagnetic radiation is emitted by such a wide variety of sources, from radio stations to electric light bulbs to X-ray machines, it is important to understand the general principles that determine whether or not any given radiation type is biologically important. The effectiveness of radiation depends primarily on whether or not it can produce chemical reactions, and whether these occur depends on the energy per quantum (not on

Figure 15-1 The spectrum of electro-magnetic radiation.

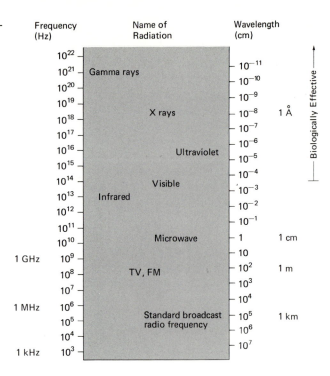

the total amount of energy) in the exposure. We know that visible light must be capable of causing some reactions; otherwise we could not see, as vision depends on a chemical reaction in the retina of the eye.

However, relatively few chemical reactions can be caused by energies lower than the quantum value for visible light; in the case of radio broadcasting or television waves, the energy per quantum is so small that there is no possibility that they can promote a chemical reaction by atomic absorption. Because the ability to produce chemical reactions is independent of the total number of quanta involved, a person standing near a radio station that produced a total power of hundreds of thousands of watts would not have to worry about the possible induction of mutations. Biological effects, such as temporary male sterility, that have been attributed to these longer wavelengths (from radio and radar transmitters) are produced by ordinary heating of the living tissue. Similarly, microwave ovens produce waves with little energy per quantum. Such sources, with low energy per quantum, can heat tissue, and it is this heating that is generally assumed to be responsible for producing eye cataracts, headaches, nausea, and other effects on humans.

PRODUCTION OF X RAYS. In principle, high-energy radiation may be generated very simply. All that is needed is an ordinary flask, with a piece of metal (target) at one end and a wire (filament) that can be heated by an electric current at the other. Then the air is removed from the flask, and a source of high voltage (tens of thousands of volts) is connected across the filament and the target (Figure 15-2), with the positive end connected to the target. When the filament is heated, it emits electrons, which, being nega-

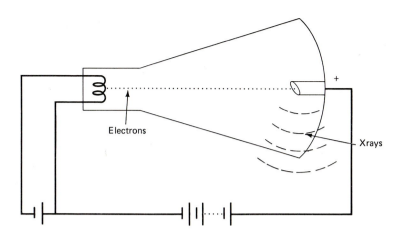

Figure 15-2 Schematic of a typical X-ray tube. The air is removed from a flask, and a hot filament is placed at one end and a solid target at the other. When a source of high voltage is connected across the two elements, X rays are generated at the target.

tively charged, are attracted to the target by its positive charge. Unimpeded in the vacuum, the electrons accelerate under the influence of the high voltage until they smash into the target. They may be stopped immediately, whereupon their kinetic energy is transformed into photons of radiant energy. An electron accelerated by 50,000 volts, and then suddenly stopped, would release 50,000 electron volts in a single photon. This much energy per photon puts it in the X-ray range.

Unstable Isotopes

Another source of high-energy radiation is the disintegration of unstable configurations of neutrons and protons in certain atomic nuclei. This more potent radiation emanating from within the nuclei of atoms is called *gamma radiation,* to distinguish it from the X rays produced by changes of electron position around the atomic nucleus. When unstable atoms disintegrate, they may also emit *beta particles* (electrons) and *alpha particles* (made up of two neutrons and two protons). The proportions of gamma rays and beta and alpha particles thrown out of the nucleus are variable, depending on the particular atom and the composition of its unstable nucleus. Atoms of the same element differing from each other by the number of neutrons in their nuclei are called *isotopes.* The isotope of hydrogen with two neutrons (*tritium*) is characterized by an instability of the nucleus: there seem to be too many neutrons present for the small positive charge of one proton, and such a nucleus may suffer spontaneous degeneration by expelling an electron, producing a rare isotope of helium.

THE HALF-LIFE OF UNSTABLE ISOTOPES. What causes some isotopes of a given element to be stable and others unstable is not completely understood; it appears that the number of neutrons and protons must bear some proportion to each other for stability. If there are too many, or too few, neutrons or protons, the nucleus spontaneously disintegrates and, during this process, ejects one or several of the types of particles making it up, along with gamma radiation, which has all of the properties of X rays, except that it is more energetic. The particles ejected are also important: virtually all the biological damage to tissue caused by the disintegration of tritium and *plutonium* comes from high-speed particles that are quite energetic but can travel only very short distances in solid matter.

Because the components of the nucleus change, there may be a change in the element that the nucleus represents, and this new nucleus may be either stable or unstable. The rate of spontaneous disintegration differs from one isotope to the next; the measure of the rate is the *half-life,* which is the time required for the disintegration of one half of the atoms in a quantity of a given isotope. Note that half of the remaining atoms will disintegrate during a second half-life interval, leaving 25 percent unaffected, and that during a third such period half of the 25 percent will disintegrate, leaving 12.5 percent unaffected. This type of reduction is called an exponential decay; regardless of how much time is involved, in units of half-lives, there will still remain a mathematically calculable amount of the original isotope unchanged. For biologically important elements, some radioactive (i.e., radiation-producing) isotopes and their half-lives are as follows:

Potassium 42(^{42}K)	12 hours
Sodium 24 (^{24}Na)	15 hours
Phosphorus 32 (^{32}P)	14 days
Sulfur 35 (^{35}S)	87 days
Calcium 45 (^{45}Ca)	164 days
Tritium (^{3}H)	12 years
Strontium 90 (^{90}Sr)	28 years
Plutonium 239 (^{239}Pu)	24,000 years
Potasium 40 (^{40}K)	1,000,000,000 years

The energy from the disintegration of radioactive isotopes is of biological importance in several ways. The explosion of an atomic bomb causes the formation of vast quantities of radioactive isotopes, which are blown into the atmosphere and released gradually over weeks, months, and years as fallout, exposing millions of humans to radiation. The production of energy by atomic fission in nuclear power plants creates large quantities of radioactive isotopes. Although virtually all of these are ordinarily contained, their potential effects on life must be seriously reckoned with. Unstable isotopes are increasingly used in industry, medicine, and scientific research.

Measurement of Radiation

When a quantum of radiant energy hits an atom, the energy may knock an electron out of its orbit around the nucleus. Such an atom, with an improper number of electrons, is said to be *ionized.*

The appropriate measurement of radiation for biological purposes must be closely connected to the amount of ionization produced by that radiation in a given volume of tissue. The increase in mutation is much the same, independent of the wavelength of the absorbed radiation or, to a lesser degree, the time span over which it is absorbed. As long as we know the total

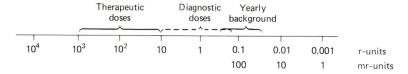

Figure 15-3 Logarithmic scale showing approximate X-ray doses used for therapeutic treatment of disease in specific organs and for diagnostic examinations, and the yearly whole-body dose from background radiation.

amount of ionization produced during the exposure, we have a good measure of the expected biological effect.

The original measurement of dose was the roentgen, or *r unit*, named for the German physicist W. C. Roentgen, who discovered X rays in 1895. In its original technical definition, it was the amount of X radiation that produced 1 standard unit of charged particles in 1 cubic centimeter of air. For our purposes, we can regard it as the amount of X radiation that produces 1.6×10^{12} ion pairs in 1 cm³ of water. An approximately equivalent unit is the *rad*, which is based on the total amount of energy absorbed, and which therefore can be used to measure the effectiveness of other biologically effective agents such as neutrons. In discussions of effects on humans, the unit commonly used is the *rem* (roentgen-equivalent-man). In any case, for gamma and X radiations, these units are roughly equivalent and may be used interchangeably for purposes of comparison without any appreciable error. In most cases, radiation applied to humans is considerably less than a rem, and instead, a thousandth of a rem, or millirem (*mrem*), is the unit applied. Because a mrem is only a thousandth of a rem, it should be very carefully noted which one is being used in any statement of dosages.

To give a feeling for the meaning of the rem as it applies to doses that human beings are exposed to, the doses for various purposes are indicted in Figure 15-3. Note that the scale is logarithmic and that each unit on the scale represents a change by a factor of 10. At one extreme are the massive doses applied in X-ray treatments to eradicate cancerous tissue, running into the tens of thousands of rems. At the other extreme are the rather small doses received regularly as background radiation, from fallout, and from nuclear reactors.

Exposures of Humans to Radiation

Although the deleterious effects of high-energy radiation on living tissue have been recognized since the turn of the century, the dangers to humans have not been appreciated by the average person until fairly recently. Accidents like that at Three Mile Island in 1978, as well as the problem of the long-term storage of nuclear wastes, have made the public acutely aware of the possibility of radiation damage, both to the individual as carcinogenic and teratogenic effects and to future generations, as mutations. On the other hand, there is generally little appreciation of the actual damage to be expected, within limits, from specific doses of radiation, although this is one of the best-researched areas of health-related biological investigation.

A Brief History

Just one year after Roentgen identified X rays in 1895, one of the first casualties from X-ray burn was an assistant of Thomas Edison, one Clarence Dally, who suffered from an inflamed and ulcerated scalp and a loss of hair after prolonged exposure to an unshielded X-ray tube. Some eight years later he died, with cancerous growths on his arms and legs.

"Radiation sickness" was described in 1897, and the first successful medical use of X rays came in 1900, when a small tumor on the nose of a patient was destroyed. In 1901 it was shown that heavy doses of X-rays were lethal to guinea pigs, even though no outward manifestations of damage appeared. A year later, skin cancer in humans was reported after X-ray exposure. The early workers with radium, A. H. Becquerel and Pierre Curie, deliberately exposed themselves to their experimental material to determine its effects. Madame Marie Curie and her daughter Irene, pioneers in the field of radioactivity, both died of leukemia, probably induced by their exposure to their research material. It has been estimated that more than a hundred persons died during the early decades of X-ray development. However, this is undoubtedly a gross underestimate, because many individuals may have died from exposures received years earlier, without the true cause of their deaths being known.

Doses of Radiation to Humans

HEAVY DOSES. Information on the effects of high radiation doses to the whole body comes from (1) industrial and laboratory accidents; (2) the accidental exposure to fallout during atomic bomb tests in the Pacific in 1954; (3) the bomb explosions in Hiroshima and Nagasaki; and (4) the medical exposure for cancer therapy or other medical reasons.

In 1946 an accident involving an atomic experiment at Los Alamos resulted in the irradiation of one of the workers, Dr. Slotin, with an estimated total of 880 rem. He survived only nine days. In 1958 a worker at Los Alamos accidentally transferred about 34 kilograms of material with a high concentration of the isotope uranium 235 (^{235}U) from storage tanks to a waste tank that had dimensions that caused the solution to go "critical." There was a miniature explosion, of no great consequence in its blast effect, but producing a high level of radiation. It is estimated that the worker received approximately 12,000 rem to his upper extremities. He became semiconscious and incoherent and died thirty-five hours later with clearly manifested brain damage.

When the first atom bombs were dropped on Hiroshima and Nagasaki, about 100,000 were killed and another 60,000 were injured. Most of the fatalities resulted from the explosions themselves—from the blast, heat, and shock. Only a small fraction of the immediate deaths can be attributed to the high radiation exposure. The survivors of these atomic explosions have been monitored very carefully over the years; they have developed leukemia and other cancers at a greatly increased rate over nonexposed Japanese populations.

SUBLETHAL DOSES. Of seven survivors of the Los Alamos accident in 1946, four received exposures estimated at between 100 and 400 rem, and another three workers received from 30 to 55 rem. In 1958 six workers in Yugoslav nuclear energy plant received an estimated 200–450 rem of whole-body radiation. They were immediately rushed to Paris, where they received intensive medical treatment, and all but one survived.

During the U.S. atomic tests of 1954 in the South Pacific, some 300 persons received an "unpleasantly large" radiation dose. Of 236 Marshall Islanders caught in the fallout when the wind shifted unexpectedly, 64 had radiation burns (Figure 15-4) and suffered from the usual symptoms of nausea, vomiting, and diarrhea, and in 39 cases loss of hair was reported. It is estimated that those receiving the highest doses got about 175 rem. Twenty-three crew members of a Japanese fishing boat, *The Lucky Dragon,* were also caught in the local fallout and suffered from burns, fever, and swelling. One died several weeks later, but it is not clear whether this death was attributable to the exposure. The average radiation dose was probably between 150 and 175 rem.

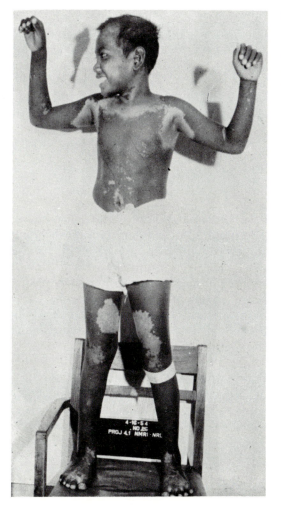

Figure 15-4 A young boy suffering from radiation burns received during the U.S. atomic tests at Bikini Island in the South Pacific, when an unexpected shift in wind direction exposed people on the Marshall Islands, about 500 miles away. (Courtesy of R. A. Conard, Brookhaven National Laboratory.)

From these unplanned "experiments," we learn that an exposure of several hundred rem will cause nausea, vomiting, bleeding, and fatigue; other symptoms include a reduction in lymphocyte and platelet counts and a high body temperature. Subsequently the victim may experience loss of hair and temporary sterility. If the dose is in the vicinity of 1,000 rem, the person usually dies; if it is only several hundred rem, the person will probably survive. The probability of survival is, of course, increased appreciably if the person is provided with continuous medical monitoring so that no serious consequences can arise from a minor infection at a time when the white blood cell count is low because of the radiation exposure.

THE MEAN LETHAL DOSE.   Because it is difficult to determine the dose at which a killing effect is just perceptible, or to determine the dose that is completely lethal, investigators in the field have adopted the convention of a 50 percent lethal dose, or LD_{50}, defined as the dose that will kill about 50 percent of a sample of exposed animals in a specified period of time, such as thirty days after treatment ($LD_{50}/30$). Clearly, the most elementary ethical considerations forbid any attempt to derive an LD_{50} for humans, and in those cases where humans have accidentally received massive doses, there is considerable uncertainty surrounding the actual doses they received. It is therefore necessary to use figures from experimental animals, such as the mouse and rat, along with some fragmentary data on the survival or death of humans after accidental exposures of the sort we have already discussed. From these it could appear that a whole-body dose of about 450 rem would probably be lethal to half of a sample of humans within a period of a few months.

Doses from Medical Technology

THERAPEUTIC DOSES.   Treatment for cancer and other pathological conditions may demand the use of many hundreds and even thousands of rem, directed specifically at the tissue involved. Whole-body doses of this magnitude would more often than not be lethal. To minimize the destruction of normal tissue, the patient or the source of radiation may be rotated during treatment so that the skin exposure is spread out as much as possible. These massive doses are justifiable when the alternative (no treatment) is likely to lead to the early death of the patient. Such doses given to large numbers of people might be expected to have a significant effect on the overall production of new mutations in the population, but, as a rule, patients with malignancies tend to be older and generally have completed their families by the time of treatment. In any case, if a decision must be made between the present welfare of a medical patient and the hypothetical future difficulties that may be caused by new mutational events, it is invariably made in favor of the welfare of the patient.

DIAGNOSTIC DOSES.   Doses from several to about a hundred rem are needed for fluoroscopy when the nature of the examination requires that the physician view an organ in action, as in gastrointestinal examination. Doses to patients are much less when X-ray film is exposed photographically than when a fluoroscopic screen is used, because of the much greater sensitivity of the film.

In discussions of radiation doses the distinction should always be made between exposure to the whole body and exposure to the organ involved, in particular the gonads, because damage to the latter would be of concern to future generations. In the case of dental X rays, for instance, when the dose to the jaw is about 100 mrem for a single exposure, the dose to the gonads, in either male or female, is less than one ten-thousandth as much. On the other hand, an X-ray photograph of the pelvis, which may require 220 mrem, delivers more than a third as much to the gonads of either sex. From these concerns comes the concept of the *genetically significant dose,* which is defined as the total gonadal dose from a given source with possible genetic effects, divided by the total number in the population, to give an average gonadal dose per person. This calculation must take into account the number of persons irradiated, the types of examinations, and the exposures received, along with the calculated accumulated dose to the gonads as well as the anticipated number of future children of the irradiated individuals. Thus in 1970 the average whole-body exposure to residents of the United States was about 70 mrem; the genetically significant dose, however, was half that: 35 mrem.

Because of the greater appreciation by radiologists of the dangers of the dose to the reproductive organs , greater care is being taken now than in the past to shield the gonads from unnecessary exposure. On the other hand, many unnecessary X-ray photographs are being taken routinely when the probability of revealing any new medically significant condition is nearly zero. It is estimated that at least one-third of the 270 million medical X rays taken in the United States annually are unnecessary. Taking superfluous X rays may be encouraged by hosital administrators and companies that ensure doctors against lawsuits, because such photographs can be subsequently used as evidence in any legal action (such as a malpractice suit brought by a patient against a doctor) to show that the physician did, in fact, provide adequate medical care. (However, it might be argued that the opposite is true, that unnecessary X-raying of patients is also reprehensible.)

Curiously, medical insurance, however commendable in other respects, also tends to increase unnecessary X-ray exposures, because the patient, who is not paying the costs directly, is less likely to question the necessity of this type of examination.

Radioactive isotopes are used quite frequently in the diagnosis and treatment of certain diseases. In the United States the estimated genetically significant dose (the average gonadal dose) per person from the use of radioisotopes applied from outside the body is calculated to be 12 mrem/year. Persons who receive radioactive iodine treatments for thyroid problems may retain the isotope to such an extent that they would be too radioactive to be accepted as baggage on a passenger airliner. They can, however, travel as regular passengers and irradiate their fellow travelers.

DOSES TO TECHNICIANS.   One large group of persons has traditionally been regularly exposed to unusually high doses: the radiologists and technicians who constantly live with high-energy radiation in their work. Although regulations provide for mandatory precautions that would prevent overexposure, there is no doubt that in many cases these rules are not observed, partly because of the lack of obvious immediate damage caused by the X rays, and partly because of the slight inconvenience that such precautions necessitate.

Studies made in the 1950s showed that at that time X-ray technicians received an average of 100–300 mrem/week, that they had a frequency of leukemia about 9 times greater than that of the rest of the population, and that the life span of radiologists was reduced by about five years. Approximately 20 percent of radiologists checked at that time were receiving more than the "acceptable" dose of 300 mrem/week. More recent figures show that medical X-ray workers now receive a *mean annual dose* of 320 mrem, dental X-ray workers 125 mrem, radioisotope workers 262 mrem, and radium workers 540 mrem. Clearly there is progressively less exposure of medical personnel.

When children are being X-rayed, the parent may be asked to hold the child still, on the reasonable assumption that it may be better for the parent to receive a single small dose than for the nurse or technician to receive the same small dose to the hands day after day. At one time, dentists would hold film in position in the patient's mouth during dental X-raying; it is now becoming common practice for them to ask the patient to hold the film in place.

The service performed for the benefit of humanity by the medical profession in applying X-ray technology for diagnostic and therapeutic purposes cannot be questioned. Although some young persons are deterred from entering this vitally important medical area because of the fear of genetic or other damage to their future progeny from the inevitable exposures incurred during such work, the figures showing the relative doses from various situations, along with the very low probability of their producing any biological effect, suggest that these technicians suffer little risk provided that care is taken to adopt standard safety precautions.

OTHER OCCUPATIONAL EXPOSURES.   Data are accumulating on groups of persons who have been exposed to doses ranging from 1 to 10 or 20 rem. Studies have been made of the deaths from cancer of nuclear workers at the Portsmouth Naval Shipyard in New Hampshire, of the deaths from leukemia of soldiers who witnessed an A bomb explosion in 1957, of deaths from lung cancer of uranium miners, and of residents of southwest Utah who were exposed by open-air tests conducted in Nevada. There have been reports suggesting that all of these people have an increased rate of disease over that of a comparable unexposed population, but, with the exception of the uranium miners, these studies have yet to reach unequivocal conclusions.

Astronauts who spend many days orbiting the earth pass through regions of high radiation density; even though their compartments are shielded, their stay in orbit must be limited. Their selection depends on, among other criteria, whether they have completed their families! From this point of view, a trip to the moon, during which the astronauts quickly pass through the high-radiation belts surrounding the earth, is safer.

Miscellaneous Exposures

Occasionally persons receive small exposures from miscellaneous sources. The radium dial on a watch gives 700 mrem/year to the wrist and 10 mrem to the whole body. Pilots of large commercial aircraft, with their panels of radium-illuminated dials, receive radiation at the rate of 1,200 mrem/year (this value would, of course, have to be adjusted to that proportion of a year during which they were at the controls); and for each transcontinental round trip that a pilot makes, pilot and passengers accumulate an additional 2–4 mrem, because of the higher cosmic-ray exposure at high altitudes. High-altitude passenger planes like the Concorde carry radiation monitors to alert the pilot to drop to a lower altitude in the event of an unusually high radiation dose. X-ray equipment used at airports to examine baggage ordinarily exposes the contents to 1 mrem; some of the more powerful devices produce 200 mrem in the area where the baggage is placed.

In other cases exposures of high doses seem to be completely unjustified. Not many years ago shoe stores used fluoroscopes to reveal the bones of the foot within a shoe. It is questionable whether any real advantage in shoe fitting resulted; the greatest effectiveness of the machine may have been a promotional one. In any case the dose delivered to the foot was about 50–150 rem/minute, and 1–10 rem/minute to the whole body. In retrospect this appears to have been a most incredible misuse of high-energy radiation. The machines were poorly shielded, had no knowledgeable operating personnel in attendance (but exposed clerks, salespeople, and bystanders in large numbers), had no time restrictions on exposure, and were readily available to small children, for whom the machines may have served as a useful entertainment and baby-sitting device. These machines are now generally prohibited by state laws.

Before 1958 some 200,000 children in the United States had been irradiated for ringworm of the scalp, some more than once. After the radiation many experienced severe headache, nausea, and vomiting, an understandable reaction, as they had received 450–850 rem to the scalp, with an average brain dose of 98–140 rem. More recently, about 16,000 children were similarly treated in Israel.

In the 1930s some gynecologists irradiated the ovaries of women in the belief that the treatment induced ovulation. The dose used for this treatment was about 70 rem. Because the rays must be focused specifically on the female germ cells to accomplish the purpose, gynecologists were severely criticized for this practice, and it can be assumed to have been stopped. For one thing, a greater sensitivity on the part of the medical profession to the genetic hazards of radiation has led to greater restraint in its use in exposing the germ cells. The possibility of an occasional defective infant attributable to the radiation (although, obviously, cause and effect in such a case could never be proved) definitely makes the procedure seem less attractive. In any case the availability of "fertility" hormones now makes it possible for gynecologists to solve more easily many problems of fertility resulting from failure of ovulation. Unfortunately in many such cases multiple ovulation results, and the consequences of the ensuing multiple pregnancy and premature birth may be fatal to the newborn.

Doses from Television Receivers

Television picture tubes as well as other high-voltage tubes in color sets can generate X rays. The energy of the radiation depends on the voltage applied across the elements of the tube; Figure 15-5 shows the X-ray intensity, in millionths of rem per hour, produced by a color television set, operating at the unusually high voltages supplied by a circuit specially designed for that purpose. Black-and-white sets operate at voltages well below those shown on the graph; their X-ray production can be considered zero. In the late 1960s measurements showed that the radiation produced by color TV sets was quite variable, but in some instances excessively high, ranging up to 12 mrem/hour, and that 20 percent (in one sample) were emitting excessive doses of radiation. In some instances, poor shielding of the internal components was responsible; in others, excessive voltages. Some technicians increase internal voltages in sets to brighten and sharpen the picture and unknowingly convert a safe set into a weak X-ray emitter. In other cases, fluctuations in the line voltage coming into the set have been multiplied by the power transformer of the set to excessively high voltages in the set components. Many brands had been recalled for alterations to reduce the external radiation.

The dose is reduced considerably simply by maximizing the distance from the set. Figure 15-6 shows the relative reduction in radiation level as the distance increases. In theory, radiation from a point source decreases in intensity as the square of the distance from that source. For distances near the set, the reduction is not as great as theory would predict because the emitting element is not a point, but at about 1 meter the square law becomes

Figure 15-5 X-ray doses in millionths of an r unit per hour after atypically high voltages have been applied to the picture tube of a color television set. Note that a typical color television set would not be capable of delivering these high voltages and that improvements in shielding, circuit design, and component construction further decrease the potential hazards of more recent color TV. (Courtesy of S. P. Wang, S. Sario, and H. Hersch, Zenith Radio Corporation.)

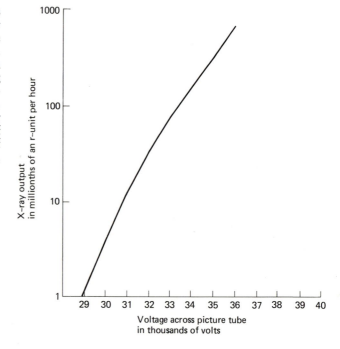

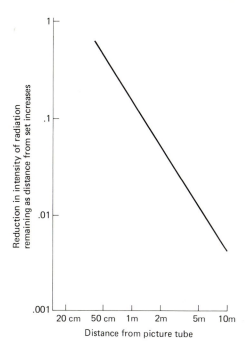

Figure 15-6 The relative reduction in radiation level from a radiating color television set as distance from the set increases. Note that at about 3 ft (or 1 m) the intensity is only 10 percent of the level at the set and that with greater distances the level drops off exponentially. (Courtesy C. P. Braestrup, Conference on Detection and Measurement of X-radiation from Color Television Receivers, 1968.)

effective. In any case the simplest way to reduce any potential exposure (other than shutting off the equipment) is to put as much distance as possible between the person and the set.

By taking into account the number of hours that the average person watches television in the United States (3 hr/day for children under fourteen and 2.6 hours for older persons), the average distance from the television receiver (somewhat more than 5 feet for children and about 8 feet for older persons), the absorption by body tissue, and so on, it is possible to calculate the dose to various organs (Table 15-1) if the set is emitting at the maximum rate of .5 mrem/hour. In fact, from studies in Washington, D.C., it would appear that the average color set now emits less than one tenth of this amount, about .043 mrem/hour, so that the values in Table 15-1 should be divided by 10. Taking into account that some persons do not watch TV at all, the genetically significant dose to the U.S. population is approximately

Organ	Children Less Than 15 Years Old (mrem/yr)	People 15 Years of Age and Older (mrem/yr)
Skin	42	18
Thyroid	30	11
Male gonads	18	8
Female gonads	5	2
Lens of the eye	34	15

Table 15-1 The estimated average dose from the front face of color television tubes radiating at the rate of .5 mrem/hr. Note that most sets radiate at a rate less than a tenth as great. (From R. H. Neill, H. D. Youmans, and J. L. Wyatt, *Radiological Health Data and Reports* **12[1]:**3, 1971.)

.5 mrem/year, or about one half of 1 percent of the average dose from natural background radiation.

All sets manufactured after June 1, 1970, have been designed in such a way that even with maximum voltages, a new *maximum permissible dose* would not be exceeded. This maximum permissible dose amounts to .5 mrem/hour at a distance of 5 cm in any direction from the set.

Background Sources of Radiation

As our discussion progresses from the heavy doses to the body to the lighter ones, we eventually reach the low, constant dosages originating in the natural environment. This is radiation from which we cannot escape, coming from the ground we walk on, and from the elements making up our bodies, as radioactive isotopes. Because of the small doses involved, the unit is invariably thousandths of a rem (millirem or mrem) per year.

COSMIC RAYS. Cosmic "rays" consist of high-velocity charged particles, originating from our sun and outer space. They enter the earth's atmosphere, collide with atoms, and yield a variety of additional high-velocity particles, as well as gamma rays. As they cascade toward the earth, they produce additional charged particles and radiation. At the same time, the particles may be slowed and part of the radiation may be absorbed during the passage through the atmosphere. For this reason, the intensity of the penetrating radiation is strongly dependent on altitude. Cosmic rays are about twice as intense at Denver, slightly more than 5,000 feet in altitude, as they are at sea level. Also, the paths of the charged particles are curved by the earth's magnetic field and so are not uniformly distributed around the earth but differ depending on the latitude. Table 15-2 shows the annual cosmic ray

Table 15-2   Approximate cosmic ray doses for a number of American cities. (From V. L. Sailor, *Population Exposure to Radiation: Natural and Man-made,* Sixth Berkeley Symposium, 1971, Berkeley. Originally published by the University of California Press; reprinted by permission of The Regents of the University of California.)

Location	Elevation (ft)	Dose Rate (mrem/yr)	Difference from Sea Level (mrem/yr)
Denver	5,280	67	39
Albuquerque	4,958	60	32
Salt Lake City	4,260	54	26
Oklahoma City	1,207	34	6
Phoenix	1,090	33	5
Atlanta	1,050	33	5
Kansas City	750	32	4
Chicago	579	31	3
San Francisco	65	28.5	.5
Baltimore	20	28	0

doses for ten American cities, along with the excess doses being received per year because of the altitude and latitude of each.

RADIATION FROM ROCKS AND SOIL. The elements contributing most to the natural radioactivity of the ground are the products of disintegration of uranium 238 and thorium 232, which give an average figure of 50 mrem/year. This average figure, however, is subject to wide variations because of the great differences in the concentration of these elements from one place to another; the range around this mean value extends from 25 to 75 mrem/year.

In some areas the soil is unusually rich in radioactive elements. About 50,000 people live over radioactive deposits in two states of Brazil, and they receive from 500 to 1,000 mrem/year. In a village in another state of Brazil, a few hundred people receive as much as 12,000 mrem/year. In an area of India occupied by about 10,000 people, the exposure from radioactive soil ranges from 200 to 2,600 mrem/year, and 70,000 Chinese live in an area where the radiation from the soil is three times the normal amount. In all the preceding cases, studies have been made to determine whether the increased exposures have any demonstrable biological effects with negative results so far. These studies continue.

Some 1,800 residents of Grand Junction, Colorado, live in houses built on "tailings," rock and soil removed from uranium mines. They may receive as much as 107 mrem/year additional radiation, and the total number of persons in the state of Colorado similarly exposed may be 10 to 20 times as great as the number of Grand Junction residents given above. Uranium miners are an unusually highly exposed occupational group: they may receive doses as high as 6,600 mrem/year to the skeleton and 500 mrem/year to the gonads.

RADIATION FROM BUILDINGS. A person inside a building may receive less radiation of cosmic or soil origin, but this may be matched or even greatly exceeded by radiation coming from the building itself. Thus, although an ordinary wooden house may reduce exposure to about 70 percent of the level outside, brick, stone, or concrete buildings may contribute 50–100 mrem/year more than the exposures received out-of-doors in the same area. Table 15-3 shows the radiation exposure inside various buildings. Although no correction has been made for the cosmic rays present at the same time, it is clear that the doses are far above the expectation for cosmic rays alone.

INTERNAL EXPOSURE. Because all the air, water, and food that enters the human body contains naturally occurring radioactive elements, the body itself accumulates a small amount of radioactivity. Most of it comes from potassium 40 (20 mrem/year); some from carbon 14, radium 222, and so on (3 mrem/year). This total is not very great; from another point of view, however, it may appear more significant because it is responsible for about 500,000 atomic disintegrations per minute within the human body.

DOSE FROM FALLOUT. At one time, the ever-increasing increment to natural radiation from the man-made product of atomic explosions seemed to present a threat of unlimited dimensions. The agreement of the major powers to halt above-ground nuclear testing has largely removed this danger,

Table 15-3 Gamma exposure rates inside various buildings, cosmic ray component included. (From V. L. Sailor, *Population Exposure to Radiation: Natural and Man-made,* Sixth Berkeley Symposium, Berkeley, 1971.)

Location	Structure	Exposure Rate (mrem/yr) Typical	Extreme
East Germany	Brick	106	1,200
New York City	Brick	79–118	
Grand Central Station	New York granite		500
United States	Wood	60	
	Concrete	130	
Aberdeen, Scotland	Granite	81	110
Cornwall, England	Granite	145	
Sweden	Wood	48–57	
	Brick	99–112	
	Concrete	158–202	

although the continued testing by certain countries (e.g., China and France) does add a small amount of additional radiation. In spite of these small additions, the *radiation from fallout* is currently decreasing. The exposure to humans depends on the degree to which the isotopes are retained in the stratosphere before deposition on the ground, as well as their fate afterward. Strontium, in particular, poses a special threat because of its involvement in the food chain, where it competes with calcium for incorporation into living tissue. The population of the northern hemisphere is currently receiving about 5 mrem/year from fallout.

Total Radiation Exposure

A rough estimate of the total amount of radiation being received by the U.S. population is about 200 mrem/year (Table 15-4). Of this total, the largest contribution is 60–170 mrem from natural background, and the next most important is about 80 mrem from medical technology (Figure 15-7).

Radiation Exposures from Nuclear Reactors

In recent years there has been an extensive debate between those individuals favoring the development of nuclear power sources and those actively opposed. The majority of people on both sides of the issue are motivated by a desire for at least a continuation and perhaps even an improvement of their present mode of life, and its extension to others they consider less fortunate. However, most do not have the scientific background for independent judgment in this area and rely on secondhand information from those they consider more knowledgeable than themselves and more reliable than others.

Natural Background	mrem/year
Cosmic radiation	25–75
Terrestrial radiation	15–75
Internal radiation	20
Medical X rays	
Diagnosis	77
Radioisotopes	14
Dental X rays	1.4
Miscellaneous	
Fallout	4–5
Television	0.5
Airline travel	0.5
Nuclear power	≪ 1

Table 15-4 Annual dose rates from significant sources of radiation exposure in the United States. (Condensed from National Academy of Science, *The Effects on Populations of Exposure to Low Levels of Ionizing Radiation: 1980.* National Academy Press, Washington, D.C.).

The Three Mile Island accident has heightened apprehensions and sharpened the controversy; the general feeling that the U.S. is failing to solve nuclear waste problems in an acceptable way has added to the sentiment that power from nuclear energy may present exorbitant costs in human terms. Development of nuclear energy in the U.S. slowed down appreciably in the latter part of the 1970s. It should be noted, however, that some European countries are proceeding with the construction of nuclear plants with unusual speed and expect to have most of their energy needs met by nuclear

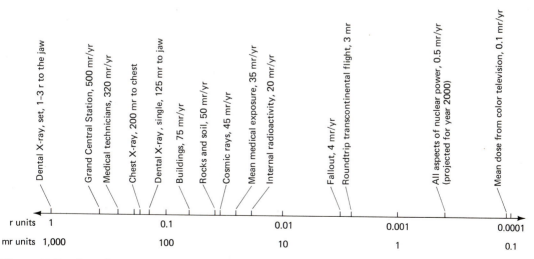

Figure 15-7 The relative exposures of the U.S. population (or the group specified) to various radiation sources, natural and man-made. Note that the scale of exposures is made logarithmic in order to include both relatively high and relatively low doses on the same plot.

power by the end of the century. However, the effluent from these reactors and the manner of waste disposal from them is a matter of international concern.

As the debate continues, other issues emerge. Besides (1) the probability of reactor breakdown serious enough to lead to a meltdown of the core, and (2) the problems of radioactive waste disposal, questions have been raised about (3) the accuracy of predictions for future energy needs; (4) the problem of a dwindling fossil fuel supply, particularly oil and natural gas; (5) the health hazards in the use of coal, with its attendant acidification of lakes and production of CO_2; (6) fears concerning sabotage and the consequent infringement of civil liberties at nuclear plants; (7) the increasing availability of plutonium that can be used for nuclear weapons either by countries or by terrorist groups; (8) the adequacy of wind, solar, or tidal power, as well as conservation, to meet part of our energy needs; and so on. Our discussion here is limited to the biological problems posed by nuclear power generation.

The important issues are primarily biological and genetic. How much damage will be wrought on the world's population if widespread radiation induces cancers, particularly leukemia, or other ill health? To what extent do we jeopardize future generations? These are important questions for which we now do have some quantitative estimates; we shall consider in some detail the probable and known human exposures from nuclear power production, as well as the biological consequences of such exposures.

HAZARDS TO THE GENERAL PUBLIC. At the outset, the possibility of an atomic holocaust, with its characteristic mushroom cloud originating in an explosion from a nuclear power plant, must be disposed of. The essential requirements for a bomb—purity of fissionable material, a precise timing device, and a system to contain the fissionable material in one compact unit long enough to achieve an explosion without premature sputtering—are missing. The serious reactor accident that might happen is the "meltdown," caused by a runaway reaction releasing excessive heat and melting all the fuel elements. Most reactors have an airtight hemispherical dome that holds all the nuclear components; the dome is engineered to contain the radioactive gases released in case of a meltdown (Figure 15-8). Whether, in fact, the dome is capable of doing this in all accidents is unknown, as no such accidents have occurred and, for obvious reasons, an actual test has not been made. The partial meltdown at Three Mile Island did involve the release of some radioactivity; the dome was not breached, but defective valves allowed some gas to escape.

Under normal operating conditions the release of radioactivity from a nuclear plant to the general public is small: in 1969 thirteen plants averaged 14 mrem/year at the plant boundaries.

DOSES TO EMPLOYEES. Workers inside a nuclear facility are allowed, by current regulation, a yearly exposure no greater than 5,000 mrem (or 12,000 mrem for workers with no previous radiation history). Most actually receive less than this: the average yearly exposure of workers at the Trojan Plant in Oregon in 1978 was 300 mrem, about the same as that for X-ray technicians.

Figure 15-8 The nuclear reactors at Three Mile Island. The four tall structures are the water-cooling towers. The arrow points to the containment building housing the reactor on the right that suffered a partial meltdown. (Courtesy of the Metropolitan Edison Company.)

However, individuals receive much higher doses from time to time, particularly during refueling operations, when occasional accidental doses of between 5,000 and 20,000 mrem are given to some workers.

NUCLEAR ACCIDENTS. Although we may dismiss the likelihood that a bona fide nuclear explosion will occur in an atomic power plant, we must nevertheless assume that other types of accidents (meltdowns, for instance) will occur at such installations and that radioactive material will be released to the outside air, and therefore to the general public.

THE THREE MILE ISLAND ACCIDENT. In 1979, a series of failures, both mechanical and human, including the turning off of the emergency cooling system, led to a partial meltdown of one of the two reactors at the Three Mile Island nuclear plant in Pennsylvania. Radioactive gases (xenon and krypton) were released, and readings at the plant gate reached 35 mrem/hour. Four employees received a total of about 4,000 mrem. It was stated that no resident living within five miles of the plant received more than 100 mrems, and the average dose to persons within fifty miles was 1.5 mrems.

PROBLEMS OF WASTE DISPOSAL. There has not yet been devised any generally acceptable plan for the unsupervised permanent disposal of high-level radioactive waste. At the present time most attention centers on the technique of imbedding the long-lived isotopes in glass, ceramic, or synthetic crystal blocks and burying these deep in the earth. It has also been proposed that such wastes might be shot into space, if space technology should improve to the point where the chance of a disastrous malfunction would be minimal.

RETENTION OF ISOTOPES IN THE BODY. When a small amount of radioactive isotope is introduced into the body (by inhalation, by swallowing, or medically by way of intravenous injection), the amount of radiation the body receives depends on the half-life of the isotope ingested and its relative accumulation and retention in various organs. Some reactor by-products, such as the noble gases xenon and krypton, are not chemically reactive and so are not retained within the body for more than a few hours, but other elements (like strontium) may be metabolized and stored for many years. The time required for the body to eliminate one-half of the dose of any substance is called the *biological half-life* of that substance. One can calculate the effective half-life easily by dividing the product of the two half-lives by their sum:

$$\text{effective half-life} = \frac{\text{biological} \times \text{physical}}{\text{biological} + \text{physical}}.$$

Thus, iodine 131 has a biological half-life of 138 days (in the thyroid gland) and a physical half-life of 8 days. Its effective half-life is therefore

$$(138 \times 8) / (138 + 8) = 7.56 \text{ days.}$$

As can be seen from the example, the effective half-life is closer to, and less than, the shorter of the two half-lives. Krypton 85 has a biological half-life of several hours; the release from the Three Mile Island plant of about 60,000 curies could be predicted to expose humans to a much smaller dose than the same amount of radioactivity of an element with a greater retention in the body, like strontium 90, with a biological half-life of 18,000 days.

ACCUMULATION OF STRONTIUM 90. Biological half-life becomes particularly important in a consideration of strontium 90. This element has many of the properties of calcium and may be metabolized in body processes that ordinarily use calcium. If the body is given equal amounts of calcium and strontium, from 25 to 50 percent as much strontium will be used as calcium. A cow feeding on equal amounts of the two elements will produce milk in which 11 percent of the calcium is replaced by strontium.

A newborn child will possess half the strontium/calcium ratio of the mother. The discrimination in favor of calcium is not sufficient to prevent considerable accumulation of strontium in the bones, and the half-lives (radioactive 10,000 days, biological 18,000 days) combine to give an overall effective half-life of strontium 90 in the body of 6,400 days. Because humans may accumulate strontium by way of contamination of the food chain (Figure 15-9), it assumes an unusual importance.

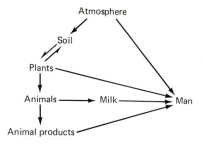

Figure 15-9   The accumulation of strontium 90 as it passes through the food chain.

Radiation Effects at the Cellular Level

When the radiation dose is higher than several hundred rem, the individual is killed because too many cells are destroyed for continued survival. This results from destruction of the intracellular components, those molecular structures necessary for normal functioning but not necessarily part of the genetic makeup. Thus a dose of 100 rem will produce about 10,000 ion pairs in a typical cell, with the result that more than 1 percent of the proteins and nucleic acids will be ionized.

ACUTE VERSUS CHRONIC EXPOSURES.   The biological impact of a given radiation dose may differ, depending on whether it is applied in a short period of time *(acute)* or is spread out over a long period *(chronic)*. A heavy radiation dose, sufficient to kill a human if given acutely in a few minutes, would have much less disastrous consequences if spread out chronically over a lifetime. On the other hand, for other kinds of damage—mutation, for instance—from which there is no comparable recovery, acute and chronic doses may be roughly equivalent, although acute doses may be several times more effective than chronic doses.

CELL LETHALITY.   If a person survives the first few weeks after exposure, there is a new hazard to be faced. Certain blood cells, which are normally characterized by rapid mitosis and which the body requires in large quantities, are destroyed. These include the cells that give rise to leukocytes, erythrocytes, and platelets. The first are essential for disease resistance, the second for oxygen transport, and the third for blood clotting. Thus, if any of these cell populations is injured so that normal mitosis is interrupted, the affected person may experience low resistance to disease, anemia, or excessive bleeding once the mature cells present in the circulatory system have run the course of their normal life span and an inadequate number of new ones are formed to replace them (Figure 15-10). The main loss of such cells comes from the action of radiation in breaking chromosomes.

Radiation-Induced Chromosome Breaks

SINGLE CHROMOSOME BREAKS.   As the radiation traverses a cell, it may break chromosomes. At the point of a chromosome break, the two newly formed free ends behave as though they were "sticky," or unsaturated, in

Figure 15-10 Counts of different cells of the blood system in five men exposed to 236 to 365 rem total body radiation in an accident at Oak Ridge, Tennessee. (Courtesy of the Medical Division, Oak Ridge Institute of Nuclear Studies, Oak Ridge, Tennessee.)

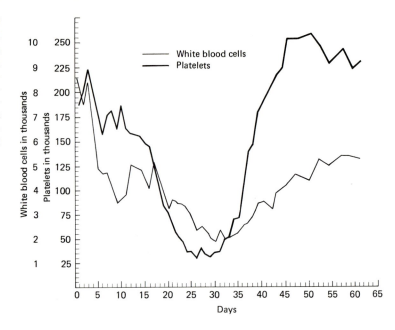

the sense that they can now rejoin with other broken ends. The simplest rearrangement that can be produced, and undoubtedly the most common, is the dicentric that is formed from a sister strand union after a single chromosome break. The consequences of this event have been described in Chapter 8. Radiation may also produce the kinds of chromosome rearrangements also discussed in Chapter 8. These include translocations, inversions, deficiencies, and ring chromosomes.

DESTRUCTION OF CELL LINES. Clearly, if enough primitive cells are destroyed in one of these ways, the body may lose an important function, such as the loss of thyroid gland activity (Figure 15-11) or the ability to fight disease if the leukocytes are affected. On the other hand, this susceptibility of cells in rapid mitosis is taken advantage of in the treatment of cancer. Here the cancerous cells are exposed to massive doses of radiation and, in many cases, can be completely eliminated. Of course, much depends on the skill of the radiologist, who must direct the radiation precisely at the site of the cancer, with minimum radiation to other organs and tissues that might be damaged. The massive doses applied to cancerous tissue, sometimes hundreds or even thousands of r-units, are not a reflection of any resistance of the tissue to radiation as much as they represent the necessity of killing all of the cancer cells, and the realization that the patient will almost certainly succumb to the disease if drastic measures are not taken.

CARCINOGENESIS. Ironically, another effect of radiation is its production of cancers, or its *carcinogenic effect*. Children who have been given X-ray treatments for such diseases as enlarged tonsils, adenoids, acne, and thymus gland problems have a 7 percent incidence of thyroid cancer. Survivors of Hiroshima and Nagasaki who received more than 20 rad died of leukemia or other cancers 3 percent more often than unirradiated persons.

Figure 15-11 Marshallese boy at age twelve suffering from thyroid deficiency as a result of exposure to fallout from atomic tests; on the right is the same boy three years later, after medical treatment. (Courtesy of R. A. Conard, Brookhaven National Laboratory.)

Adult cells that are completely differentiated and engaged in specialized tasks may be induced by radiation to revert to a primitive condition of rapid mitosis. How radiation induces cells to behave in this way is not known. Perhaps in some cases the radiation destroys an intracellular mechanism that suppresses or inhibits mitotic divisions. Perhaps in other cases it decreases the cell's normal resistance to invaders and allows a carcinogenic virus to enter the cell. More likely, different cancers may be produced by any one of several mechanisms, including the preceding. In any case the probability that any one person will develop an induced cancer is low even when that person has been given heavy radiation doses; otherwise, diagnostic or therapeutic exposures would not be acceptable.

SPECIFIC RADIATION EFFECTS ON BLOOD TISSUE.   The decrease in blood cells (Figure 15-10) seen after a minor whole-body radiation exposure comes from the destruction of the cells that have suffered chromosome breaks and are prevented from completing normal mitosis and differentiating into normal blood cells. As the mature leukocytes die off over a period of several weeks, and their normal replacements do not appear, the total count is diminished, giving rise to a decreased number of leukocytes, or *leukopenia.* It is at this time that danger from external invaders becomes most acute; death can result from infections that the immune system could take care of quite effectively under ordinary circumstances. However, with the passage of time the unaffected leukoblasts, those cells giving rise to mature leukocytes, may proceed through sufficient normal mitoses to compensate for the deficiency, and the person's recovery, from this aspect at least, will be complete.

A long-range hazard comes from the induction of a blood cancer called *leukemia.* In this disease the cells are changed so that they go through rapid uncontrolled mitoses, without subsequent differentiation, so that they cannot function normally. In contrast to leukopenia, in which the white cell count decreases, in leukemia it increases. Treatment follows the same pattern as in other malignancies; an attempt is made to kill the mitotically superactive cells. Leukemia differs from leukopenia also in its greater lethality and the greater delay between the exposure to an inducing agent such as radiation and its appearance in the human body. Observations of survivors from Hiroshima suggest that those who developed leukemia after receiving 500 rem manifested the disease some three years later, whereas the disease did not appear in persons who received only 200 rem until five years later.

EMBRYONIC AND FETAL DAMAGE.   Another kind of somatic effect is that of interference with the normal development of an embryo or a fetus. If a woman is X-rayed during the first few months of pregnancy, there is some chance that vital tissue or organ development of the fetus will be stopped, or misdirected, so that the child may be born with severe defects. X-ray radiation is a potent teratogen. In mice, doses of 200 rem have produced abnormalities (defects of the vertebrae, ribs, skull, and limbs) in 100 percent of offspring, when the dose was applied during a critical period early in development. The possibility of teratogenic effects has led some states to prohibit X-raying pregnant females before the sixth month of pregnancy, unless immediate critical problems of the mother's health are involved. In some cases women are X-rayed for therapeutic or diagnostic reasons and subsequently discover that they are pregnant. If the dose to the embryo or fetus exceeds 10 rem, the probability of a defective offspring is high.

Children have been exposed *in utero* during prenatal measurements of pregnant women's pelvic dimensions and other observations of the state of fetuses during the last trimester of the gestation period. When such exposed children have received a dose of 2 rem, they are about twice as likely as other children to die of a malignancy before their tenth birthday; it is estimated that perhaps 400 deaths per year in this age group in the U.S. can be attributed to prenatal X-ray examinations. The obstetrician making use of such an examination might counter with the argument that without the information provided by such examinations, unexpected problems arising during childbirth might cost more than 400 lives per year. It is undoubtedly true, however, that this type of examination may not be required for all pregnancies in which it is made and that it should not be used routinely. One estimate made several years ago was that such examinations were made 5–10 times more often than necessary.

Surveys of the Japanese survivors at Hiroshima and Nagasaki between the second and sixth months of intrauterine life at the time of the blast showed a reduced average head size and an increased incidence of mental retardation. Rough calculations of the relationship between the degree of effect and the dose suggest that the amount of mental retardation with reduced head size is 10 percent per 100 rem. Thus during this period the fetus in the uterus may be much more susceptible to the induction of mental defect than to leukemia. In the past some pediatricians exposed infants to bimonthly fluoroscopic examinations for the first two years of life, as an aid in monitoring childhood development, with whole-body doses of more than 100 rem. Such examinations should be viewed with some concern.

Mutational Changes

INDUCED GENE MUTATION. Prior to the discovery of the effects of radiation on gene mutation, its deleterious action on living systems was considered much the same as the deleterious action of chemical poisons. At very low concentrations poisons ordinarily have no effect, there being some minimum concentration (the *threshold*) necessary before any biological consequences can be noted. If this pharmacological concept were applicable to radiation damage, a person receiving a small dose day after day (as a careless X-ray technician might) would not have to be concerned because the threshold dose would not be reached on any one day. It was also assumed that if a person recovered from an exposure that had perceptible deleterious effects, these effects would wear off over a period of time and he or she would be in the same condition as if the exposure had never been received.

It is not possible to perform experiments at very low radiation doses to determine if there is a threshold, or if the dose–effect curve is linear. The noise of spontaneous mutation, occurring erratically, would mask any slight induced effects. Instead, we make the following simple argument: Radiation, when absorbed, manifests itself in ion production—roughly a million million in each cubic centimeter of tissue for each rem. Any one ion could be responsible for a genetic change, so that if we received a dose of only a millionth of a rem, there would still be a million ions produced in each cubic centimeter of tissue with the potential of damaging the DNA. Reducing the dose simply reduces the probability of a biological change. Consequently we assume that any dose, however small, carries with it its proportional amount of mutational damage (Figure 15-12).

Furthermore, *genetic damage accumulates.* If a mutation occurs during one year of early life, and a second mutation during another year, the individual will have the equivalent of the sum of the two events. The reason is

Figure 15-12 The linear relationship between radiation dosage and induced mutation rate, as suggested by numerous studies in *Drosophila*. Although it is not possible to make an accurate determination of the rate at very low doses because the erratic spontaneous mutants are indistinguishable from the induced ones, there is every reason to believe that the line is straight, even at low doses; that is, there is no threshold.

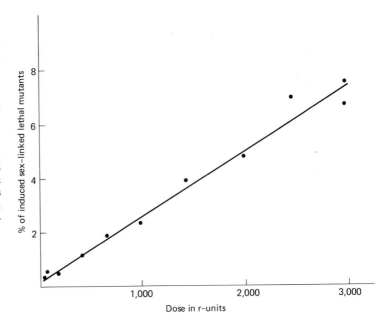

Dose in r–units

that even if many cell divisions should occur between the two events, the first mutational change will perpetuate itself by cell division just as the nonmutated cells do—and to the same extent—so that the relative proportion of cells with the mutational change will not be altered by cell division. A second mutation, whenever it occurs, will simply be added to whatever has happened before.

From this point of view, then, the total amount of radiation reaching the gonads during the individual's reproductive lifetime is important. Indeed, even doses applied to the mother during an individual's embryonic and fetal life should be added in as well.

In summary, the *mutational* damage caused by radiation has two properties of utmost significance different from those of most other types of biological damage. The first of these is that the dose–response relationship is linear, without any threshold, and the second is that doses are strictly cumulative. These two points are especially significant in estimations of genetic damage to the population under present conditions. Because there is no threshold, each small dose makes its proportionate contribution to the pool of new mutations; thus a minute amount of radiation (e.g., from fallout) applied to the population of the world can result in significant numbers of mutations. Similarly, because these exposures, although very low-level, may be continuous throughout the life of an individual, the cumulative effect may add up to a significant exposure even in a single individual.

EFFECTS ON MUTATION RATE. It is clearly not possible to design an experiment to measure the increase in mutation rate caused by radiation to humans. In fact, studies of the children of the survivors of Hiroshima and Nagasaki have failed to uncover any increase in new mutations. It can be stated further that there is little possibility of attributing any specific instance of mutation to radiation exposure because, regardless of the intensity of exposure, there can be no absolute assurance that any new mutation (e.g., a rare dominant) resulted from that exposure and not from some spontaneous event.

If we know (1) how often, on the average, a locus is mutated by radiation, (2) how many loci a human has, and (3) what dose of radiation people are exposed to, then we can easily calculate the number of mutations expected by multiplying these together. We do not, of course, have any way of knowing how many mutations are induced with any dose of radiation in humans, but we do have such information from extensive experiments with mice. From these experiments we estimate that with an exposure of 1 rem, a specific locus will mutate about 2.5 times out of every 100 million exposed. Thus our first factor (1) becomes 2.5/100 million, or 2.5×10^{-8} per locus per rem.

With respect to our second factor (2), we do not know how many loci a person has, but the figure of 40,000 (4×10^4) is an estimate that cannot be very far off.

Finally, let us assume that our population receives a total dose of 10 rem, to give us (3). Then, if we multiply these (2.5×10^{-8} mutations/locus rem) $\times$ (4×10^4 loci $\times$ 10 rem) $= 10 \times 10^{-8+4+1} = 10^{-2}$ mutations. As $10^{-2} = .01$, or 1 percent, we estimate that in a human population that receives 10 rem, there will be about one person per hundred with a new mutation caused by that radiation.

It has been shown that acute doses are approximately 4 times more effective in inducing mutation than chronic doses. This may be the result of overloading the DNA repair system. If the preceding calculations were based on acute rather than chronic doses, the net result would be 4 times as great, even though the total dose was the same in both cases.

THE DOUBLING DOSE. In an equally simple fashion we can make an educated guess about the amount of radiation that it would take to produce the same number of mutations as are produced spontaneously. Spontaneous mutation rates have different values for different loci, but a good overall average can be taken as 1×10^{-6} mutations per locus. With an induced mutation rate of 2.5×10^{-8} mutations per locus per rem, how many rem would it take to make the induced rate equal the spontaneous rate? We can arrive at this simply by solving the following equation:

$$\text{Dose required in rem} \times \frac{2.5 \times 10^{-8}}{\text{locus, rem}} = \frac{1 \times 10^{-6}}{\text{locus}}$$

The two sides of the equation are equal when the dose required is 40 rem. For this reason we can refer to 40 rem as the *doubling dose*.

Of course, such a calculation is based on figures that themselves are rough estimates, and therefore the values calculated here should not be taken too literally. Each investigator who attempts a calculation of this sort, based on rates that are themselves open to question, will come up with a final answer different from the answers of others. Most estimates generally agree, however, in placing the doubling dose somewhere between 20 and 200 rem; some more recent estimates suggest an even higher doubling dose—in the several hundreds of rem.

Two aspects of induced mutation should be kept in mind. First, whereas beneficial mutations must form only a minute fraction of all spontaneous mutational changes, they must be even rarer among radiation-induced mutations, because radiation breaks chromosomes, with attendant damage to genes, loss of chromosome segments, and so on. On the other hand, in most cases these mutations are recessive. They are expressed only in the homozygous state, produced by chance some generations after the initial induction, when two persons heterozygous for the defective allele happen to produce progeny. When such recessive alleles do become homozygous, they are often lethal to the embryo at an early stage, so that society does not suffer from the deleterious effects of those particular mutations. Others, however, are viable and are responsible for gross physical defects, a reduction in physical vigor, premature death, or other "semilethal" conditions.

NUMBERS OF MUTATIONS PRODUCED BY RADIATION EXPOSURES. Using the preceding information, we can readily make some quick calculations showing what to expect from any dose of radiation. Suppose that each person should receive 170 mrem per year, a dose called the *maximum acceptable dose* for the population at large. Over a thirty-year period (one generation), the accumulated dose will amount to 5 rem. If 20 rem is the doubling dose, then the radiation will be responsible for 5/20, or 1/4, as many mutants as occur spontaneously. On the other hand, if 200 rem is the doubling dose, the additional number of mutants will amount to only 1/40 of the spontaneous number.

RELATIVE FREQUENCIES OF NEW MUTATIONS. The calculations for relative frequencies of new mutations and other congenital defects are summarized in Table 15-5. For autosomal dominant traits there are now about 10,000 cases per million live births, of which about 2,000 are new mutants (the other 8,000 being born of affected parents). If the doubling dose is 20 rem, the effect of 5 rem will be to multiply the above figures by 25 percent. Thus we can expect 500 new cases in the first generation and, as these new mutants also produce progeny over time, 2,500 cases per generation. On the other hand, if the doubling dose is 200 rem, the effect of 5 rem will be only one fortieth as great. In that case 50 new cases will appear per generation, which would eventually make a total of 250 cases (again, per million live births) attributable to the radiation.

A somewhat similar calculation can be made for X-chromosome traits where from 1/4 to 1/40 of the current number of new mutants will be attributable to radiation. On the other hand, for recessive traits the newly mutated alleles will not appear until, by chance, they become homozygous, so that it is not possible to make any estimate of the number that will appear within the near future. Calculations based on the known incidence of congenital anomalies, aneuploidy, and abortions are also given in Table 15-5.

THREE MILE ISLAND EXPOSURE. We can make a similar calculation for the radiation effects at Three Mile Island (TMI). A convenient unit for expressing the exposure of groups of people is the *person-rem*, the total of all doses received by all individuals. This is easily calculated by multiplying the average dose per person by the number of persons receiving that dose.

Table 15-5 The estimated effect of 5 rem per generation on a population of 1 million live births, compared with the current incidence. When two estimates are given, the first assumes a doubling dose of 200 rem, the second a doubling dose of 20 rem. (Data from the 1972 Report of the Advisory Committee on the Biological Effects of Ionizing Radiation, National Academy of Science, Washington, D.C.)

Type of Defect	Current Incidence per Million Live Births	Effect of 5 rem per Generation	
		Immediate	Eventual
Specific genetic defects			
Autosomal dominants	10,000	50–500	250–2,500
X-chromosome traits	400	0–15	10–100
Recessive traits	1,500	Very few	Very slow increase
Congenital anomalies			
Chromosome rearrangements	1,000	60	75
Aneuploidy	4,000	5	5
Recognized abortions			
Aneuploidy and polyploidy	35,000	55	55
X0	9,000	15	15
Chromosome rearrangements	11,000	360	450
Total	71,900	545–1,010	860–3,200

In the case of TMI, 2,200,000 persons living within fifty miles of the reactor are said to have received an average of 1.5 mrem, for a total of 3,300 person-rem. Table 15-5 shows the incidence of immediate genetic defects (middle column) when 1 million individuals have received 5 rem, which equals 5 million person-rems.

Because 3,300 rem is .066 percent of 5 million rem, we can make a rough estimate of the mutational damage from this accident by multiplying the overall frequency of immediate effects, given in Table 15-5 as between 215 and 1,010, by .066 percent. The results are less than 1, suggesting that, assuming our underlying assumptions are true, perhaps one new mutation was induced.

MUTATIONS FROM MEDICAL X RAYS. Next to natural background radiation, the largest contribution to our radiation load is delivered to us by members of the medical profession, who expose the average U.S. citizen to 35 mrem/year. If the population of the United States is 280 million then the total dose is

$$280 \times 10^6 \times 35 \times 10^{-3} \text{ rem} = 98 \times 10^3 = 9.8 \times 10^6 \text{ person-rems.}$$

Of this dose, one third is generally considered to be unnecessary: This amounts to 3.27×10^6 person rems. Table 15-5 is based on 5×10^6 person-rems. Therefore we can make a quick estimate of the mutational damage contributed from medical X rays by multiplying each of the two right-most columns of entry in Table 15-5 by 65 percent (3.27/5).

Between 140 and 660 defective children can be attributed to the immediate effects of this radiation, and from 560 to 2,090 each year in the future, assuming that the amount of unnecessary medical radiation continues at the same rate as in the past.

Malignancies Induced by Radiation

A more immediate effect of radiation is the induction of malignant cell growth, particularly leukemia. Unlike the situation for gene mutations (where it can be reasonably assumed that there is a linear dose–response relationship so that any dose, however small, will have its proportional effect, and any mutations so produced at different times will persist to give a cumulative effect), for the induction of carcinomas the situation is not so clear.

CANCER AND LEUKEMIA INDUCTION IN HUMANS. Innumberable studies on humans receiving high radiation doses show an increasing frequency of induced cancers with increasing dose. One such study involves the survivors of the atomic explosions of Hiroshima and Nagasaki. Figure 15-13 shows the relationship between the doses (estimated) that the bomb survivors received and the numbers subsequently stricken with leukemia.

MALIGNANCIES FROM NUCLEAR ENERGY. The problem of malignancies induced by nuclear energy is clearly most important. If there exists a threshold below which radiation damage has no effect, as some nuclear proponents argue, then we can disregard the very low doses that the production of nuclear energy is responsible for (under normal operating conditions). On the

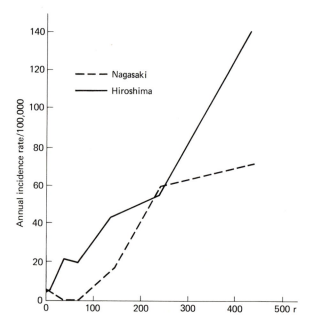

Figure 15-13 The incidence of leukemia per 100,000 of the exposed survivors of Hiroshima and Nagasaki, plotted as a function of the dose estimated to have been received by those survivors. (T. Ishimaru et al. Leukemia in Atomic Bomb Survivors, Hiroshima and Nagasaki. *Radiat. Res.* **45**:216–33, 1972.)

other hand, if the relationship is the same as for the gene mutations discussed earlier, then, however small the dose, a corresponding effect occurs. In any case, for our purposes we do not wish to underestimate any potentially disastrous biological damage to the population; we will therefore assume that there is no threshold and that the linear relationship holds. This means that we can add all doses together, however small any one individual dose may be, and calculate the resulting cancer rate from the total of person-rem.

An estimate of the effectiveness of radiation in producing cancer is that 10,000 person-rem will produce one case of cancer. If the total exposure after TMI was 3,300 person-rems, the expectation is that less than one person $(3,300/10,000 = 1/3)$ will eventually die of cancer from this exposure. This low estimate has been revised upward by a factor of 50 or more by those who believe, first, that the dose to the public may have been higher than the official estimate indicates and, second, that cancer may be more readily induced by low doses of radiation than is usually assumed. Thus, if it is assumed that 138 person-rem will cause 1 case of cancer, then the population of 2,000,000 exposed at Three Mile Island will, over the years, experience 3,300/138, or about 24 additional cases.

MALIGNANCIES FROM MEDICAL X RAYS. We can make a similar calculation for unnecessary medical X-rays, which we have calculated to be about 3.3×10^6 person-rem per year in the United States. If we divide this by 10^4 (the number of person-rem necessary to produce 1 malignancy), we have a total of 3.3×10^2, or 330 malignancies. Furthermore, because this is a continuous dosage each year, that number of deaths will also occur each year.

SUMMARY. In this chapter the radiation exposures to humans from a wide variety of sources have been evaluated, and simple calculations have been introduced that show how the amount of mutational dosage and malignancy is calculated from known population exposures. Although these estimates are very crude and are open to criticism from those who believe that radiation effects are either very much greater or very much less, they do give some measure of the degree to which radiation may be affecting the health and well-being of the population.

Additional Reading

BARNABY, F. 1977. Hiroshima and Nagasaki: The survivors. The reckoning. *New Scientist,* **75:**472–475.

GOFMAN, J. W. 1975. The cancer hazard from inhaled plutonium. Committee for Nuclear Responsibility, Inc., Yachats, Oregon, CNR Report 1975-1-R. 36 pp.

Health survey in high background radiation areas in China. 1980. (From Wei Lüxin, Laboratory of Industrial Hygiene, Ministry of Public Health, Beijing, China). *Science,* **209:**877–880.

HOLDEN, C. 1979. Low-level radiation: A high-level concern. *Science,* **204:**155–158.

KERR, R. A. 1979. Nuclear waste disposal: Alternatives to solidification in glass proposed. *Science,* **204:**289–291.

LAND, C. E. 1980. Estimating cancer risks from low doses of ionizing radiation. *Science,* **209:**1197–1203.

LA PORTE, T. R. 1978. Nuclear waste: Increasing scale and sociopolitical impacts. *Science,* **201:**22–28.

LINOS, A., et al. 1980. Low-dose radiation and leukemia. *New England Journal of Medicine,* **302:**1101–1105.

MARSHALL, E. 1979. A preliminary report on Three Mile Island. *Science,* **204:**280–281.

MARSHALL, E. 1979. The crisis at Three Mile Island: Nuclear risks are reconsidered. *Science,* **204:**152–153.

MARX, J. L. 1979. Low-level radiation: Just how bad is it? *Science,* **204:**160–164.

MAUGH, T. H., II. 1979. Burial is last resort for hazardous wastes. *Science,* **204:**1295–1298.

More cancer links to low-level radiation. 1978. *Science News,* **113:**117.

NEEL, J. V., et al. 1980. Search for mutations affecting protein structure in children of atomic bomb survivors: Preliminary report. *Genetics,* **77:**4221–4225.

ROTBLAT, J. 1977. Hiroshima and Nagasaki: The survivors. The puzzle of absent effects. *New Scientist,* **75:**475–476.

TORREY, L. 1979. The accident at Three Mile Island. *New Scientist,* **84:**424–428.

Questions

1. Ultraviolet light and X rays are both powerful mutagenic agents when bacteria are exposed to them. Why is ultraviolet much less mutagenic in its action on mammals?

2. Describe the electromagnetic spectrum and indicate the point at which the energy per quantum becomes great enough to exert a mutagenic effect. Does this consideration have any bearing on the kinds of radiation we are exposed to daily, such as microwave ovens and radio and TV transmitters?

3. How does an X-ray tube generate radiation?

4. What is an isotope? An unstable isotope? When does an isotope produce radiation? If the half-life of an unstable isotope is ten years, how long would it take for all but one sixty-fourth of the total to disintegrate?

5. What are the units by which radiation is measured? Do they differ appreciably?

6. How do we get information on the effects of high radiation doses on humans?

7. What is meant by the LD_{50}? What is its value for humans?

8. There are two general types of medical radiation exposures: diagnostic and therapeutic. Which exposes the body to the higher dose of radiation? Can you suggest why?

9. Define the *genetically significant dose (GSD)*. What factors must be

taken into account in estimating it? What is the contribution to the GSD from medical X rays?

10. Can you give any reasons for the fact that probably as many as one third of all medical X-ray exposures are actually unnecessary?

11. Who are the persons who regularly receive radiation in the course of their occupations? Approximately how much radiation do they receive?

12. What is the dose that adults and children in the United States receive from color television sets?

13. What do we mean by *background sources* of radiation? How much radiation do we receive from each of these sources?

14. Can you list some of the considerations both pro and con that bear on the use of nuclear energy as a major source of power in the United States?

15. How does the radiation dose received by workers in a nuclear plant compare with that received by those who work with radiation in the medical profession?

16. How much radiation were residents living near the Three Mile Island nuclear plant exposed to by the accident of 1979 (according to official sources)?

17. Which is more important in determining the damage that a radioactive isotope might do to the human body: the biological half-life or the physical half-life?

18. Why do we make a distinction between acute and chronic doses of radiation?

19. What is the difference between leukemia and leukopenia?

20. Mutational damage to the DNA in a human cell has two properties of utmost significance in a consideration of the long-range effects of radiation on populations. What are they?

21. How do we estimate that 10 rem applied to a human population will result in one new mutation per hundred persons exposed?

22. Calculate the doubling dose. Precisely what is meant by this expression?

23. How do we make estimates of the mutational damage produced in an exposed population, such as the residents living near Three Mile Island or persons who receive medical X-ray treatments?

16

Prenatal Development

Normal Embryonic Development

During the nine-month period between the time of fertilization of the egg by a sperm and the birth of the infant, a series of most remarkable developmental changes takes place. The total number of cells increases phenomenally from one to countless billions (about 10^{14}), which then become transformed into hundreds of different types, one to become a heart muscle cell, another a nerve cell, and so on. In addition, these cells take up special positions, in some cases by migrating from one spot in the developing embryo to another, and in other cases by undergoing extra mitoses at those sites. In its final form the newborn child is a creation of extraordinary complexity, with a set of highly integrated systems that normally enable it to survive and cope with an almost infinite variety of situations for its long life span.

To accomplish this complex development, two basic forces are at work. The first is mitosis, which allows for the precise reproduction of all the essential contents of a single cell into two daughter cells. As we have seen in Chapter 2, our knowledge of the mechanism of mitosis is fairly complete. Also, as development proceeds, different cell lines take on new and characteristic structures and functions; this process is called *differentiation*. Although differentiation has been studied in great detail from the single-cell zygote to the adult in a wide variety of animals, our knowledge of the process is, by comparison with our knowledge of mitosis, quite primitive. We are only now beginning to understand why one of the descendants of the original fertilized egg becomes a white blood cell and another descendant of that same egg becomes a color-detecting cell in the retina of the eye. At this point, we will consider only the gross morphological changes occurring prior to birth.

FERTILIZATION.   At intervals of slightly less than a calendar month, at approximately the middle of the menstrual cycle, a fertile woman of childbearing age experiences *ovulation*, the discharge of a mature egg or *ovum* from an ovary into the corresponding *Fallopian tube*. Here it may survive for as long as a day, and then it degenerates unless it is fertilized by a *sperm*, which, having a longer life span, may have been present in the female genital tract for several days. When one sperm succeeds in fertilizing the egg (Figure 16-1*A*, *B*, *C*), the egg becomes refractory and will not accept any additional sperm, with only rare exceptions.

The human egg is about 0.1mm (1/200 of an inch) in diameter and is barely visible to the human eye; the sperm is much smaller, 0.06mm (1/500 of an inch) long, the majority of its length consisting of a long tail. Together these minute cells contain all of the instructions necessary for the production of a new human being.

THE CLEAVAGE DIVISIONS.   Some thirty hours after the union of sperm and egg, the fertilized ovum (the *zygote*) divides into two (Figure 16-2*A*) and after another ten hours into four (Figure 16-2*B*). These early mitoses of the fertilized egg are referred to as *cleavage divisions* (Figure 16-2*A*, *B*, *C*). These divisions continue as the cells travel down the Fallopian tube to the uterus; the resulting clump of cells takes on the appearance of a little mulberry, from which it gets its scientific name, the Latin *morula* (Figure 16-2*D*). A cavity is formed within the cells and fluid accumulates within; this is now the *blastocyst* stage (Figure 16-3). At about six or seven days after fertilization, the blastocyst attaches itself to the uterine wall in a process known as *implantation*. Implantation is one of a long series of hazards that the fertilized egg must confront in its progression toward becoming a fully formed fetus; perhaps as many as a half of all fertilized eggs fail to implant and are lost.

The blastocyst changes shape when a quarter or so of its cells concentrate on one side to form the *embryo* proper, while a cavity forms between these cells and other cells that are responsible for attaching the embryo to the uterine wall. At the same time, the maternal tissue reacts to the presence of the embryo by forming enlarged blood vessels at the site of implantation; later these will provide oxygen and other nutrients to the developing child. On occasion this sudden growth of blood vessels may cause bleeding, and because this bleeding occurs about thirteen days after ovulation, or approximately twenty-eight days after the last menstrual period, it can be confused with ordinary menstrual bleeding, leading to an error of a month in the estimate of the date of conception.

LATER EMBRYONIC DEVELOPMENT.   At two weeks the embryo has become two-layered, and at three weeks the nervous system and the main organs begin to form. At this point the heartbeat begins. During the period from

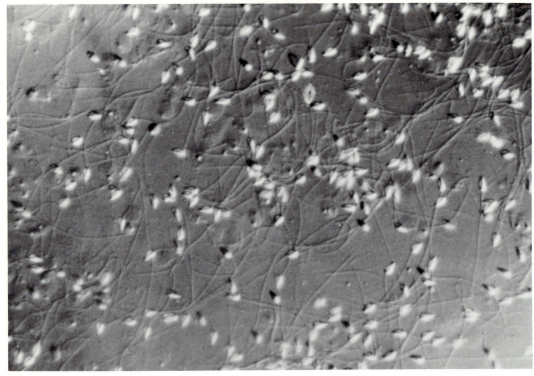

A

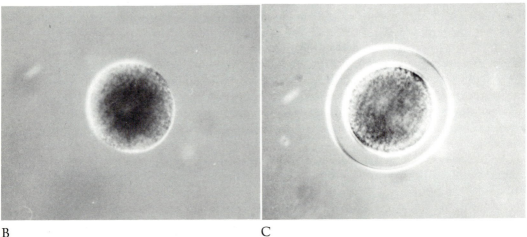

B   C

Figure 16-1 The gametes and fertilized egg of the sea urchin of the Pacific Northwest. The structure of the gametes and the appearance of the egg during early divisions are so similar in virtually all animals that good illustrative material, such as this, can be used equally well to illustrate the gametes and early development in humans. *(A)* The sperm, with its head showing up clearly under special lighting conditions and high magnification, and the thinner long tail. *(B)* The egg prior to fertilization. *(C)* The egg shortly after fertilization showing how the production of a "fertilization" membrane protects the egg against entry by a second sperm. (Photographs of sea urchin development courtesy Oregon Institute of Marine Biology, University of Oregon.)

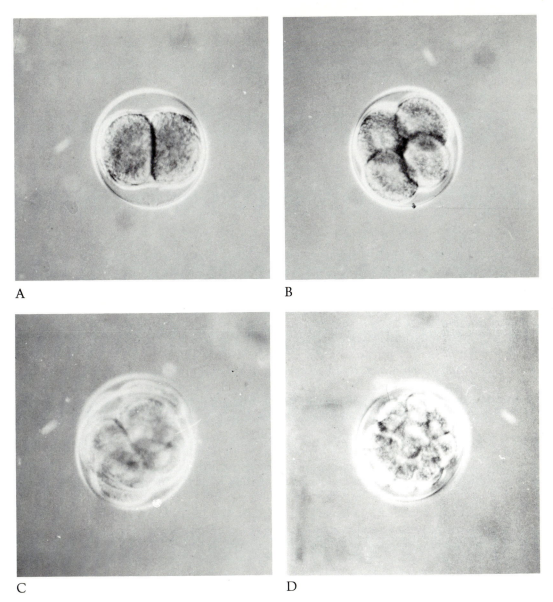

A

B

C

D

Figure 16-2 The progressive division of the fertilized egg of the sea urchin into the two-cell stage *(A)*, the four-cell stage *(B)*, and the more advanced stages *(C* and *D)*, the last of which now resembles a morula.

four to eight weeks, the change from an unrecognizable mass of cells to a new individual is apparent (Figure 16-4). At the end of the second month of development, the embryo, now less than an inch long, has progressed to the point where it is recognizable as a human being, with the major features to be possessed by the newborn infant present, and its embryonic phase is at an end. From the beginning of the third month on, it is referred to as a *fetus,* from the Latin meaning "a young one."

Figure 16-3 The transformation of the morula into a mass with a hollow center, the blastocyst. The larger mass of cells at one end becomes the embryo proper and the other cells become the fetal membrane known as the *placenta*. (A. T. Hertig, J. Rock, and E. C. Adams, *Am. J. Anat.* **98:**435–93, 1956.)

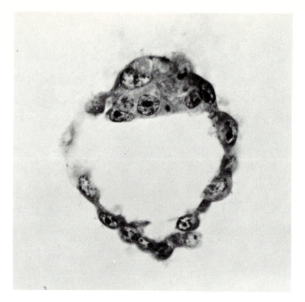

FETAL DEVELOPMENT. Except for additional growth and changes in the relative dimensions of the body parts (the head, for instance, still being abnormally large), the basic elements of the complete human being have been formed. During the early fetal stages the sexual development is neutral, with primitive elements of both sexes being present. At the end of the tenth week of gestation the structures forming the internal male organs start to form (if the fetus is male), and somewhat later those of the female differentiate (if the fetus is female). By the end of the third month the external genitalia of the fetus may be distinctive, making it possible to classify the sex. The fetal heartbeat is detectable from outside the mother's abdomen, and at about four months fetal muscular activity may be felt by the mother. The most pronounced changes during this time are those of weight and size: at the beginning of the third month when the embryonic period has ended and the fetal period has begun, the fetus weighs only three grams (about a tenth of an ounce).

Abnormal Development

The series of events starting in the fertilized egg involving rapid mitosis, cell migration, and differentiation, to produce the complete infant, represent such an extraordinarily complex process that it may seem nothing short of miraculous that the vast majority of children are normal at birth. Irregularities of development can lead to observable defects in the individual; the study of such abnormalities is known as *dysmorphology*. The serious disturbances of normal development can lead to abnormalities at birth that are known as *congenital defects*. The most common are such viable anomalies as *cleft lip, cleft palate,* and *club foot*—defects that can, in many cases, be repaired by surgery. Fortunately most of the serious defects that occur dur-

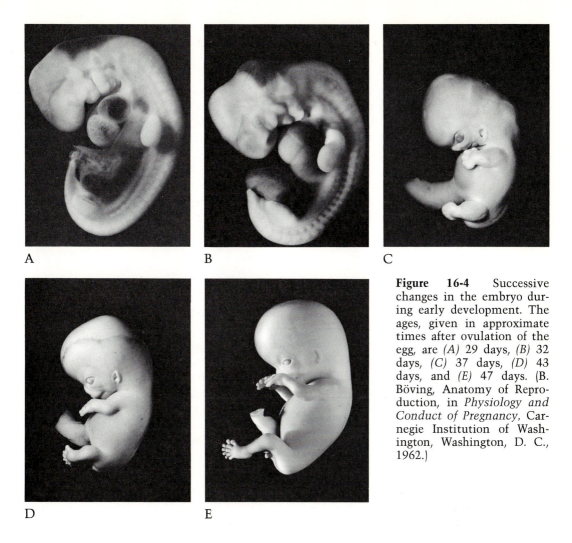

A   B   C

D   E

Figure 16-4 Successive changes in the embryo during early development. The ages, given in approximate times after ovulation of the egg, are *(A)* 29 days, *(B)* 32 days, *(C)* 37 days, *(D)* 43 days, and *(E)* 47 days. (B. Böving, Anatomy of Reproduction, in *Physiology and Conduct of Pregnancy,* Carnegie Institution of Washington, Washington, D. C., 1962.)

ing early development give rise to an inviable embryo, which is lost in early pregnancy by *spontaneous abortion* (miscarriage).

Teratogenic agents produce their effects by interfering with mitosis, cell migration, or differentiation and can cause profound deviations from normality. However, their effect is not produced by an alteration of the genes or chromosomes, and for this reason persons born with congenital defects induced by teratogens can expect to have unaffected offspring.

Statistics on the overall incidence of congenital defects vary widely with the country from which they come, with whether they come from hospital birth records or from births at home, and with the age of the child at the time the observation is made. The frequency from birth certificates seems to be less than 1 percent, and from hospital records somewhat greater. The frequency is greatest when detailed examinations are made by pediatricians. Pediatric examinations in the United States have given a figure of almost 9 percent, whereas in Germany the figure is as low as 2 percent. As an overall average it can be estimated that from 2 to 3 percent of all newborn infants show congenital malformations at birth, and after one year this figure is twice as high because of the appearance of malformations not obvious at birth.

CENTRAL NERVOUS SYSTEM MALFORMATIONS.   Spina bifida, the abnormal development of the spinal column, and anencephaly, incomplete development of the head, are two comparatively common congenital defects of the central nervous system, together occurring with a frequency of about 1 in 250 births. Anencephalic children die at birth; spina bifida cases may, after radical surgery, survive as grossly incapacitated cases. It has been found that the concentration of a certain protein, alpha-fetoprotein (AFP), is higher in the amniotic fluid that surrounds the fetus before birth when it has such severe developmental defects than it is in normal pregnancies. This phenomenon makes it possible to monitor any pregnancy suspected of such complications.

GERMAN MEASLES (RUBELLA).   Because the embryo and the fetus are so well protected—both from outside influences and from toxic agents in the mother's blood stream,—by the fetal membranes and the amniotic fluid in which the developing infant rests, it had been assumed for a long time that the embryo was relatively immune to external damage. In 1941 an Australian eye specialist, Gregg, called attention to an extraordinary increase in the frequency of eye cataracts in newborn children. Of 78 children born with opaque lenses, 68 had mothers who had contracted German measles during early pregnancy. Gregg suggested a cause-and-effect relationship between the *rubella* virus responsible for German measles and the cataracts.

Under ordinary circumstances, only 4–8 pregnant women out of every 10,000 in the United States are afflicted with rubella. During the year 1964, however, this country experienced an epidemic of German measles and the incidence went as high as 220 per 10,000. There were approximately 20,000 cases of congenital malformations among newborns as a result. In addition to blindness, other defects such as heart disease, deafness, and mental retardation, as well as an extremely small size at birth, were observed among the children of infected mothers.

Stillbirths and spontaneous abortions are from 2 to 4 times more likely in pregnancies accompanied by rubella than in uncomplicated pregnancies. In addition to the congenital malformations resulting from the 1964 epidemic, 30,000 stillbirths and miscarriages were attributed to this cause.

The kind of defect caused by rubella depends on the stage of embryonic development at which it occurs: infection during the sixth week of pregnancy may cause cataracts, whereas deafness follows infection during the ninth week. Heart defects may result from infection between the fifth and tenth weeks. The overall probability of malformations in newborns is about 50 percent when the infection occurs during the first month of pregnancy, and it decreases to 20 percent when the infection comes between the fifth and eighth weeks. It is less than 10 percent thereafter. These percentages would increase if inspection for defects that cannot be diagnosed immediately at birth were made somewhat later.

Many women of reproductive age have been immunized by an exposure to this virus earlier in their lives. It is now the practice to immunize childern (both girls and boys) against the virus to minimize later problems. However, vaccination immediately before or during the early stages of pregnancy is not recommended because there is a slight chance that the vaccine itself might cause congenital defects.

OTHER VIRUSES.   However, it is not an invariable rule for all teratogens that the danger to the embryo or the fetus is greatest during the early stages of development. Women infected early in pregnancy with hepatitis B virus rarely produce defective fetuses, but if the infection occurs during the middle of pregnancy, the risk to the fetus is about 10 percent, and if it occurs toward the end, the danger increases to 76 percent.

Another virus that has been positively shown to produce congenital malformations is the *cytomegalovirus.* Unfortunately it has no visible effect on the infected mother and so usually goes unrecognized in pregnant women. In fact, the only way of knowing that the virus has this capability is by the observation that a stillborn child sometimes carries large quantities of the virus, visible microscopically. The disease is often fatal to developing infants, and survivors can be severely mentally retarded. About 30,000 infants per year in the United States are born infected with this virus, of whom 4,000 may be mentally impaired.

Other maternal infections, such as Asian flu, mumps, polio, and chicken-pox, have been reported on occasion to have preceded the births of malformed children, but the data at the present time conflict and suggest that these infections play little, if any, role in producing birth defects. It was thought at one time that syphilis might be a major cause of congenital defects. At present there is some doubt that it produces as many defects with as great a frequency as was once believed, although it has been repeatedly observed to lead to congenital deafness and mental retardation as well as to defects of other organs of the developing fetus. *Herpes simplex*, a type of virus one of whose forms is responsible for cold sores, leads to brain damage, deafness, blindness, and other malformations of the nervous system. One third of infected newborns die, another third have defects, and the remainder are unaffected.

Medical Teratogens

THALIDOMIDE.   During the period from 1957 to 1961 a very effective antinauseant and sleeping pill called *thalidomide* was produced by a German pharmaceutical company and was made available throughout the world. Its properties made it very valuable in combating morning sickness and other discomforts experienced during early pregnancy. Within four years it became apparent that large numbers of children were being born with *phocomelia,* a condition involving flipperlike arms and legs, usually very short and sometimes entirely missing (Figure 16-5). Both in Germany and in Australia the correlation was made between the mother's taking thalidomide and the birth of a defective child. Before the drug was withdrawn from the market, at least 5,000 affected babies were born in Germany and

Figure 16-5 The failure of the long bones of the arms to grow normally in phocomelia, the congenital defect most commonly resulting from the ingestion of thalidomide by the mother. (Courtesy of D. Smith, University of Washington Medical School.)

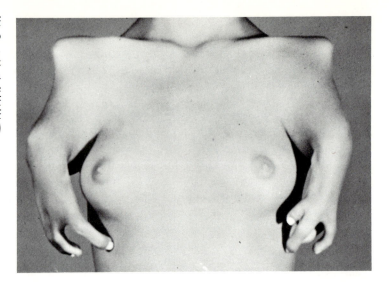

another thousand in other countries throughout the world. Because of stricter laws regulating the sale and distribution of new drugs in the United States, thalidomide was kept off the market in this country by the action of a concerned official of the Food and Drug Administration; nevertheless some women were able to obtain the drug from outside the country, with occasional defective births.

A study of the range of defects that thalidomide causes showed that the most disastrous effects occurred if it was taken by the mother during the third and fourth weeks of the embryo's life. The timing is consistent with the observation that the limbs develop during this embryonic period. Lesser effects were noted when the drug was taken subsequent to this time. In addition to deformities of the long bones, heart abnormalities and defects of the ear and the intestinal tract have been recorded.

DES. In the 1950's, the synthetic female hormone *diethylstilbestrol* (DES) was prescribed to about 5 million women in the United States to prevent miscarriage; their daughters, who were exposed *in utero*, showed at a young age an increased frequency of abnormal tissue growths in the reproductive tract. These daughters have proved to have a higher incidence of miscarriages and stillbirths after they become pregnant.

Of male mice whose mothers had been treated with DES, 60 percent proved to be sterile, apparently the result of DES-induced changes in the gonadal tissue during embryonic development. Of human males who had DES-treated mothers, 25 percent have also been reported to have abnormalities of the genital tract.

Adding DES to livestock feed decreases by about 10 percent the amount of food necessary to achieve a given animal weight; in addition, the meat from animals so fed is leaner than usual. Persons eating the meat of such animals ingest a small quantity of the hormone. However, because the amount is less that 1/2,500 the normal concentration of estrogens in a woman's body, it is debatable whether this ingestion has any substantial deleterious effect on humans.

OTHER MEDICATION. Oral contraceptives have been accused of producing congenital defects; however, the data are contradictory and the increase in frequency caused by these compounds must be slight, if it exists at all. Hormone pills used in pregnancy tests can cause cleft palate, nerve defects, limb malformations, and other birth defects, according to a study in England and Wales. When taken during the first four months of pregnancy, progestin, a female hormone prescribed for women with menstrual disorders, may increase the risk of heart defects or deformed arms and legs in the fetus.

Sedatives containing phenobarbital, when given to pregnant female rats in doses so small that the rats are not put to sleep, affect sexual development of the fetuses so that both male and female progeny tend to be sterile. A similar effect has not been noticed for humans; the long delay between the time of taking the drug by the mother and any effect in her grown offspring, perhaps twenty or more years later, make it difficult to correlate cause and effect, if it exists.

Pregnant women are particularly prone to excessive use of drugs (defined in the broadest sense, including vitamins, iron, pain relievers, tranquilizers, and anti-morning-sickness remedies). One study of 168 pregnant women in Florida showed that all took at least two drugs, almost 95 percent took five or more, and the average number taken was eleven.

RADIATION. Of pregnant Japanese women who were exposed to heavy doses of radiation at the time of the atomic bomb explosions at the end of World War II, 25 percent produced children who died during their first year of life, and another 25 percent gave birth to infants with severe congenital anomalies.

Embryos developing in women who must be given pelvic radiation treatments in early pregnancy are particularly at risk because of the high doses of radiation used in such medical treatments; as many as 50 percent of the offspring of such treated women may be abnormal. These abnormalities include *microcephaly* (an unusually small head) accompanied by severe mental retardation, skull defects, blindness, cleft palate, and defects of the limbs. As with most other teratogenic effects, defects are most common and most severe when the mother is X-rayed during the first two months of pregnancy and decrease markedly thereafter. Other effects of radiation, such as the production of changes in the chromosomes and genes, were discussed in Chapter 15.

Other Teratogens

ALCOHOL. Some 6,000 babies born per year in the United States may be mentally or physically damaged because their mothers drink alcohol to excess while pregnant. Two glasses of wine or one and a half pints of beer per day may double the chances that a child will be stillborn, have low birth weight, or suffer from some other congenital defect. If a pregnant woman drinks heavily, the child may (in perhaps as many as 20 percent of all cases) suffer malformations, especially of the head and face (Figure 16-6), be retarded mentally and show slow development in walking. Affected children are said to have the *fetal alcohol syndrome (FAS)*. The offspring of more

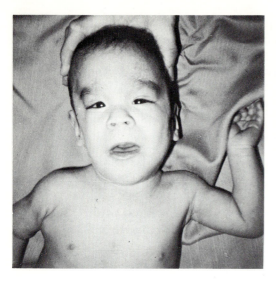
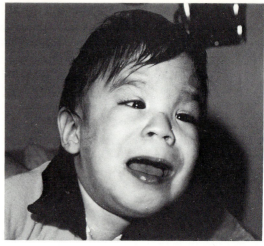

Figure 16-6 A child with some of the symptoms of fetal alcohol syndrome, at an early age and several years later. The diagnostic features include mental retardation, somewhat closed eyelids, reduced head diameter, delayed physical development, and abnormal palmar creases. (Courtesy of D. Weaver, University of Indiana Medical School.)

moderately drinking mothers may be tremulous and sluggish and may also have an impaired intellect.

In another survey, it appeared that a woman who drinks 2 ounces of a 100-proof whiskey a day has only a very small chance of producing an abnormal offspring, but if she drinks between 2 and 4 ounces, the risk goes up to 10 percent, and an intake of 10 or more ounces per day increases the risk to 50 percent. One study in Seattle suggests that a number of different drinking patterns can lead to FAS, ranging from moderate constant tippling to the infrequent blast.

Alcohol can apparently disrupt development of the central nervous system during the first three months and lead to behavioral problems in the child when taken in later pregnancy. In experiments with rats, progeny showed effects even when the supply to an "alcoholic rat" was cut off immediately before pregnancy. The reason for this toxic effect of alcohol on embryonic development is unknown. It has been shown, however, to be independent of both malnutrition and poor physical condition, both of which also occur in alcoholics.

Experiments with mice show that different strains differ in their ability to metabolize alcohol and so produce different degrees of abnormalities in their offspring. Because humans show a similar individual and group variability, it is reasonable to conclude that the risk of a newborn with FAS cannot be accurately estimated simply from the amount of alcohol the mother consumes during pregnancy.

CAFFEINE. Caffeine is a mildly addictive drug found in small quantities in tea and coffee; about half as much is present in cocoa and a third as much in cola drinks. It is also used in many popular pain relievers. If experimental animals are given very high doses of caffeine during pregnancy, an amount equivalent to a human's drinking about twenty-four cups of coffee per day, they produce a higher proportion of defective offspring. However, in these experiments, previous exposure to the drug reduced its effect in increasing congenital malformations. There is some evidence that ordinary tea or coffee drinking (four or five cups a day) may increase the risk of congenital abnormalities, such as heart defects and cleft palate.

SMOKING. Babies of smokers weigh an average of 200 grams (almost half a pound) less than those of nonsmokers and show a slower rate of mental and physical growth. The dangers of miscarriage and fetal death are increased; it has been estimated that as many as one seventh of all premature deliveries may be attributed to smoking. The ill effect of smoking tobacco is thought to be caused by the production of carbon dioxide during smoking, which binds hemoglobin and so prevents the normal amount of oxygen from getting to the tissues of the fetus, as well as by a reaction of the nervous system which constricts the blood vessels supplying the uterus, futher decreasing the oxygen available to the developing fetus.

2,4,5-T. The weed- and brush-killing agent known as *2,4,5-T,* used extensively in the United States since 1948, contains a minute fraction of a potent class of toxic contaminants called *dioxins,* formed during the manufacture of the herbicide. Dioxin was also present in the defoliant known as *agent orange* used by the American Armed Forces in Vietnam; the use of this substance was stopped in 1970 after it was shown that there was an increased incidence of birth defects in mice and rats exposed to it. It was subsequently reported that there was an increase in the frequency of miscarriages in Vietnam. Furthermore many of the men who were involved in the spraying have reported a variety of medical problems that they attribute to the agent.

Since 1970, 2,4,5-T has been banned in household products and its use is prohibited in parks, lakes, and recreational areas. However, it is still used to inhibit the growth of brush competing with trees in the forests of the Northwest. The threefold increase in the number of miscarriages and other complications of pregnancy among women living in areas of Oregon so treated has led to increased concern about the use of this herbicide.

In 1976 a chemical plant explosion in Seveso, Italy, released toxic vapors, including dioxin, over a wide area. Animals and vegetation were killed; only a few people have shown any serious effects, but the long-range effects will not be known for years. In any case, it has been shown that minute amounts of dioxin in monkeys, mice, rats, and hamsters have caused miscarriages and birth defects. There is no reason to believe that the human body should not be similarly susceptible to the toxic effects of this chemical.

HEXACHLOROPHENE. At one time *hexachlorophene,* a potent bactericidal agent, was used in over-the-counter health-care products, such as deodorant soaps and acne medicines. It is able to penetrate unbroken skin and migrate into cells of the blood, fat tissue, brain cells, liver, and so on. Swedish nurses who worked in geriatric wards (and therefore washed their hands with bactericidal soap frequently but were less likely than other nurses to be exposed to other potential teratogens, such as anesthesia gases) had children with congenital defects 50 times as often as would be expected. Of 460 children, 25 had severe defects and 46 had minor deformities. None of the children born to a matched control group had severe defects, and only 8 had minor problems. The normal incidence of severe congenital defects in Sweden is about 1 per 1,000 births.

GENERAL COMMENTS. There are several difficulties in proving that any particular chemical is teratogenic. Congenital defects are relatively infrequent after the ingestion of known teratogens; the manifestation of defects may not only be erratic but may also be variable with respect to the specific organs affected. Further, it is obviously not possible to test potential teratogens deliberately by feeding them to pregnant women. Experimental work with animals usually involves feeding them excessive amounts of the material, and very often different species of animals differ in their responses to these chemicals. Even different strains or races within a given species may manifest different reactions to a given teratogen. Simple prudence would suggest that no mother expose herself to any unusual chemical during her child's prenatal life, especially during the very sensitive period of early embryonic development. Unfortunately women are not always aware they are pregnant during this critical early period.

Additional Reading

The bendectin battle. 1980. *Science News,* **118:**395.

CHRISTOFFEL, K. K. and I. SALAFSKY. 1975. Fetal alcohol syndrome in dizygotic twins. *Journal of Pediatrics,* **87:**963–967.

HANSON, J. W., A. P. STREISSGUTH, and D. W. SMITH. 1978. The effects of moderate alcohol consumption during pregnancy on fetal growth and morphogenesis. *Journal of Pediatrics,* **92:**457–460.

HERBICIDE is "real-risk" to birds and people. 1981. *New Scientist,* **89:**400.

KENWARD, M., and L. MCGINTY. 1980. Mother's drugs affect baby's hormones. *New Scientist,* **85:**64.

LEVINE, C. 1979. An act of greed. (Book report on "Suffer the Children: The Story of Thalidomide," by the Insight Team of Sunday Times of London. New York: Viking.) Hastings Center Report, pp. 43–44, June 1979.

MARX, J. L. 1978. The mating game: What happens when sperm meets egg. *Science,* **200:**1256–1258.

PATTEN, B. M. 1968. *Human Embryology.* New York: McGraw-Hill.

SIEGEL, M. 1973. Congenital malformations following chickenpox, measles, mumps and hepatitis. *JAMA,* **31:**1521–1524.

Teratogens acting through males. 1978. *Science,* **202:**733.

Questions

1. In rough outline, give the sequence of events from the fertilization of the egg to the birth of an infant.
2. What is the distinction between an embryo and a fetus?

3. What is your personal opinion about the time at which human life begins? Can you give any arguments that would support a different time?
4. What are the most common congenital defects? What is the approximate frequency of congenital defects in the newborn?
5. How is it possible to monitor pregnancies for abnormalities of the central nervous system, like spina bifida?
6. Name several viruses that have been implicated in congenital defects. Is there a relationship between the type of congenital defect produced and the teratogenic agent involved?
7. Which medication was responsible for a high incidence of phocomelia in the late 1950s? Are there any other medications that are suspect?
8. What types of developmental abnormalities does radiation induce?
9. Approximately how much alcohol taken by a pregnant woman may cause the newborn to appear with fetal alcohol syndrome (FAS)?
10. List some additional substances that a pregnant woman might be advised to avoid if at all possible. (This list might be supplemented by compounds whose effects are described in Chapter 14.)
11. A chemical plant in Washington hires only infertile women for jobs that could involve exposure to chemicals implicated in birth defects. A woman who was told that she would have to undergo a hysterectomy before being hired for such a job filed a complaint with the State Human Rights Commission. If you were a lawyer, whose side would you prefer to represent?

17

Twinning

One of the commonest criticisms directed at the science of genetics by social scientists is that the biological basis of human attributes is grossly oversimplified by geneticists. This criticism is, to a large extent, justified. During the early days of genetics such characterisitics as honesty, virtue, and criminality were commonly depicted in pedigrees as the consequence of single dominant or recessive genes. A substantial proportion of social scientists, and others as well, embrace a philosophy that denies any substantial contribution to the behavior and psychology of normal individuals from their genetic composition. They maintain that behavioral and psychological differences stem predominantly, if not entirely, from environmental rather than genetic variables.

Certainly it seems self-evident that the important attributes of humanness, to whatever extent they are genetic, are not determined by single loci acting independently of all other loci. The most obvious physical characteristics of height, weight, and bodily structure, as they appear as variants in the normal population, are not controlled by simple genetic factors, and one might imagine any genetic basis (to the extent that it actually exists) to be equally complex for behavior, intelligence, and other more important mental and emotional qualities. Where it appears that alleles at many loci make contributions to one charateristic, the genetic system is said to be *polygenic*.

352   Of course, the illustration of the simple rules of inheritance, an important

first step in understanding the transmission of genetic traits in humans, must always start with simple, clear, and unambiguous traits. This oversimplification is deliberate; it allows us to avoid any discussion of the complicated interactions that occur among large numbers of loci during development. However, now that we have discussed principles of Mendelian transmission as well as the essential details of embryonic development, it is appropriate to consider these more complicated aspects of genetic determination.

The genetic analysis of any characteristic is difficult when there are large numbers of different alleles at different loci making variable contributions to the phenotype. It is desirable, whenever possible, to make the genetic background uniform. In the lower organisms, and particularly in agriculturally important plants and animals, it is possible, by successive generations of inbreeding, to produce strains in which individuals are homozygous at almost all loci, with all members of the strain having essentially the same genetic composition; that is, they are *isogenic*. Regular variability that then appears among such isogenic individuals can be attributed to environmental causes. The progress of experimental genetics would be seriously hampered if genetic variation could not be kept under control in some way. However, it is obviously not possible to produce isogenic strains of humans. This problem can be circumvented to some extent by the study of twins.

Use of Twins

Each person has a unique genetic constitution, not only different from every other living person, but also probably different from that of any other person who has ever existed. The exception to this rule is the case of identical twins, two different persons who are both the product of a single fertilized egg and therefore have the same genetic content. Consequently such twins provide a natural experiment that yields data not available from any other source.

GALTON'S CONTRIBUTION.   The possibility of using twins in assessing the influence of inheritance was suggested by Francis Galton, a cousin of Charles Darwin. Galton was a pioneer in the application of mathematics to biology. His basic philosophy seems to have been that virtually anything could, and should, be measured. He attempted to ascertain the existence of a Supreme Being by comparing frequencies of shipwrecks of vessels with and without missionaries aboard, on the assumption that the latter might be shipwrecked more frequently. They were not. Along a similar line, he checked on the effectiveness of prayer by examining the longevity of royalty, who might be expected to live longer because their subjects prayed for their health and safety. They did not.

In his researches on twins Galton obtained most of his information by correspondence with friends and the friends of friends who were twins, or who had twins in their families. He recognized in a general way the existence of the two types of twins, now known as *identical* and *fraternal*. Of the 94 pairs of twins he studied, 35 of whom were probably identical and another 20 almost certainly fraternal, he properly assessed their potential value in distinguishing between genetic and environmental contributions to specific traits. He came to the conclusion that twins who were very much alike in childhood remained so throughout their lifetimes, whereas those who were quite different were no more or less different in later life than ordinary sibs. He also noted that in many cases twins developed the same kinds of mental illness.

Characteristics of Twinning

TYPES OF TWINNING.   There are two fundamentally different kinds of twins: those who arise from two zygotes, *dizygotic* (DZ) or *fraternal* twins, and those who arise from a single zygote, *monozygotic* (MZ) or *identical* twins. Fraternal twins are of no greater genetic relationship to each other than sibs born at different times; about half the time they are of the same sex. Identical twins, on the other hand, come from a single fertilized egg that has divided during the course of early development into two cells or two cell masses, each forming a new individual. Such twins are genetically identical to each other, with rare exceptions to be noted later.

INCIDENCE OF TWIN BIRTHS.   The frequency of twinning in the white population of the United States is slightly more that 1 twin pair per 100 births, with about twice as many fraternal twin births as identical. Older mothers tend to bear more twins, and the recent decline in the frequencies of twin offspring may be the consequence of a reduced reproduction rate of older women. However, in other population groups in which older women still produce children, the twin rate is high. Thus in Ireland, where women tend to marry at an older age, the incidence of twinning is the highest in Europe, 14 per 1,000 births.

There are other differences between population groups. In the black population of the United States the rate is about 1 twin pair per 73 births, or about 14 per 1,000. In one African tribe, the Yorubas, the frequency has been reported to be as high as 1 pair per 30 births. Among Asians on the other hand, the Japanese and the Chinese have a very low rate, with an average of only 1 twin pair per 160 births. It is interesting that these groups differ only in frequencies of fraternal twinning; identical pairs appear to occur with about the same frequency in all populations (Table 17-1). The ratio of frater-

Table 17-1   Twinning rates per thousand births by race of the mothers. (From tables by N. E. Morton, C. S. Chung, and M. P. Mi. *Genetics of Interracial Crosses in Hawaii*, S. Karger, AG, Basel, 1967; D. Hewitt and H. Stewart, *Acta Genet. Med. Gemellol.* **19**:84, 1970.)

	MZ	*DZ*
Indians of Arizona, New Mexico, and Oklahoma	2.0	5.0
Orientals	4.5	2.5
U.S. whites	3.8	7.4
U.S. blacks	3.9	11.8
Johannesburg blacks	4.9	22.3
Nigerian blacks	5.0	39.9

nals to identicals is about 8:1 among the Yorubas, but the direction is reversed among the Asiatics, where the identicals are twice as frequent as the fraternals.

EMBRYOLOGICAL ASPECTS OF TWINNING. At the time of a twin birth the obstetrician usually makes a preliminary determination of the type based on the distribution of the fetal and the maternal tissues that make up the afterbirth. These include the *amnion,* the *chorion,* and the *placenta.*

Early in development of the largely undifferentiated mass of cells, an amniotic cavity is formed within which the embryo will develop. The membrane surrounding the cavity becomes the amnion, which will be the innermost of the several birth membranes (Figure 17-1). Part of the outermost layer of the cell mass forms a chorion, which as development proceeds becomes thinner and forms a second cell layer around the fetus. The placenta is of dual origin, formed at the juncture of the fetal chorion and the maternal tissue. Figure 17-2 shows how these three types of tissue may individually surround a fetus during the prenatal period.

Because both monozygotic and dizygotic twins may develop as separate embryos, they may develop with seperate amnions, chorions, and placentas (Figure 17-3A). MZ twins may have a single chorion (DZ twins never do), and within that chorion may have either double (Figure 17-3D) or single (Figure 17-3C) amnions. Thus it is clear that the identification of twins as monozygotic or dizygotic is not unambiguous unless there is a single chorion. The frequencies of the combinations of single and double chorions are given in Table 17-2. The determination of zygosity is subject to some error (unless the twins are unliked-sexed), and very often twins misclassify themselves, primarily on the basis of the obstetrician's diagnosis at the time of their birth. One study shows that about a quarter of all MZ twins believe they are dizygotic, whereas a fifth of all DZ twins consider themselves monozygotic.

ALGEBRAIC ANALYSIS OF TWINNING FREQUENCIES. It is useful to be able to analyze twin data as to zygosity on a mathematical rather than a biological basis. This makes it possible to extract information from crude birth records and to come to some conclusions about the frequencies of the different kinds of twin births in the overall population.

This approach was suggested by the German physician W. Weinberg many years ago. Basically it is very simple. Because the ratio of male to female births is about 50:50, DZ twins stand a 50 percent chance of being of unlike sexes. On the other hand, MZ twins must be like-sexed. Therefore, if we look at any birth data on twins, we should see an excess of like-sexed twins (coming from the MZ class), and we should be able to make use of this excess to estimate the numbers of the two types of twins.

Suppose, for instance, that in a population of 1,000 twin births there were 200 unlike-sexed twins and 800 like-sexed twins. What proportion of MZ and DZ births would this distribution represent? The answer is very simple. Corresponding to the 200 unlike-sexed (boy–girl) twins, which must be of DZ origin, there should be another 200 twins of like sexes (girl–girl and boy–boy) also of DZ origin. If this total of 400 DZ twins is then subtracted

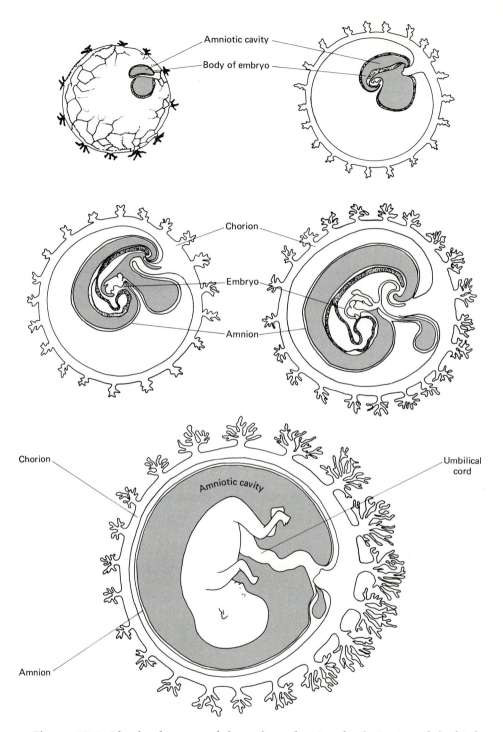

Figuure 17-1   The development of the embryo showing the derivation of the birth membranes. The amnion originates as the lining of the small amniotic cavity and eventually comes to surround the fetus. The chorion surrounds the amnion. (B. M. Patten. *Foundations of Embryology*, McGraw-Hill, New York, 1958.)

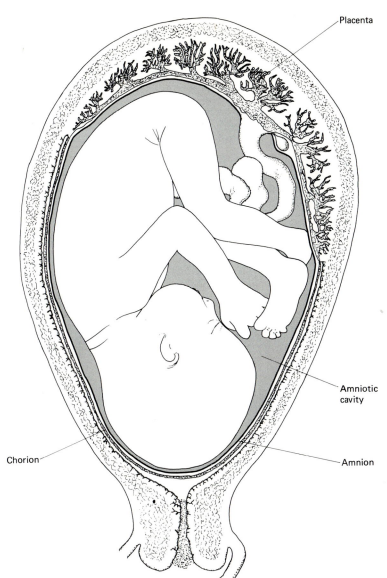

Placenta

Figure 17-2 A more advanced fetus, showing the relationship of the placenta, chorion, and amnion. (B. M. Patten. *Foundations of Embryology*, McGraw-Hill, New York, 1958.)

Amniotic cavity

Chorion

Amnion

from the total of 1,000, we are left with 600 like-sexed twins who must represent the MZ twins. We would therefore conclude that in this particular population of twins, 60 percent are monozygotic and 40 percent dizygotic.

It is possible to look at census birth data and calculate the probable frequencies of MZ and DZ twins, knowing no more than the number of like- and unliked-sexed twins. This has been done by a number of workers to show the change in the frequency of the different kinds of births in the black and white populations in the United States. One such analysis is shown in Figure 17-4. From this analysis it appears that the frequency of MZ twinning in both races does not change appreciably as the age of the mothers changes, but that the frequency of DZ twinning increases markedly as the

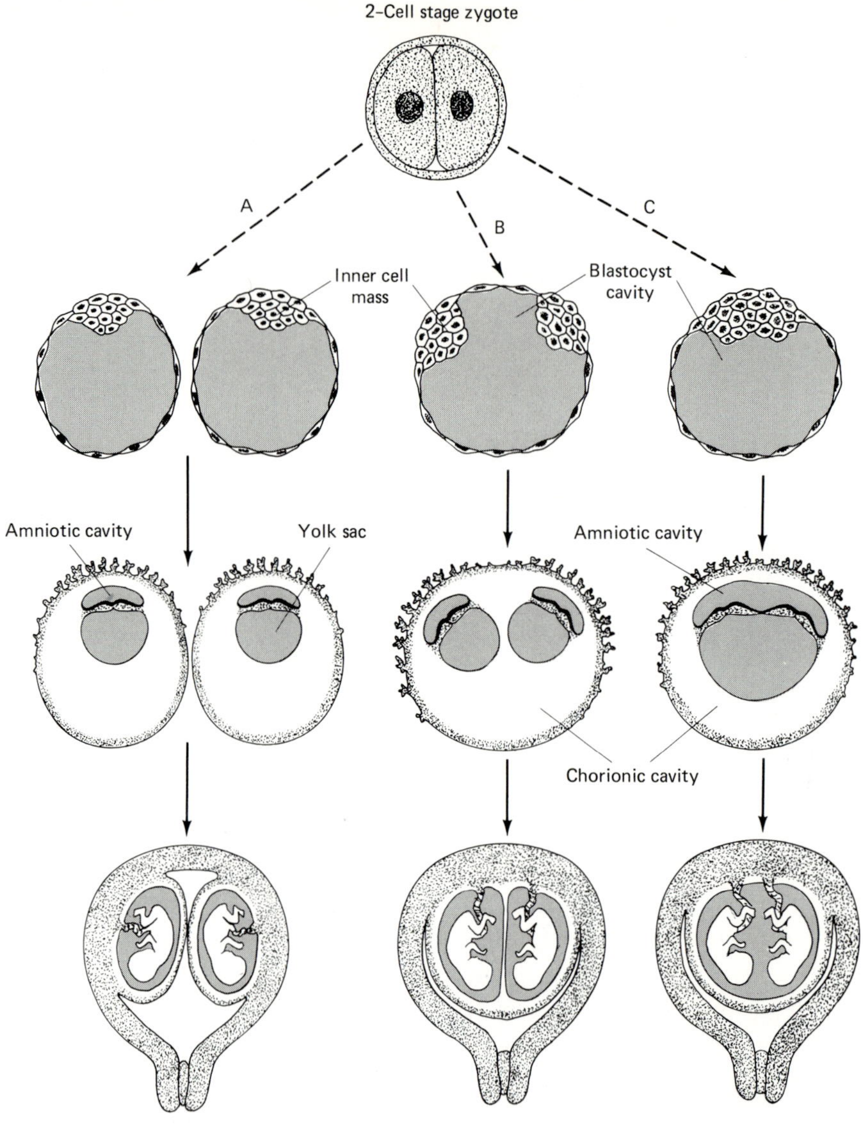

2-Cell stage zygote

Figure 17-3 Schematic diagrams showing the distribution of the birth membranes in monozygotic twinning. *A.* The separation of the two cell masses is complete, so that each embryo has its own amnion, chorion, and placenta. Sometimes the two placentas fuse to form a single one. This sequence is also characteristic of dizygotic twinning. *B.* The two inner cell masses separate, but within a single blastocyst cavity. Each embryo then has its own amnion, but both are within a single chorion. *C.* The inner cell mass does not separate until later, so that both embryos are included within a common amnion and chorion. (J. Langman. *Medical Embryology,* Williams & Wilkins, Baltimore, 1969.)

mother reaches middle age, more than quadrupling in frequency from age seventeen to age thirty-eight, with a small drop thereafter. At the highest level, the percentages of twin births are almost 1.3 percent for the white population and 1.2 percent for the black population. Because DZ twins

Table 17-2 The frequencies with which MZ and DZ twins are found within a single or a double chorion. This table shows that the type of twinning can be definitely established only if there is a single chorion, that twinning type being MZ. (From J. H. Edwards, in *Birth Defects: Orig. Art. Ser.*, ed. D. Bergsma. *New Directions in Human Genetics.* Published by Williams & Wilkins Co., Baltimore, for March of Dimes Birth Defects Foundation, White Plains N.Y., [2]:79, 1965.)

| | **Number of Chorions** | | |
Type of Twinning	Single	Double	Totals
MZ	20%	10%	30%
DZ	0	70%	70%
			100%

occur much more frequently at older mothers' ages but MZ twinning is relatively independent of age, a shift in the age distribution at which mothers produce children will change the frequency of DZ twins but leave the MZ unaffected. For this reason, in recent years in developed countries, there has been a decline in the frequecy of DZ, but not MZ, births.

The algebraic approach to the partitioning of twin births into MZ and DZ categories can be very useful because it makes possible analyses of bulk twin data, unclassified with respect to zygosity. A necesary refinement is to adjust the probabilities of the two sexes to some value other than .5, one

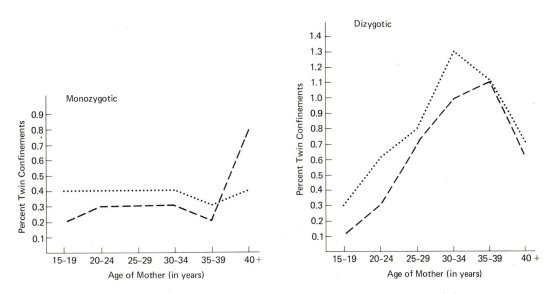

Figure 17-4 The change in frequency of twin births as a function of the age of the mother. The MZ levels seen in blacks (dotted line) and whites (dashed line) appear constant, but the DZ frequencies rise sharply to about age 35. (N. C. Myrianthopoulos. An Epidemiologic Survey of Twins in a Large, Prospectively Studied Population. *Am. J. Hum. Genet.* **22**:611–29, 1970 © 1970 American Society of Human Genetics. Published by The University of Chicago Press.)

obtained directly from the data, usually about .515 for boys and .485 for girls.

Although this algebraic approach does yield valuable information and, for the analysis of crude birth data of the sort collected by bureaus of vital statistics, may be the only way to estimate the relative frequencies of MZ and DZ births, it is nevertheless open to several criticisms. For one thing, it has been discovered that among twins who are known to be dizygotic on the basis of differences in blood groups or chromosome polymorphisms, there is an excess of like-sexed twins.

In addition, such data underestimate the total number of twin pregnancies because the frequency of miscarriages and stillbirths of one or both members of a twin pair is higher than that of single fetuses, and it is not known whether twins classified either as MZ or DZ, or as like- or unlike-sexed, are represented equally in such losses.

INHERITANCE OF THE TWINNING TENDENCY.   Studies of the Yorubas in western Nigeria, with their high twinning frequency, have made it possible to get additional information about the relative contributions of the two parents to the twinning tendency. Because polygamy is practiced in this group, it is possible to compare the frequency of twinning among the various wives of one man. These studies show (1) that mothers who have previously had twins do not have any greater tendency toward twinning than those who have not; (2) that fathers do not contribute to the twinning tendency (because the other wives of their polygamous union do not have any more twins than average); and (3) that mothers who are themselves twins do not have a higher tendency toward twinning than those who are not twins.

The total information on the inheritance of twinning is diffuse and vague. There do exist pedigrees with a concentration of twin births that cannot be attributed to chance alone. On the other hand, the sibs of parents of twins show only a very slight increase above average in the probability of their having twins also. Because DZ and MZ twinning have completely different biological causes, one might expect that a genetic predisposition to twinning would result in a clustering of one or the other but not both types within the same family pedigree. In those few cases in which there is an indication of a slight clustering of twins, they appear to be randomly distributed among the two types.

FETAL LOSS IN TWINNING.   It would be expected that the frequency of twinning would go down in a population in which there is a high rate of fetal loss because, for a twin pair to be born, both members must survive. All populations carry recessive lethal genes that are brought together more often under conditions of inbreeding, resulting in embryonic or fetal death. Because the loss usually occurs very early, the pregnancy may not even be noticed. If two populations are compared, one with an embryonic loss of 10 precent for genetic reasons and the other with an embryonic loss of 20 percent, also from genetic causes, the difference could go unnoticed because there is now no good method of measuring early embryonic loss, and in any case, these losses are only a fraction of the total zygotic losses found in all populations. However, the chance of two DZ twins both surviving in the first population would be .9 × .9, or .81, and .8 × .8, or .64, in the second. Thus the frequency of DZ twinning, with its dependence on the genetic constitution of two separate embryos, would appear to be about 27 percent higher (.81/.64 = 1.27) for the first population than for the second.

It should be pointed out, however, that such loss, of genetic origin, should affect the frequency of DZ twinning only. Because MZ twins have identical genotypes, they would both be affected or not, in the same proportions as single children, with no net change in the MZ twinning frequency relative to total births. Perhaps the quite variable DZ but relatively constant MZ birth frequencies from one population to another (Table 17-1) are a reflection of some fetal loss due to genetic factors. American Indians have a relatively high rate of inbreeding compared with other groups, so it is possible that their lower twin frequency may be explained in this way. Similarly it has been observed that there is a low frequency of trisomy-21 in twin births. An embryo with trisomy-21 has only a one-third chance of surviving to birth, so the chance of two identical twins surviving simultaneously is very much reduced. The data show that among twin pairs with at least one affected by trisomy-21, there are far fewer identical pairs than would be expected.

Diagnosis of Zygosity

PHYSICAL SIMILARITIES. If one were challenged to make a list of specific physical characteristics—facial features, for instance—with a genetic basis, it would be difficult to name half a dozen. It might appear that the total number of such features is quite limited. However, a comparison of the features of two identical twins, often so alike that their parents and sibs may confuse them, leads to the opposite conclusion: that the genetic composition of a person is effective in determining even the most trivial aspects of physical appearance.

It should come as no surprise, then, to discover that measurements of physical attributes show MZ twins to be much more alike than either DZ twins or nontwin sibs. (It should be noted that for all comparisons of MZ twins with either DZ twins or sibs, the latter two must be limited to like-sexed pairs to avoid the obvious difficulties in comparing boys with girls. In all subsequent discussions of comparison it should be taken for granted that both DZ twin and sib data come from like-sexed individuals, unless otherwise stated.) Thus the average difference in height between sibs as they reach the same age is 4.5 cm, or about 2 inches; for DZ twins the difference is about the same, 4.4 cm; but for MZ twins it is only 1.7 cm (Figure 17-5). Fingerprint patterns of MZ twins are quite similar, and on the basis of this criterion alone zygosity can be diagnosed correctly in about 86 percent of all cases. Other physical characteristics show varying degrees of likeness. Weight, which might be expected to be more dependent on environmental influences, is not much more alike in MZ than in DZ twins.

PHYSIOLOGY. Virtually all aspects of physiology are more similar in MZ twins. Blood pressure, breathing rate, changes in heartbeat and breathing after exercise, and electroencephalographic patterns are strikingly similar.

Figure 17-5 A pair of twins, superficially somewhat dissimilar, particularly in height, but alike in all other criteria to the extent that the probability of monozygosity is practically one. This is a nice illustration of the danger of dogmatic generalizations based on appearances only. (S. Pruzansky, M. Markovic, and D. Buzdygan. *Acta Genet. Med. Gemellol.* **19:**225, 1970.)

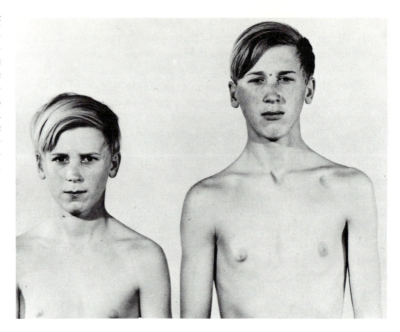

DZ twin girls show, on the average, a difference of about a year in the time they begin menstruating whereas MZ twins differ by less than three months. Curiously, left-handedness, although generally more frequent in twins, appears more often in fraternal than in identical twins.

INDEX OF SIMILARITY. When twins are alike with respect to having a given characteristic, they are said to be *concordant;* if different, they are *discordant.* A convenient index, the frequency of concordance, is calculated by dividing the number of twin pairs in which both have the characteristic by the number in which at least one has the characteristic. Thus, if 83 MZ twin pairs are examined for the presence of club feet and 50 pairs show it not at all, 20 pairs have one member affected, and 13 both, the frequency of concordance in this hypothetical example would be 13/(20 + 13), or .39.

Frequencies of concordance for some common developmental defects are given in Table 17-3. The fact that MZ twins are not completely concordant shows that these defects are not entirely under genetic control, that is, that there is a substantial nongenetic influence. On the other hand, the higher concordance of MZ twins over DZ twins is strong evidence for some genetic component.

BLOOD GROUPS. In addition to the information on the birth membrane and physical similarity (including sex), other criteria may be weighed in the judgment of zygosity. The blood groups are particularly helpful when the parents are so constituted as to produce a variety of phenotypes in their progeny. Clearly if twins are of group O and both their parents are O, the similarity is of no use in determining zygosity, as *all* of the children in that sibship will be of group O. On the other hand, if one of the parents is A and the other B, then the similarity of O children would be helpful because the

Table 17-3 Estimated rates of concordance for some congenital malformations in MZ and DZ twins. (S. Hay and D. Wehrung, *Am. J. Hum. Genet.* **22:**6, 1970.)

Congenital Malformations	MZ Pairs			DZ Pairs		
	Total Pairs	Concordant	Percentage Concordant	Total Pairs	Concordant	Percentage Concordant
Cleft lip, with or without cleft palate	51	9	17.6	84	2	2.4
Cleft palate	15	6	40.0	42	2	4.8
Central nervous system malformations	112	6	5.4	110	2	1.8
Congenital heart disease	69	5	7.2	60	2	3.3
Positional foot defects	99	19	19.2	192	4	2.1
Polydactyly	47	20	42.6	106	6	5.7
Limb reduction deformities	27	2	7.4	32	0	0.0
Down syndrome	5	4	80.0	46	0	0.0

parents, who must be heterozygous AO and BO to produce any O children at all, could produce A, B, and AB as well as O children. Dissimilarity with respect to any one of the blood groups (ABO, MN, Rh, and so on) would lead to the conclusion of dizygosity immediately; identity with respect to all the blood groups would suggest monozygosity with a very high probability. All in all, there are eighteen different loci (mostly blood group loci) that are polymorphic and can be represented by different alleles in DZ twins but, of course, must be the same in MZ twins. Using all of these loci makes it possible to determine zygosity with a probability of misclassification lower than 1 mistake in every 500 cases.

CHROMOSOME POLYMORPHISMS. The new staining techniques that reveal chromosome bands and other longitudinal differentiation have revealed widespread permanent structural variations in otherwise normal chromosomes. When these variants are found in some but not all members of a sibship, they may be used as genetic markers like any other inherited trait, and identical twins should be concordant in this regard also. In one study involving forty like-sexed twin pairs, the determination of zygosity on the basis of chromosome variants agreed in every case with the determination based on the blood groups. Taking into account the ten or so chromosomes of the human that may show polymorphisms, along with their frequencies in the population, it can be calculated that in only one time out of twenty-five should two members of a DZ pair have identical karyotypes.

BIOCHEMICAL IDENTITY.   One interesting test for identity based on subtle biochemical differences has been reported in a reputable medical journal. A trained bloodhound can readily distinguish between sibs and DZ twins but confuses the scents of MZ twins. Unfortunately this sort of test, apparently about 100 percent effective, might be difficult to implement in the average run-of-the-mill genetics clinic. In any case such extreme measures may not be necessary; an experienced observer can make a correct zygosity diagnosis with an accuracy of 90–95 percent on the basis of physical features alone.

ANTIGENIC SIMILARITY.   Other tests for zygosity depend on differences in antigen specificities likely to be found in DZ twins but not in MZ twins. One of these is the mixed-lymphocyte test, described in Chapter 12. If lymphocytes from MZ twins are mixed in a test tube, there should be no effect on mitosis because they should carry the same antigens, whereas the lymphocytes from DZ twins should induce mitosis, just as a mixture of those from sibs does. A more critical test, quite informative but more difficult to perform, involves making reciprocal skin transplants between two sibs. Skin, one of the most difficult of tissues to get to take, is rejected by the nonidentical host in almost every case but is usually accepted when the graft is from an identical twin. Finally, the precise determination of the alleles of the major histocompatibility complex—that is, the HLA-A, HLA-B, HLA-C, and HLA-D factors—is a powerful means of determining genetic identity.

Unusual Twin Events

CHIMERAS.   Very often biologically important observations are made by chance during routine clinical tests. An illustration comes from a blood-typing laboratory in England.

When a certain Mrs. K. in Sheffield was blood-typed, it was found that she had two types of blood cells, about three fifths of type O and the rest A. These results prompted the investigators to ask a simple question: Was she a twin? Her answer was yes, but it was given with some surprise, as her twin brother had died twenty-five years earlier.

Why did the investigators ask her this question? Because the most likely explanation for her condition was that she was one of a twin pair and that during embryonic development, cells of the other twin (group A) became incorporated with her blood-forming system, which happened to be of group O. In fact, by further testing of the fraction of her cells that were A, it was possible to describe the antigenic properties of eight additional blood groups of the long-deceased twin brother. Interestingly he happened to be more similar in these particular groups to an older sister than he was to his twin. The invasion of A cells could not have been very extensive, except for erythroblastic tissue, because those cells must have had a Y chromosome (as they came from a male twin). However, there were no evidences of the masculinization of Mrs. K., expected if the migrating Y-bearing cells had had any developmental impact. In fact, Mrs. K. had married and had had a child.

Several such cases have since been found in which both twins were still alive. Each one had some cells of the blood groups of the other twin and appeared to have normal sex development in cases of unlike-sexed twins. It is likely that the interchange of blood cells occurred rather late in development by way of a fusion of blood vessels. This event is well known in cattle, in which DZ twins interchange cells in 90 percent of the cases; when the sexes are different, the hormones from the male cause the female to develop into a sterile, somewhat malelike type called a *freemartin.* Such an effect of one sex on the other in developing twins is unknown in humans.

When cells of two demonstrably different origins combine to produce one individual, as in the case of Mrs. K., the result is called a *chimera.* This is a distinctly different phenomenon from mosaicism. A mosaic consists of two different types of cells, originally the same at the stage of the zygote, but becoming genetically different subsequently by some event such as mutation, by nondisjunction in one cell line, or by X-chromosome inactivation.

In another case in Austria, a mother of blood group AB had a child of group O. On further investigation of additional blood groups, the HLA alleles, and other biochemical markers, it was shown that her other three children also had alleles inconsistent with her genotype. The staff of the obstetric department at the hospital at which all four children were born ruled out the possibility of any kind of mix-up with other infants at or near the time of birth.

In this case there was no twin sibling, and all her erythrocytes and lymphocytes had the single genetic composition inconsistent with those of the children. However, when the woman's parents were typed, they were found to have the puzzling alleles that the children had. The simplest explanation is that this woman was a chimera, with a different gonadal cell population from that of her blood cell population. Presumably the woman originated from two different fertilized eggs, which, as they multiplied, fused to form a chimera.

How common chimerism is in humans is unknown. Studies of twin pairs for evidences of cell interchange between them suggest a very low frequency; however, it is possible that some cases classified as mosaicism, and explained by mutation or nondisjunction, may in fact be chimeras that have arisen from the fusion of two cell masses. For instance, there are persons some of whose cells appear to be XX and others XY. Possibly there existed two twin zygotes initially, one XX and the other XY. After one twin acquired cells from the other, the second twin embryo might have failed to become a viable fetus or, for that matter, might not have been formed in the first place if the two cell masses fused to give a single embryo. Such a single birth might not be recognized as a case of chimerism, and if subsequent inspection showed two different kinds of cells to be present, the person might simply be referred to as a mosaic.

ABNORMAL DEVELOPMENT.   In some cases, separation of the dividing cells into two different groups to produce twins occurs rather late in development. The embryos may then have some structural feature in common, to produce what have been called *Siamese twins.* Depending on the nature of the common elements the twins may (or may not) be separated safely by surgery. All such twins are monozygotic.

SUPERFECUNDATION. Among the unusual circumstances surrounding the births of twins are those few cases in which it has been shown that the twins must have had different fathers. One such case involved a mother with twin son and daughter. It was shown that the girl twin was of blood group A and MN, inconsistent with the groups of the mother (O and M) and her husband (B and M). However, the boy twin was B and M, consistent with those groups. A boarder at the residence proved to be A and MN, consistent with the groups of the girl twin but not the boy. The twins, then, seem to have had different fathers. This phenomenon, the production of members of the same multiple birth by different fathers, well known in domestic animals, is known as *superfecundation*.

MZ TWINS OF DIFFERENT PHENOTYPES. It is to be expected that if one member of an identical twin pair carries some unusual genetic trait, the other will as well. This expectation is implicit in the term *identical*. Identical twins, both with Down syndrome, or both with Klinefelter syndrome, or both with Turner syndrome, have been reported, as well as several instances of twins who had both Down and Klinefelter syndromes simultaneously.

In rarer instances twins who satisfy all the criteria for monozygosity are strikingly different. Sometimes one appears to be male, the other female. As an illustration of how such cases can be simply explained, consider the following course of events. Mitotic nondisjunction may occur in an XY zygote during early development (Figure 17-6), giving rise to two cell lines, one XY and the other XO (having lost the Y). Later separation of the cell lines into two different masses may occur in such a way as to produce two embryos, one a normal male and the other a female with Turner syndrome, but both of MZ origin. In other cases, MZ twins are like-sexed but one is normal and the other chromosomally abnormal. For instance, suppose that the zygote was XX and that during the early cleavage divisions, an X chromosome was lost during one of the mitoses. A twin pair might result, a female from the XX cells and a Turner from the XO (Figure 17-7). This is actually the most common type of chromosomally unlike MZ pairs. Because the mitotic

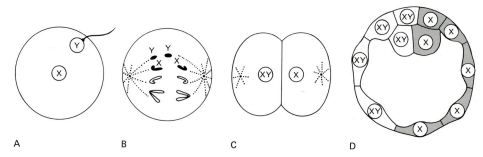

A   B   C   D

Figure 17-6 A hypothetical scheme to account for the origin of a mosaic embryonic mass of cells by chromosome loss during the first cleavage. A Y-chromosome is lost during mitosis so that some of the resulting cells of the XY embryo are XO. If this clump of cells now separates into two groups, one of constitution XY and the other XO, two monozygotic twins may develop, of opposite sexes. (From *Principles of Human Genetics*, Third Edition, by Curt Stern. W. H. Freeman and Company. Copyright © 1973.)

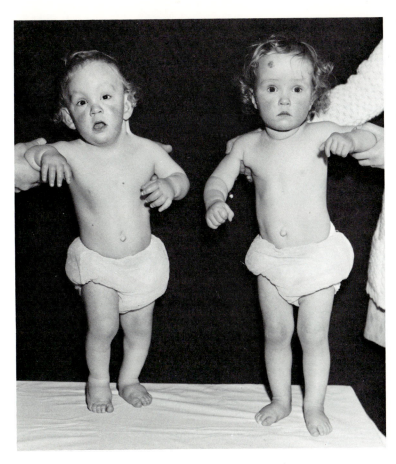

Figure 17-7 A pair of monozygotic twins of different phenotype. The one on the left has the Turner syndrome; the one on the right appears normal. Both girls in fact have both XX and XO cells and are mosaics. The likely method of origin is shown in Figure 17-8. (A. M. Potter and L. S. Taitz, Acta Paediat. Scand. 61:473–476, 1972.)

event gives rise to a clone of cells that tends to be located together, the cell masses separating later preferentially include either normal or abnormal cells, although it would not be surprising to find (as is, in fact, the case) that such different MZ twins are sometimes mosaics (Figure 17-8). Curiously one of the first unlike MZ pairs to be described consisted of a male with short stature whose lymphocytes were XO and a Turner female who was a lymphocyte mosaic for XO and XY cells. It can be surmised that the male was also a mosaic, with some cells also of XY composition. In this case the cells observed almost certainly did not include all the types actually present in the mosaic.

In half a dozen cases MZ twin pairs have been described in which one was normal and the other had Down syndrome. The course of events in this case is probably the following (Figure 17-9): After one or more normal mitoses of the zygote, a nondisjunctional event in a dividing cell gives rise to two different cell lines, one with three chromosomes 21 and the other with one. There are then three different cell lines present. The monosomy-21 cell type reproduces slowly or not at all and is eventually lost. The splitting of the cell masses made up of normal cells and trisomy-21 cells then occurs, with the result that one of them has primarily, if not entirely, disomy-21, and the other trisomy-21.

Figure 17-8 A schematic diagram showing the origin of the phenotypically different but monozygotic twins of Figure 17-7. An X chromosome was lost during the first cleavage division to give two products, one normal XX and the other XO. These both divided by mitosis to produce a group of cells made up of both types. The cell mass then split to form MZ twins, one an XX/XO mosaic with mostly XX cells and the other a similar mosaic with mostly XO cells.

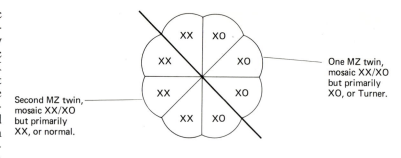

Second MZ twin, mosaic XX/XO but primarily XX, or normal.

One MZ twin, mosaic XX/XO but primarily XO, or Turner.

INFECTIOUS DISEASE.   Comparisons of degrees of concordance (similarity) in MZ versus DZ twins for various disease conditions give some indication of the degree of genetic cause of those diseases (Table 17-4). In these comparisons DZ twins are divided into two groups: like-sexed (DZ_1) and unlike-sexed (DZ_2). In this table the first three pathological conditions show only a slight increase, if any, in MZ over DZ twin types, from which it can be concluded that the genetic contribution to the disease is far outweighed by other (environmental) factors. For the remaining conditions, even those that are known to be caused by infection (such as tuberculosis), there is a definite increase in concordance rates in MZ over DZ twins, suggesting a genetic predisposition toward that disease.

ALLERGY.   For allergic disease a clear genetic component is indicated (Table 17-5). Out of some 500 twin pairs, 76 were found to have at least one affected member. When these were then classified according to concordance and zygosity, most of the discordant pairs were dizygotic and most of the concordant pairs were monozygotic. Furthermore, when the pairs with an allergic member were examined for the incidence of migraine headache, 23 cases were found, compared with only 1 case in a control set of 76 pairs of nonallergic twins. This result strongly implies a basic connection between migraine headache and allergy.

CANCER.   Although there are many relatively rare genetic diseases that are responsible for cancerous growths (neoplasms) affecting specific organs or areas of the body, the possible inheritance of a predisposition to cancer as a more general and widespread disease is a particularly vexing question. There is no doubt that a few families, as rare exceptions in the medical literature, show high incidences of cancer with an inheritance pattern that appears to be that of a simple single-locus effect. Furthermore the probability of a second occurrence in a family in which one member has already been affected is often doubled or trebled over the chance of a new occurrence in a family

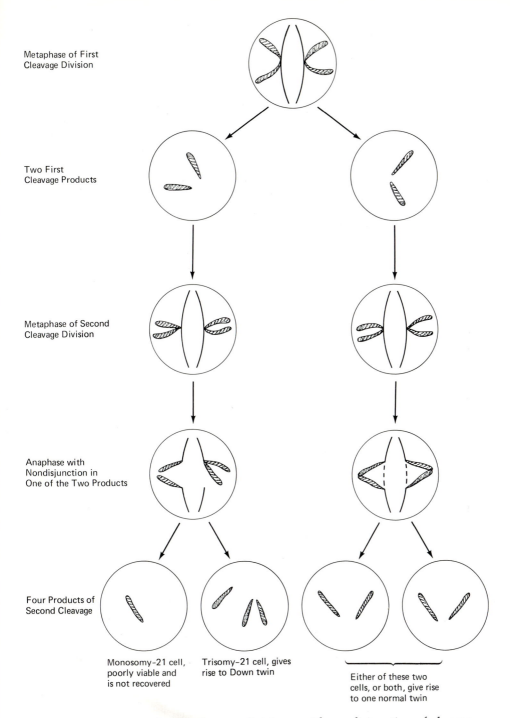

Metaphase of First
Cleavage Division

Two First
Cleavage Products

Metaphase of Second
Cleavage Division

Anaphase with
Nondisjunction in
One of the Two Products

Four Products of
Second Cleavage

Monosomy-21 cell,
poorly viable and
is not recovered

Trisomy-21 cell, gives
rise to Down twin

Either of these two
cells, or both, give rise
to one normal twin

Figure 17-9 A series of early cleavage divisions, with nondisjunction of chromosome 21, giving rise to two MZ twins, one with trisomy-21 and the other normal.

Table 17-4 Concordance rates for various disease conditions in monozygotic twins (MZ), like-sexed dizygotic twins (DZ₁), and unlike-sexed dizygotics (DZ₂). (B. Harvald and M. Hauge, *Genetics and Epidemiology of Chronic Diseases*, H.E.W., PHS Publ. No. 1163.)

Type of Disease	MZ		DZ₁		DZ₂	
Verified cancer, same site	8/160	.050	8/335	.024	4/321	.012
Verified cancer, all sites	17/160	.106	43/335	.128	36/321	.112
Coronary occlusion	20/102	.196	24/155	.155	10/133	.057
Peptic ulcer	30/144	.208	24/208	.115	18/243	.074
Arterial hypertension	20/80	.250	10/106	.094	4/106	.038
Cerebral apoplexy	22/90	.224	16/148	.108	22/179	.123
Tuberculosis	50/135	.370	42/267	.157	36/246	.146
Rheumatic fever	30/148	.203	12/226	.036	14/202	.069
Rheumatoid arthritis	16/47	.340	2/71	.028	8/70	.114
Bronchial asthma	30/64	.469	24/101	.238	22/91	.242
Diabetes	36/51	.706	12/55	.218	10/55	.182

previously unaffected. Indeed, in the case of breast cancer, the chance may increase by as much as ninefold when other members of the family have had the disease. If this seems like a striking increase, it should be remembered that the overall incidence in the population is quite low and that even a ninefold increase does not actually amount to a substantial risk for any one individual. In any case, in line with the evidence from concordance in identical twins (Table 17-4), it can be stated with some assurance that the genetic predisposition to cancer is relatively weak and that the major cause must be elsewhere, a conclusion we reached for other reasons in Chapter 9.

Additional Reading

BULMER, M. G. 1970. *The Biology of Twinning in Man.* Oxford: Clarendon.

LANGMAN, J. 1969. *Medical Embryology.* Baltimore: Williams & Wilkins.

MITTLER, P. 1971. *The Study of Twins.* London: Penguin.

TERASAKI, P. I., et al. 1978. Twins with two different fathers identified by HLA. *New England Journal of Medicine,* **299:**590–592.

TURPIN, R., and J. LEGEUNE. 1969. Monozygotic twinning and chromosome aberrations (heterokaryotic monozygotism). In R. Turpin and J. Lejeune, eds., *Human Affliction and Chromosomal Aberrations.* Oxford: Pergamon.

Table 17-5 Breakdown of twin pairs, in which one member suffers from allergy, by concordance and zygosity. (M. Milani-Comparetti, *Acta Genet. Med. Gemellol.* **19:**240, 1970.)

	MZ	DZ	Total
Discordant	8	30	38
Concordant	28	10	38
Total	36	40	76

Questions

1. Sociologists and psychologists often criticize human geneticists for oversimplifying the biological basis for human traits. To what extent is this criticism justified?

2. How do inbred lines in lower animals serve a useful function in genetic analysis? What common human phenomenon serves as a substitute? Who first recognized the usefulness of twins in analyzing human characteristics?

3. What is the relative frequency of monozygotic and dizygotic twins? With what variables does this relation change?

4. Are twins almost always aware of the particular class of twins they belong to? Why not? How often are they wrong?

5. How does the number of the different kinds of membranes surrounding the fetus differ in the two kinds of twins?

6. Suppose that 1,000 pairs of twins attend a twin convention and that there prove to be 900 pairs of like sex and only 100 pairs of unlike sex. What would your estimate of the frequency of the two types of twins be in the sample, assuming a sex ratio of 1:1?

7. If there is a set of twins in your sibship, and you also have a pair of twin cousins, does this fact make it likely that you yourself will have twins among your children? Would your answer be different if it is specified that the twins were monozygotic rather than dizygotic?

8. Why are there fewer twin pairs with trisomy-21 than might be expected?

9. What kinds of characteristics of twins can be used to make a rather positive determination of the type of twinning?

10. How can a chimera for the blood groups be detected? What is the difference between a mosaic and a chimera?

11. Can twins who are identical by all of the usual criteria still be grossly different physically, even to the extent of being of different sexes? How can this happen?

12. What does the study of concordance in twins tell us about the genetic basis of cancer?

13. A few cases have arisen in which twin pregnancies are revealed at the time of amniocentesis. If one fetus should be grossly abnormal (trisomy-13, for instance), and the other perfectly normal, the parents may be faced with a painful dilemma. What would your decision be in this circumstance?

18

Behavior and Intelligence

Genotype and Behavior

Many persons regard the human at birth as being virtually devoid of behavioral characteristics, except, possibly, for a few innate drives like maternal concern for offspring and interest in the opposite sex. Variations from the norm, according to this idea, are entirely environmental in origin. This point of view is consistent with the idea that "all people are created equal" and that the maximization of opportunity for all and improvement of social conditions will inevitably produce a society free of ignorance, crime, and mental illness. Such a view denies the existence of any substantial genetic contribution in accounting for the differences between individuals. For this reason the burden of proof might rest with those proposing any substantial effect of the genotype. In this section we bring together evidence concerning different behavioral characteristics, some more convincing than others, in favor of a genetic contribution.

There is no difficulty in finding specific cases in which the genotype of an individual affects important behavioral characteristics. A wide variety of genetic conditions lead to obvious differences in the behavior of the affected person. At one extreme are those cases in which the behavior of an individual is profoundly affected by genetically simple conditions, such as phenylketonuria (caused by homozygosity for a single recessive gene) or trisomy-21 (the addition of a single chromosome). In such cases development may be so defective that the deficiency of mental ability affects the individual's ability to perform satisfactorily throughout an entire lifetime.

There are other clearly genetic conditions having effects not quite so drastic but still having a profound influence. Hereditary deafness is one that, as a secondary effect, may reduce language development, academic capabilities, and normal interaction with other people. Other examples may not be so clearly genetic in origin but may be genetic in part. For instance, infantile autism is a severe behavioral defect in which children are highly withdrawn and have severe speech difficulties. Of 11 pairs of monozygotic and 10 pairs of dizygotic twins, of whom at least one was affected, 17 pairs were discordant; the remaining 4 concordant ones were all MZ. None of these concordant sets had any previous history of brain damage, whereas 12 of the 17 from the discordant pairs, both MZ and DZ, had.

Other inherited characteristics operate at a more subtle level to lead to a diminished participation in normal activities. Any genetic condition that leads to a physical disease, lowered vitality, a reduced capability of performing physical endeavors, or an inability to perform adequately in academic life (for instance, poor eyesight) will affect behavior in the long run. It might even be argued that the sex of an individual is a genetic determinant of behavior. These may not seem to be good examples of a genetic component of behavior, however, because they are either too abnormal or too much a part of the normal aspects of ordinary human life.

Instructive behavioral examples can be readily found in domesticated animals. The relative docility of animals bred for human consumption (e.g., cattle and sheep) may result from generations of selection for such characteristics, because unruly members of a herd are removed from the breeding population at an early date. The slavish obedience of the dog to its master must have undergone similar intense selection. Furthermore, in the relatively short period of time (historically speaking) that dogs have been intensively inbred, we have been able to produce different strains with quite distinct behavioral characteristics, and certain strains are known for their specific psychological traits: the terrier is known as an aggressive type, the sheep dog as having a natural instinct for herding, and the German shepherd as a good watchdog, and behavioral differences among strains could be extended to include most, if not all, strains of dogs (Figure 18-1). We would have difficulty, however, identifying such widely differing innate characteristics in the human population, if for no other reasons than that intensive selection for specific characteristics has not taken place and that the intermixing of different groups of people over a long period of time has diluted any tendency toward the concentration of specific characteristics in any one group that might theoretically have been possible.

Differences attributed to nationalities, such as the greater industriousness of one or the stronger emotional makeup of another, are undoubtedly more of a fictional than a factual origin (and are likely to be due primarily to nongenetic factors to the extent that they exist). For evidence of a genetic basis for behavioral characteristics, we must look elsewhere.

Except for the pathological conditions of the sort noted above, giving rise

Figure 18-1 Differences in response of five different strains of dogs to obedience training. From the top, the strains include the cocker spaniel *(CS)*, white-haired fox terrier *(WH)*, beagle *(BEA)*, Shetland sheepdog *(SH)*, and basenji *(BA)*. (J. P. Scott and J. L. Fuller. *Genetics and the Social Behavior of the Dog,* University of Chicago Press, Chicago, 1965 © 1965 by The University of Chicago Press. All rights reserved.)

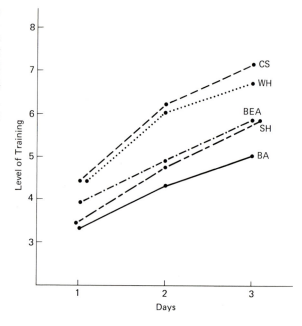

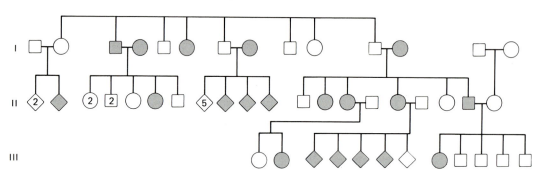

Figure 18-2 A pedigree published in the early days of genetics purporting to show the inheritance of violent temper, with the following explanation:

"Family 27. The patient, III,8, aged 21 years, early showed wilfulness and 'quick temper'. Placed in an institution her 'outbursts of temper' gave trouble. She gets depressed, and longs for the fascination of the life of the demimonde. None of her four brothers shows her temper.

"The father began early to go on sprees; when drunk he is so violent that he is like a madman and has often had to be locked up. One of his sisters is of quiet disposition and so is probably one brother, but three sisters have a violent temper. Of one sister four of the five children have quick tempers and of another sister one of the two daughters has a bad temper.

"Of the father's father not much is known. Two of his sibs had violent tempers; and two of them married women of violent temper and had children with bad tempers. A sister, (I,2) of unascertained morals and temper has one child (one out of three) who has a quick temper.

"The patient's father's mother was quick-tempered.

"The mother is gentle, overlenient, easily led and her parents were easy-going."

(C. Davenport. The Feebly Inhibited. I, Violent Temper and Its Inheritance, *J. Nerv. Ment. Dis.* **42**:593, 1915.)

to severe deficiencies in behavior, there are no known genes that are specifically responsible for honesty or dishonesty, activity or lethargy, violent or calm temperament, or other such traits, even though the early literature on "human genetics" consisted almost entirely of such pedigrees (Figure 18-2).

Abnormal Chromosome Constitution

When an individual is characterized by a deviation from normal diploidy, the first effect to be noticed is a depression of intelligence. The depression is less extreme when the deviation involves additions and subtractions of the X chromosomes, but there is a special explanation in terms of X-chromosome inactivation. Deviations from the balanced autosomal state are responsible for other characteristics that affect an individual's behavior. Because these are in the range of pathological types, we shall not consider them further in this connection.

THE XYY MALE.   Of more interest is the controversial effect of additional Y chromosomes on the behavior of males. For each study showing a correlation of this chromosomal condition with abnormal aggressive tendencies, another can be found denying such a correlation. At this stage it is possible only to present evidence on both sides of this unresolved question.

Studies on XYY individuals suggest that these males are more likely than XY males to be involved in behavioral problems related to the impulsive, exaggerated performance of a deed that is carried out without regard for the long-range consequences. This behavior very often leads to intrusions on other people or their property, followed by arrest and imprisonment. Comparisons of XYY males with ordinary males, however, show that among those institutionalized, the XYY prisoners have convictions more frequently for crimes against property and less frequently for crimes of violence against individuals. It should be noted also that although most non-institutionalized XYY males are indistinguishable from normal males, many exhibit an antisocial behavior pattern that goes well beyond the normal range, and they exhibit this pattern more frequently than do XY males.

Although the incidence of XYY newborn infants has been variously estimated from about 1 in 500 to 1 in 1,000, the incidence of XYY individuals in some institutions for the criminally insane and mentally defective delinquents, based on surveys from Europe, the United States, Canada and Australia, is about 1 in 50. On the other hand, a recent study of 2,538 inmates of all types of institutions in England has not revealed a significantly greater incidence of XYY males than can be expected on a random basis, and a study of 13 XYY males in Denmark also failed to show any greater tendency toward aggressiveness than in ordinary males, although they were incarcerated more often than XY males.

THE XXY MALE.   Males with the Klinefelter syndrome (XXY), of a birth frequency of 1 in 500, are found in psychiatric institutions at a frequency of 1 in 100. They show an increased incidence of mental retardation and are found as mentally retarded institutionalized males with a frequency of 1 in 50. They are therefore much more likely to be characterized by some major behavioral disability than the average person, and when they are classified according to the kind of misdeed responsible for their incarceration, their defect is often one of failure or inability to accomplish some task, a pattern different from that of an XYY male.

EFFECT OF ABNORMAL Y CHROMOSOMES.   In several instances unusually large Y chromosomes have been reported, and at the same time it has been suggested that an aggressive or violent behavioral pattern of the males carrying them is associated with this Y (Figure 18-3). The obvious implication is that perhaps this large Y has been formed by the joining of two smaller normal Ys, either by translocation or by an insertion of one Y segment into another, thus making these individuals effectively XYY. Various morphological Y chromosome types have been described. With quinacrine staining techniques, some are large and stain intensely, whereas others are small and stain hardly at all. In general it can be said that there is no definite association of either the size or the staining properties of the large single Y chromosome in a male with the characteristics of those individuals having extra Y chromosomes. Of course, it is always possible that each of these altered Ys has a different composition, and, irrespective of size or staining, it is in those in which certain loci on the Y chromosome are effectively doubled that we would get the equivalent of two Ys in the same cell.

It should be emphasized that in none of these cases of Y chromosome variations are the tendencies described necessarily expressed. In the vast majority of instances the characteristics of individuals with these abnormal

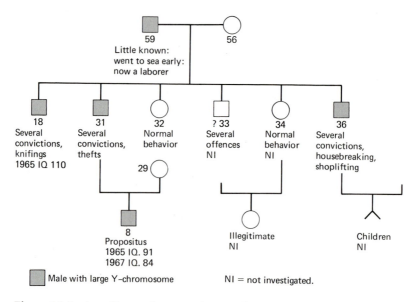

Figure 18-3 A pedigree of recent vintage showing a concentration of antisocial behavior in one kindred. It has been possible to examine the chromosomes of the males indicated by solid symbols, and all have unusually large Y chromosomes. (P. W. Harvey, S. Muldal, and D. Wauchob. Antisocial Behavior and a Large Y-Chromosome, *Lancet* **7652**:888, 1970.)

chromosomes appear to be quite within the normal range, and they do not run afoul of the law or in any way appear grossly different from their peers. In the first place, it must always be kept in mind that a specific genetic condition depends for its final biological expression on the other genes in the genotype, many apparently unrelated, and these are quite different from one zygote to the next. Then, there are innumerable uncontrolled variables that can modify the course of development and switch it from one course to another. Finally—and this is particularly true of behavioral traits—the impact of other persons, social conditions, and so forth during early life must have a profound influence on how a genetic condition potentially affecting behavior will be expressed, or on whether it is expressed at all. It is invariably a gross oversimplification to maintain that a human with a certain genotype will inevitably show some characteristic behavior pattern. Therefore, rather than simply concluding that an extra chromosome may be responsible for abnormal behavior of specified types, we might say that the genetic trait predisposes the affected individual to behave in a certain way, provided that appropriate stimuli are present at crucial points during that person's life. This is a point implicit in all subsequent discussions of behavioral traits.

Psychoses and Socially Deviant Behavior

Many students of gross behavioral abnormalities in humans are of the opinion that mental illnesses are functional, caused by a traumatic disturbance in the individual's life that triggers a mental breakdown, and that the susceptibility is based entirely on the person's environment and not on his or her genetic constitution. Others have suggested that there is an underlying genetic predisposition to psychosis, making certain persons more susceptible than others to mental illness as the result of stressful circumstances. We shall consider the kinds of evidence that have been interpreted in favor of a genetic predisposition toward psychoses.

MANIC-DEPRESSIVE PSYCHOSIS. Manic-depressive psychosis is one in which there exist alternating moods of depression and elation, with periods of normal behavior interspersed; it may consist of cases in which either the manic or the depressive periods predominate. It occurs with a frequency of about 3 per 1,000.

Studies of MZ and DZ twins provide substantial evidence that this illness has some genetic component. Data from six different studies show a concordance of 70 percent for MZ twins, but only 13 percent for DZ twins. In addition, twelve pairs of MZ twins who were raised separately had a concordance rate of 67 percent. Thus, although the lack of complete concordance of MZ twins can be interpreted to mean that environmental factors may play a role in this psychosis, the higher concordance of MZ twins than DZ twins, particularly when raised separately, is good evidence for a substantial genetic component.

Adopted children also provide information on this point. In an analysis of data from Belgium, it was found that when an adoptee had manic-depressive psychosis, in only 16 percent of the cases did the adoptive parents have any evidence of a psychosis, but 32 percent of the natural parents did.

It has been suggested that in some cases this mental disease may be caused by a dominant gene on the X chromosome, in agreement with the observation that it occurs about twice as frequently in women as in men (as women have two X chromosomes and therefore have twice the chance of carrying a rare dominant). A more generally accepted view is that no single genetic explanation adequately accounts for all of the twin and family data, and that the genetic contribution to the illness has a multifactorial basis, with different thresholds in the two sexes.

SCHIZOPHRENIA.   Schizophrenia is a personality disorder in which the affected individual has difficulty distiguishing between reality and the products of the imagination. It occurs with a frequency of about 1 in 100 persons and accounts for about one quarter of all hospitalizations in the United States. When the rates of schizophrenia in the biological parents and close relatives of adopted children who later developed the illness were compared, they proved to be more than 4 times higher in the biological parents than in the adoptive parents.

There is a much higher concordance in monozygotic twins than in dizygotics (25–40 percent versus 4–10 percent). Those who advocate a simple genetic basis that might lead, under stressful circumstances, to a manifestation of the disease have suggested a large variety of mechanisms, including a simple dominant, a simple recessive, and multiple genes, all with variable penetrance. It is also considered quite likely that the environment plays a variable role in determining the onset of the disease, in many cases being the predominant factor.

The enzyme monoamine oxidase, which is an important enzyme in nervous system chemistry and is also present in blood cells, proves to be present in less than normal levels of activity in schizophrenics. That this low enzymatic activity has a genetic rather than an environmental basis is shown by the fact that identical twins are concordant with respect to the reduced amount. In one study of thirteen pairs of genetically identical twins, in which only one of each twin pair was known to have schizophrenia, all had low levels of this particular enzyme. Other biochemical compounds as well have been reported to be deficient in schizophrenics.

In all of these cases a real criticism can be raised with respect to the higher concordance of monozygotic twins because of the greater psychological and cultural similarities of monozygotes over like-sexed dizygotes. Of some help in this connection are those cases of monozygotic twins separated shortly after birth and raised by different families. In fifteen such cases where there was at least 1 affected twin, 10 were concordant and 5 discordant, once again suggesting a genetic basis for the predisposition toward schizophrenia.

In another study of about fifty children who were removed from their schizophrenic mothers before one week of life and put into foster homes, about half developed schizophrenia or suffered from other mental disorders. In a control group of about the same number, only one in five showed any mental disturbance.

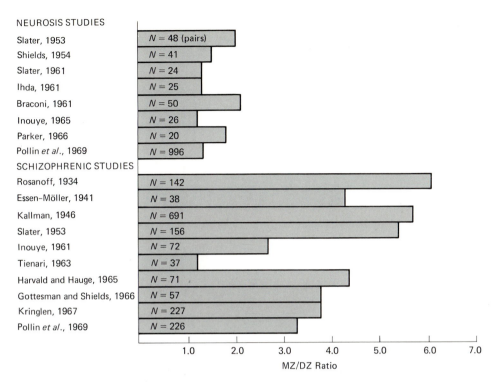

Figure 18-4 The ratio of the concordance rate for MZ twins over DZ twins in a number of studies throughout the world for schizophrenia and other psychoses. (W. Pollin. The Unique Contribution of Twin Studies to the Elucidation of Nongenetic Factors in Personality Development and Psychopathogenesis, *Acta Genet. Med. Gemellol.* **19**:301, 1970.)

A large number of studies have been made of schizophrenia and other psychoses; these agree in showing a much higher concordance rate in monozygotic than in dizygotic twins (Figure 18-4).

CRIMINALITY. Although it may be true that in cases of aneuploidy a person's behavioral characteristics, particularly decreased intelligence, may make that person more likely to become involved in an antisocial act, it seems nonsensical to suggest that such a vaguely defined characteristic as *criminality* might be determined by a simple dominant or recessive gene. Yet this is precisely the suggestion that was made by proponents of the eugenics movement in this country, as well as in certain Western European countries, shortly after the rediscovery of Mendel's work. Mendelian principles were applied directly to humans, with socially undesirable traits unequivocally described as being determined by simple dominant or recessive factors and attributed to the "lower classes" (the immigrants and the poor), to other nationalities, and to other religious and ethnic groups. It is probably not a harsh judgment to conclude that this point of view may have been motivated by a desire to justify the supremacy of the wealthier white Anglo-Saxon Protestants of the period.

Two families, the Jukes and the Kallikaks, are now considered the classic examples of this distortion of genetic principles (Figure 18-5). From their

Figure 18-5 A historically interesting pedigree of part of the Jukes family, illustrating the way in which a number of socially reprehensible characteristics were attributed to the "poor genes" in this family. (Arthur H. Estabrook. *The Jukes in 1915*, Carnegie Institution of Washington, Washington, D. C. 1916.)

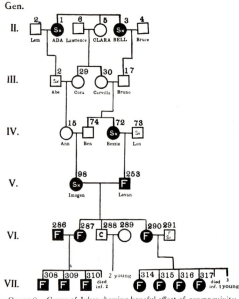

CHART 8.—Group of Jukes showing baneful effect of consanguinity, in a defective germ-plasm.

Sx, licentious.   F, feeble-minded.   C, criminal.

pedigrees it could be seen that not only criminality but also alcoholism, "licentiousness," and other unpleasant characteristics were considered simple dominants or recessives and therefore appropriate for pedigree analysis. It hardly needs to be pointed out that missing from this kind of analysis is the question of the environment in which these less fortunate individuals found themselves and the effects of dire poverty on the human condition. Fortunately pedigrees of this sort are now largely forgotten, except as illustrations of the extremes of the early days of eugenics.

It might be concluded that it would be difficult to find evidence for genetic predisposition to "criminality." This is not entirely true. Table 18-1 lists eight different studies comparing the concordance of criminality in monozygotic and dizygotic twins, where, in each case, one member of a twin pair was known to have been convicted of a crime. In each of the studies there is a greater frequency of identical twins similarly involved than of fraternal ones.

In a study involving about 6,000 pairs of twins born in Denmark in the period from 1891 to 1910, in which both twins survived at least to age fifteen, there was a correlation with respect to conviction for a crime or for a minor offense. If one of the male twins was convicted of a crime, the chance of a monozygotic twin's also being convicted was .53, but if the male twins were dizygotic, the chance was only .22. If one of the male MZ twins was convicted of a minor offense, the probability that the second twin would be similarly convicted was .235, but only .08 in DZ twins.

HOMOSEXUALITY.   Innumerable social and psychological studies have concentrated on the many forces operating early in life that might lead a person to prefer a sexual orientation toward his or her own sex. The possibility that

Table 18-1 Comparison of monozygotic twins and like-sexed dizygotic twins with regard to behavioral difficulties, delinquency, and criminality; data taken from eight different studies. (E. Slater and V. Cowie, *The Genetics of Mental Disorders*, University Press, London, 1971.)

		MZ Pairs		DZ Pairs		Concordance Rate as a Percentage	
		Conc.	Disc.	Conc.	Disc.	MZ	DZ
Rosanoff, Handy, and	Males	80	14	32	36	88	47
Plessett (1941)	Females	39	4	29	35	91	45
Adult criminality, sexes undifferentiated							
Lange (1929)		10	3	2	15	77	12
Le Gras (1933)		4	—	—	5	100	0
Kranz (1936)		20	11	23	20	65	53
Stempfl (1936)		11	7	7	12	61	37
Borgström (1939)	Adult	3	1	2	3	75	40
Yoshimasu (1965)		14	14	—	26	50	0
Hayashi (1967)	Juvenile	11	4	3	2	73	60

there could be a contribution, however small, from a person's genetic constitution is usually not considered. Yet one twin study strongly suggests such a genetic contribution. Of 44 MZ male twin pairs of whom one was homosexual, the other was also; the concordance was 100 percent. On the other hand, of 51 male DZ twin pairs, 13 were concordant and 38 discordant: the concordance was 25 percent. Furthermore there was similarity in the degree of homosexuality. Of the MZ twin pairs, 70 percent had similar degrees of homosexuality, whereas only 15 percent of the DZ twins did. Clearly such concordance can be explained on the basis of the greater association of MZ twins. However, the twins themselves denied that they developed their homosexuality in collaboration with their co-twins and maintained that the sexual behavior developed independently.

It should be noted, however, that some other studies of isolated twin pairs have not shown a similar remarkable concordance. In any case, it is difficult to see how alleles predisposing a person to homosexuality could maintain themselves in the population since such alleles would suffer a progressive decrease in frequency from one generation to the next.

Monozygotic Twins Raised Apart

Analyses of this sort raise the perplexing problem of the extent to which identical twins share not only the same genetic composition but also a common set of childhood experiences, possibly developing an unusual affinity for each other during the period of infantile and childhood development that is then accentuated by overt and subtle tendencies of their parents, peers, and others to regard each pair as a single individual rather than as two separate entities.

Furthermore, when the MZ twins are described as having been raised "separately," it is very often the case that they were raised in different households by members of the same family, or by family friends. In many cases, they attended the same school and were quite aware of each other's existence as a twin. For this reason the case studies of human interest that we now present are of twins who were separated within a month after birth and did not meet until adulthood, when they were brought together by the investigators who wished to compare them.

The similarities in the lives of identical twins raised separately are in many cases so close as to verge on the unbelievable. In the first extensive analysis, nineteen pairs of MZ twins were studied in Chicago in 1936. As one example, Ed and Fred were separated in early infancy and grew up in different cities unaware of each other's existence. At the time that they were reunited, it was found that they were both repairmen at the same telephone company, each had married in the same year to rather similar women. And each had a fox terrier named Trixie!

In a more recent study conducted at the University of Minnesota, two English MZ twins, who were unaware of each other's existence, having been separated at two weeks of age, were brought together for the first time after thirty-eight years. At their first meeting each wore seven rings and had two bracelets on one wrist and a watch and a bracelet on the other (Figure 18-6A). Each had a son and a daughter; the sons were named Richard Andrew in one case and Andrew Richard in the other; one daughter was named Catherine Louise, the other daughter was Karen Louise (and the name Karen was chosen only to honor a relative—the first choice of the parents had been Katherine!).

Then there are the twins, both named Jim, who were adopted as month-old infants by different sets of Ohio parents. Upon being reunited at age thirty-nine, they discovered a number of remarkable similarities. They both had law-enforcement training and worked as part-time deputy sheriffs. Each had a dog named Toy. Each had married twice; the name of the first wife in both cases was Linda and the second wife, also in both cases, was Betty. One had a first son named James Allan; the name of the other's first son was also James Alan. Both drove Chevrolets and vacationed in St. Petersburg, Flordia. They both chewed their fingernails excessively. And they both suffered from hemorrhoids and a combination tension headache and migraine, which started in both twins at eighteen years of age.

The above are just a few of the many anecdotal items about identical twins raised separately. There are many other interesting similarities—the twins who were both great gigglers, those who read magazines from back to front, dipped buttered toast in their coffee, and flushed the toilet before (as well as after) using it, and those who think it humorous to sneeze in a group of strangers. Surely, to a good extent, such a collection of similarities must be selected from an almost infinite number of possible concordances. Nevertheless one is left with the impression that many of these similarities must have some explanation other than chance. It has even been suggested that perhaps they are joined by some mystical bond that only they experience and that makes them inclined to a high degree of concordance (Figure 18-6*B*).

It is generally true that when evidence bearing on human psychological traits is based on twin studies, it is always possible to make an interpretation based on one's own preconceptions and personal philosophy. In fact, it is common to find that otherwise serious and objective scientists become highly emotional and colorful in their choices of words in controversies of just this sort, when unambiguous evidence is minimal.

Intelligence

There are many who believe that all humans are equally endowed mentally at birth and that differences apparent later in life result almost entirely from environmental variables. Others maintain that mental capabilities are entirely determined by the individual's genetic constitution, environmental influences playing a minor role at best. This is not a trivial question. It forms the basis for divergent educational philosophies, which in turn determine the nature of the education considered appropriate for young children. This problem of maximizing educational opportunities in the face of wide variations in academic aptitude has occupied the attention of educational psychologists for many years, and considering the widespread debate on this issue, it can hardly be viewed as resolved at the present time.

The controversy over the extent of the genetic component in determining behavior discussed previously (such as in the case of the XYY male) is mild compared with that surrounding discussions of the genetic basis for intelligence. The difficulty begins with the definition of *intelligence* itself, as it is regarded differently by people in different fields, and even within a given field. In any case it must be a complex, multidimensional aspect of human capability and behavior, with a large number of interrelated factors. Surely certain of these factors must be more important than others—more important in some cultures than in others, more important to some people than to others, and even more important at one time in a person's life than at some other times.

EFFECT OF CHROMOSOME CHANGES. That some persons may have depressed intellectual capabilities because of their chromosomal makeup has been demonstrated repeatedly. One quarter of all patients in a New Jersey school for the mentally retarded possess a chromosome abnormality (mostly trisomy-21), whereas only .5 percent of newborns in the general population show such abnormalities. The mentality of persons with extra X chromosomes (e.g., XXX and XXY) has already been commented on. One of the most frequent types, if not the most frequent type, of inherited mental retardation occurring in the male is found in the Renpenning syndrome. Cultures of leukocytes show that some of their X chromosomes have "fragile sites," points at which the chromosomes may break.

In general it can be stated that whenever there is any deviation from the normal diploid condition, as either a duplication or a deficiency for a whole chromosome or for part of one, that abnormality has a negative effect on

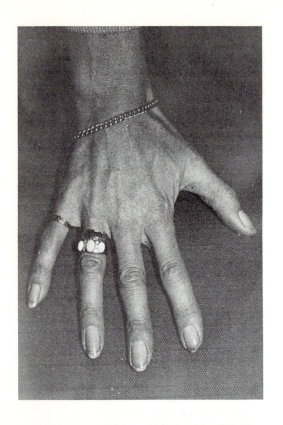
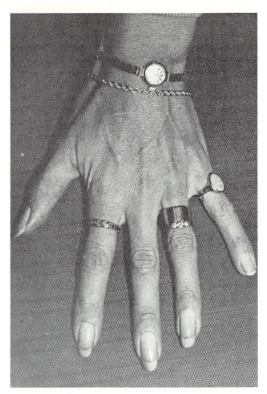
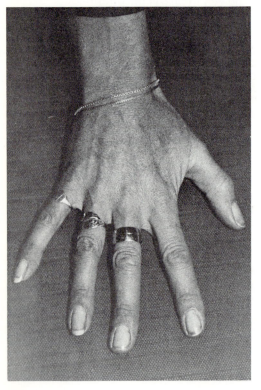
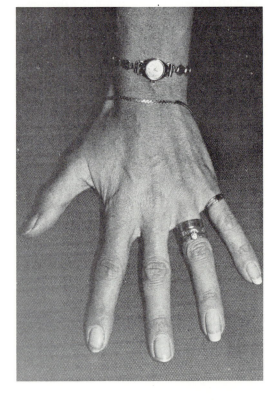

Figure 18– 6   A

B

Figure 18-6 Unusual concordance in identical twins. *A.* On the facing page, hands of two twins (upper vs. lower) separated at two weeks of age, shown at the time of first meeting 38 years later. Each has seven rings, three bracelets and one watch. (With permission of T. J. Bouchard and D. T. Lykken, University of Minnesota.) *B.* The physical similarity of the twins, being genetic, is expected, and so is the similarity of dress, which is environmental. The striking feature to this photograph, however, is something else—without having been given any instructions about how to hold their hands, each twin pair has unconsciously put them in a characteristic position. (Courtesy of K. Fredga, University of Lund.)

mental ability, which appears to be the most finely adjusted of all human attributes.

EFFECT OF DEFECTIVE ALLELES. A detailed analysis of data on mental defect in Colchester, England, shows that 33 percent of all the cases studied could be accounted for by simple genetic mechanisms. Inbreeding, especially incest, was more frequent in the parents of dull and retarded children than in the population at large. Using the degree of inbreeding as a measure of homozygosity for defective genes in the population, the workers concluded that there may be alleles at about 350 different loci that can cause mental defect and that, on the average, each person has about one such allele. It goes without saying that these alleles are recessive and that the person with one such recessive is a heterozygote who is normal!

The Arab population of Israel has a high rate of consanguineous marriages, about 34 percent. Furthermore double first-cousin marriages, where the brother and sister of one family marry the sister and brother in another, and where all four have a common set of grandparents, amount to about 4 percent of the marriages in the sample studied. This high frequency is explained on economic grounds: When a brother and sister of one family marry a sister and brother of another, the "bride price" due the parents of the groom cancels out.

In a series of eight different intelligence tests of children of these two types of consanguineous matings, along with tests of children of unrelated parents, the children of first-cousin marriages scored lower (as a group) than the unrelated children, and the children of the double first cousins were in all eight tests the lowest of all. Because children of first cousins have, on the average, one sixteenth of all their loci homozygous, and those of double first cousins two sixteenths (or one eighth) of theirs homozygous, these results can be considered a strong argument in favor of the existence, in people in general, of recessive alleles that, when homozygous, depress the intelligence somewhat.

In some cases persons heterozygous for specific recessive alleles may appear to have slightly depressed intelligence. This result has been shown in a study of women heterozygous for the PKU allele. In addition, the heterozygous children of heterozygous mothers may also show mental retardation. Males with an X-chromosome allele defective in producing the enzyme ornithine transcarbamoylase are seriously affected, cannot metabolize protein properly, and suffer vomiting, coma, and death. Females heterozygous for this allele have depressed intelligence compared with their sibs who lack the defective allele.

Evidence for the existence of factors that, in some families, are responsible for a generalized depression of intelligence is shown in Figure 18-7, which gives the relative intelligence of the sibs of individuals who are somewhat retarded and the sibs of those who are severely retarded. The sibs of the severely retarded tend to be quite normal; apparently the severely retarded carry some specific chromosomal or recessive defect usually not found in another sib. The mildly retarded, however, tend to have sibs similar to themselves, suggesting that this degree of retardation may in some cases be characteristic of the kinship and so may be genetically determined. An environmental cause, however, cannot be excluded.

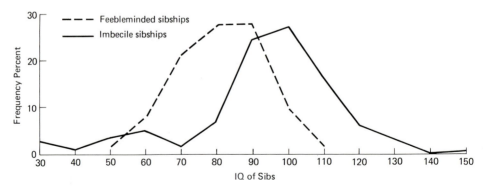

Figure 18-7 A comparison of the relative intelligence of the sibs of the mildly retarded and the severely retarded, with the curious result that the sibs of the more severely retarded are more normal. (J. A. F. Roberts. The Genetics of Mental Deficiency, *Eugen. Rev.* **44**:71, 1952.)

However, discussion of the extent of a genetic contribution to intelligence might better avoid such obvious abnormal conditions and focus on the variations in the population generally considered "normal."

THE INTELLIGENCE QUOTIENT.   The idea of testing large numbers of people for various characteristics originated with Frances Galton, who was the first to make such tests on a large scale, in a systematic manner, and with a definite purpose in mind: statistical analysis. It is Galton whom the school-child must thank for the innumerable batteries of tests that are given (often without warning) throughout the school year. Galton collected data on a large number of physical and psychological characteristics in humans. He examined the biographies of those persons in England considered eminent and discovered that many of them were close relatives. He thought that characteristics leading to eminence had a biological basis and were geneti-cally inherited. He did not seem to be as concerned as we might be today with the possibility of inheritance by way of class or wealth (rather than genes) in nineteenth-century England.

During the early part of this century a French educator named Alfred Binet constructed a test for the purpose of predicting the success of French schoolchildren as they proceeded through academic life. Because the tests he designed were based on the French elementary-school curriculum, they proved, as might have been expected, to be excellent predictors. From these tests came the concept of an *intelligence quotient,* or *IQ:* the age level at which the student performed on the test, divided by his or her actual age, then multiplied by 100. Thus, if a student of age 10 performed at the level expected of a 12-year-old, the IQ would be 12/10 × 100, or 120. At the present time the IQ value of a given test is based on a statistical calculation dependent on the number of persons in that age group with better (or worse) scores.

The IQ is a simple numerical score that is supposed to reflect a person's "intelligence," and, it goes without saying, high intelligence is a character-istic greatly desired by many people and richly rewarded by society. Con-comitantly, low intelligence is considered an inferior attribute. For these reasons, in any discussion of the nature of intelligence a person is likely to accept or reject data or arguments not on the basis of their validity, but on personal prejudices and preconceptions.

Characteristics of living things that show a continuous distribution around an average—height of human males is a good example—tend to be found in a "normal" distribution (Figure 18-8). The greater the deviation from the average, the less frequent the class. This bell-shaped curve of the relative frequencies of the different classes in a typical population with vari-ation applies when there are a large number of chance factors combining to produce the end result. Consider, for instance, the results of tossing a coin a hundred times. While the most likely result is 50 heads and 50 tails, there is a considerable but less likely chance of getting 49 heads and 51 tails, or vice versa, and still less of getting 48 of one and 52 of the other. Should we perform this simple 100-toss test a million times and plot the frequency of the different results, we would come close to obtaining a *normal curve.* The normal distribution is a mathematically sound concept; the science of sta-

Figure 18-8 The normal curve. If every person in the world tossed an unbiased coin 100 times, and the results were recorded on graph paper, the distribution would be much like this. The average number of heads (and tails) would be 50, and the expected frequencies of the other possibilities are given by the relative heights of the bars on the graph.

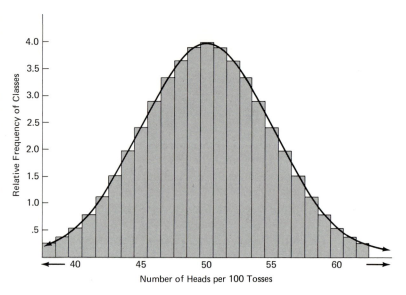

tistics, as it developed, was based primarily on the properties of the normal distribution.

Galton argued that this distribution should also hold for psychological characteristics, because they are based on many complex physiological factors acting in different directions, in the same way that height and weight are determined. We might have some reservations, however, about the meaning of a unit on the IQ scale, as it may not have the same significance as ordinary units of measurement, like pounds or feet. In what sense are two persons who each score only 80 on an IQ test the equivalent of one who can score 160? The data obtained from the tests fit a normal distribution only because the test scores for any age group are adjusted to conform to that distribution, and the initial definition of the IQ has been changed so that it now indicates the extent of the deviation from the mean of this normal curve. In other words, one unit of score at the low end of the scale might represent something completely different from a similar difference at the upper end of the normal distribution. For this reason, arithmetical manipulations of IQ test scores—even such a procedure as taking simple averages—may be open to question, and this qualification should be kept in mind later in this discussion when comparisons are made by means of averages of test scores.

ENVIRONMENTAL FACTORS AFFECTING IQ TEST PERFORMANCE. The scores from IQ tests, the scores obtained prior to any manipulation, may be affected by many factors. Most people would agree that the environment in which a child grows up must have a great influence. Families in which there is a strong emphasis on reading, book learning, arithmetic, and general conversation produce children much more competent to handle IQ tests than do families in which these educational advantages are absent. It is not surprising that in general the children of professional workers score higher on IQ tests than children of other socioeconomic groups. It has been found that seven-year-olds in Scotland, for instance, are an average of eleven months ahead of other British children with respect to reading ability; Scottish parents more often read to their children than do parents elsewhere in Great Britain, and Scottish teachers introduce the elements of reading earlier than do English or Welsh teachers. Therefore, in every discussion of IQ performance, the question of environmental background becomes a crucial one.

Educational psychologists have always been painfully aware of the validity of the criticism that a child's performance on these tests is inevitably modified by the environment and may depend on many extraneous factors. Attempts have been made to construct tests that would be independent of the culture in which the child is found, so-called culture-fair tests (Figure 18-9). Whether such tests completely eliminate the effect of the child's cultural background may be open to question, but they undoubtedly have many advantages over the usual written tests.

MULTIPLE-BIRTH AND BIRTH-ORDER INFLUENCES. It has been found in one investigation that the mean verbal reasoning scores are 100.1 for 48,913 persons of single births, 95.7 in 2,164 cases of twins, and 91.6 in 33 sets of triplets. In addition, when two identical twins have different birth weights, the smaller of the two usually has the lower IQ in early childhood. This difference disappears as the twins grow older. These observations suggest that retarded prenatal development or immature condition at the time of birth is responsible for these low performances. On the other hand, when

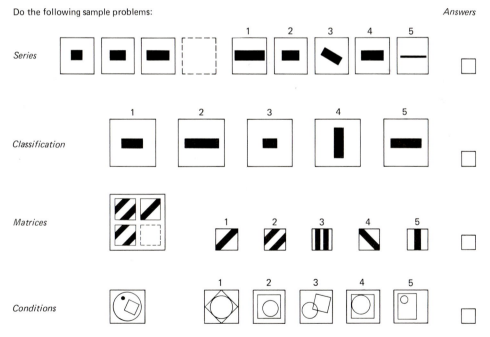

Figure 18-9 A "culture-fair" test, one designed for use by children of all languages and cultures. (From the Culture Fair Intelligence Test, Scale II, © 1949, 1957 by the Institute for Personality and Ability Testing, Champaign, Illinois. Reproduced by permission.)

one of a twin pair is stillborn or dies shortly after birth, the remaining twin scores just about as high as individuals from single births. Perhaps the lower scores of twins may be explained by the observation that they are limited in vocabulary and use primitive sentence construction, a characteristic that becomes more acute as the children age from two to five years, although at later ages they tend more toward the norm for their ages, possibly as a result of schooling.

The performance on IQ tests appears to be highest for the firstborn and to drop with succeeding births (Figure 18-10). This result was found in a study of 400,000 men nineteen years old born in the Netherlands from 1944 to 1947, who were tested primarily for the effects of the Dutch famine of 1944–1945 on mental and physical development. Interestingly the prenatal life famine conditions for many of these men produced no measurable adverse effects on their later intellectual performance.

One explanation for the decline in test performance as sibship size increases is that a child's mental development during the formative years may depend on the degree of close association with adults, as a result of which she or he may more closely conform to adult standards of vocabulary, logic,

Figure 18-10 The drop in IQ performance in successive births; data from men in the Netherlands born during World War II and immediately thereafter. (L. Belmont and F. A. Marolla. Birth Order, Family Size and Intelligence. *Science* **182**:1097, 1973. Copyright 1973 by the American Association for the Advancement of Science.)

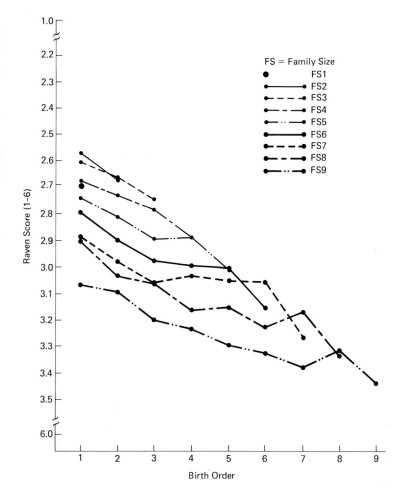

behavior, and performance. However, when the birth is multiple, or when there are several children in the family, the children learn to play and communicate with each other and to satisfy their needs at a less sophisticated level, so that their intellectual development, based on adult standards, may be somewhat retarded. This effect may be accentuated for twins, whose identical age would increase the empathic bond, lessening their dependence on verbal communication.

This steady drop in IQ test performance in younger children was not found in a study of African-Asian children. The explanation here is that in these families, the parents were, on the average, rather poorly educated and that the younger children learned primarily from their older sibs. Because those older sibs were better educated than the parents, the younger children profited more from their sibs than the older sibs had from the parents.

Similarly a study of French children showed a rise, rather than a decline, for younger children. The explanation is that in France the birth rate was low, so that there were greater intervals between births. The older sibs of a young child were therefore much older in France than for the sample studied in Holland, where the birth rate was high. Older and therefore more highly developed sibs would have a positive effect on the young child's intellectual development.

Another attractive hypothesis advanced to account for decreasing test performance as sibling size increases is that it is actually a reflection of the fact that sibship size is greater, on the average, for families with fewer financial, educational and other resources.

EVIDENCE OF GENETIC FACTORS FROM ADOPTED CHILDREN. When children are adopted into foster homes, their performance on IQ tests is, as expected, largely determined by the economic standing and professional orientation of the adopting family, so that children adopted into the higher-income categories do better on these tests than those adopted into the lower-income categories. Children who are adopted into the higher-income categories, however, do not reach the level of aptitude of those who are the natural children in those families. One interpretation of this finding is that the natural children are genetically constituted to do better than the adopted ones, because adopted children represent a more randomly selected (and therefore average) group than the natural children of the professional group. On the other hand, the adopted children may have undergone psychological trauma with the change of parents early in life. Another interpretation, of course, is that the adopted children may be subject to a certain amount of discrimination by their foster parents, who may unconsciously favor their natural children in their allocation of time and attention. Curiously it appears that children who are adopted by the lower economic groups tend to perform somewhat better than the natural children in those same groups. A simplistic interpretation of this effect is that the capabilities of the average adopted child are somewhat superior to those of the natural children, and that even in the less academically inclined environment they are able to do better than the natural children.

Perhaps a more convincing demonstration than the preceding is found when the correlation of the IQs of adopted children is made in a different

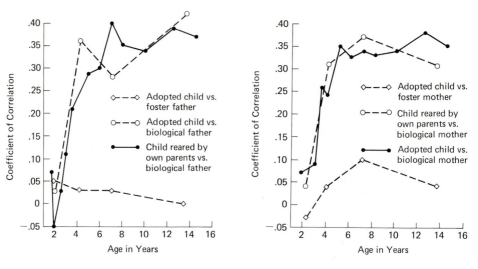

Figure 18-11 Correlation of the IQ of children with the educational level of actual and foster parents, showing the higher correlation with real than with foster parents. (M. P. Honzik. Developmental Studies of Parent–Child Resemblance in Intelligence, *Child Dev.* **28**:215–28, 1957.)

way, by a comparison of the adopted children with their natural parents and foster parents. If the IQ is primarily environmentally determined, the highest correlation should be between the child and the adoptive parents with whom the child grows up. This, however, is not the case (Figure 18-11).

A correlation has been made of IQs of adoptive parents and the natural mother with adopted and natural children, involving 342 women who gave up their children for adoption within three days after birth. The adults were given the Beta IQ test, which primarily measures performance, and the results were correlated with both the full-scale IQ (FSIQ) and the performance IQ (PIQ) of the natural and adopted children. From Table 18-2 it can be seen that the correlation of the adoptive parents with their natural child is higher than the correlation with the adopted child (lines 1 and 2 vs. lines 3 and 4), but that the correlation of the biological mother with her own child, with whom she has had no contact, is higher than that of the adoptive parents with that child (line 5 vs. lines 3 and 4). The investigators who arrived at these results conclude that there is an appreciable genetic component in IQ performance.

EVIDENCE OF GENETIC BASIS FOR IQ PERFORMANCE FROM TWIN STUDIES. The comparison of MZ twins with like-sexed DZ twins can be applied in this case. This comparison, however, leaves much to be desired because, as has been shown earlier, it can be argued that as MZ twins grow up, they may behave alike for reasons related to their obvious similarity. Parents might very well treat MZ twins differently from like-sexed DZ twins, accentuating their likenesses and providing them with more similar environments (dress and so on). Such influences might produce the greater similarity that MZ twins show over DZ twins. It is interesting, therefore, to compare twins

Table 18-2 Correlations of natural mother or adoptive parents with natural or adopted children[a]. (From J. M. Horn, J. C. Loehlin and L. Willerman, *Behavior Genetics*, 1979, **9**:177–192.

	Verbal IQ		Performance IQ		Full-scale IQ[b]	
	r	N	r	N	r	N
Adoptive father and						
biological child	0.36	144	0.38	144	0.42	162
adopted child	0.14	403	0.13	403	0.17	457
Adoptive mother and						
biological child	0.27	143	0.16	143	0.23	162
adopted child	0.15	400	0.17	398	0.19	455
Unwed mother and						
her child	0.34	53	0.25	53	0.32	53
other adopted child						
in same family	−0.07	40	0.15	40	0.07	40
biological child in						
same family	0.19	28	0.13	28	0.11	31

[a]Ns refer to the number of pairings (= the number of children)—the same parent may enter more than one pairing. In the case of twins, the second twin was excluded from the unwed mother = other child comparisons.
[b]Includes 19 biological and 44 adopted children with Binet IQs.

who are misclassified both by themselves and by their parents. Such studies show that the IQs of twins thought to be fraternal (but who were later shown to be identical) were much more alike in IQ scores than those who were thought to be identical (but who were actually fraternal). A particularly valuable source of information might be cases of identical twins who were separated shortly after birth and brought up under different family conditions. The comparison of such twins with like-sexed DZ twins who were brought up within their natural families would minimize the influence that identical twins might have on each other.

Such studies agree that MZ twins, even when raised separately, show an unusually high concordance, in most cases surpassing the similarity of like-sexed DZ twins raised together. In Table 18-3 it can be seen that even when MZ twins are raised separately, they show more similarity than DZ twins raised together. (Some data from these twin studies have been omitted from this table because the validity of the studies has been questioned.)

Nevertheless the basic argument can be approached in a completely different way, with essentially the same conclusions. If the mother of a sibship is an identical twin, then those sibs are as closely related genetically to their aunt as to their mother (Figure 18-12A). On the other hand, if the identical twin parent is the father, then the genetic relationship of the children to the aunt is no greater than their relationship to a person taken at random from the population (Figure 18-12B). A parallel set of comparisons can be made between the progeny of a twin with their uncle, when that uncle is a member of the twin pair (Figure 18-12C) and when he is not (Figure 18-12D).

Table 18-3 The correlations in IQ of various pairs of family members. The data on MZ twins exclude some questionable cases. (From Bouchard, T. J. and McGue, M. *Science*, 1981, **212**:1055–1058.)

	Number of Pairs	Average Correlation
MZ twins together	4,672	.86
MZ twins apart	65	.72
DZ twins together	3,670	.61
Siblings together	26,473	.47
Siblings apart	203	.24
Half-siblings	200	.31
Adopted versus natural children	345	.29

Because MZ twins may have experienced quite similar environments as children, some of the similarity later seen in the children of a twin with their twin aunt or uncle may be environmental in origin. The extent of this congruence may be estimated from a comparison of the children of a male twin or the children of a female twin with the nontwin uncle (Figure 18-12*B*). Since such children share neither genes nor a common environment with the nontwin aunts or uncles, any similarities make it possible to estimate any common environmental component in the traits being measured.

To check on the reliability of such a scheme, researchers studied the fingerprint patterns of each person under investigation. These patterns are known to be quite similar in identical twins, and, as they are determined about twelve weeks after conception and are essentially unaffected by environmental influences, they can be regarded as having a strong genetic basis. The similarity of the offspring to the average of the parents (midparent) was greatest, and the similarity of the nephews and nieces to their genetically unrelated uncles and aunts was least. There was, however, a substantial correlation between sons or daughters and mothers or fathers, which would be expected, and there was an almost equal relation between nephews and nieces and their twin aunts and uncles.

When the process was repeated for an IQ test (the Block Design Test, a nonverbal intelligence test), the results were similar. Calculations of the proportion of variability in the scores that can be attributed to a genetic

Figure 18-12 Simplified pedigrees showing the relation of the offspring of identical twins to their twin and nontwin aunts and uncles. Comparisons of the degree of similarity on various tests between the individuals in such families make it possible to estimate the genetic contribution to the characteristic being studied.

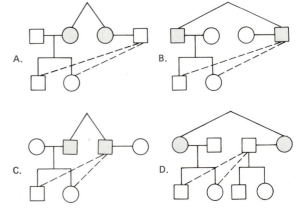

component ranged from 40 to 60 percent. Those who did this work are careful to point out, however, that it is possible to postulate some type of environmental influence that would parallel a hypothetical genetic contribution and would be indistinguishable from it.

This conclusion is sometimes expressed in terms of *heritability,* that is the extent to which the genetic background makes a contribution to the variability of a given characteristic within a specific group. Thus, if identical twins are always completely concordant with respect to some ordinarily variable characteristic (as they might be for eye color), then we would say that the heritability is 1, or 100 percent. On the other hand, if they are discordant as often as two people taken at random from the overall population might be (as in the case for identical twins being affected by certain kinds of cancer), the heritability would be considered 0. The information from MZ twins leads to the conclusion that the heritability of IQ performance in the samples tested is about 45 percent.

However, in view of the variability in testing procedures, the uncertainties in the mathematical treatment of the results, and the cultural and environmental differences separating socioeconomic and ethnic groupings, a serious question can be raised about the general applicability of this figure, except, perhaps, to indicate that there is some substantial genetic component in performance on IQ tests.

Additional Reading

BASHI, J. 1977. Effects of inbreeding on cognitive performance. *Nature,* **266**:440–442.

BELMONT, L., and F. A. Marolla. 1973. Birth order, family size, and intelligence. *Science,* **182**:1096–1101.

BELMONT, L., Z. STEIN, and P. ZYBERT. 1978. Child spacing and birth order: Effect on intellectual ability in two-child families. *Science,* **202**:995–996.

BODMER, W. F., and L. L. CAVALLI-SFORZA. 1970. Intelligence and race. *Scientific American,* **223**:19–29.

BOUCHARD, T. J. 1980. Identical twins reared apart. *Science,* **207**:1323–1328.

CARTER-SALTZMAN, L. 1980. Biological and sociocultural effects on handedness: Comparison between biological and adoptive families. *Science,* **209**:1263–1265.

DAVIS, D. J., S. CAHAN, and J. BASHI. 1970. Birth order and intellectual development: The confluence model in the light of cross-culture evidence. *Science,* **196**:1470–1472.

GERSHON, E. S., et al. 1976. The inheritance of affective disorders: A review of data and of hypotheses. *Behavior Genetics,* **6**:227–261.

GOTTESMAN, I. I., and J. SHIELDS. 1972. *Schizophrenia and Genetics: A Twin Study Vantage Point.* New York: Academic.

GREENBERG, J. 1979. The schizophrenic brain: Rewriting the chapter. *Science News,* **116**:26–27.

HESTON, L. L. 1970. The genetics of schizophrenia and schizoid diseases. *Science,* **167**:249–256.

LEWIN, R. 1977. "Head-start" pays off. *New Scientist,* **76:**508–509.

LUDMERER, K. M. 1972. *Genetics and American Society: A Historical Appraisal.* Baltimore: Johns Hopkins Press.

MEDAWAR, P. B. 1959. *The Future of Man.* New York: Basic Books.

MENDLEWICZ, J., and J. D. RANIER. 1977. Adoption study supporting genetic transmission in manic-depressive illness. *Nature,* **268:**327–329.

MONEY, J. 1975. Human behavior cytogenetics: Review of psychopathology in three syndromes—47,XXY; 47,XYY; and 45,X. *Journal of Sex Research,* **11:**181–200.

MORTON, N. E., et al. 1977. Colchester revisited; A genetic study of mental defect. *Journal of Medical Genetics,* **14:**1–9.

RISTOW, W. 1978. IQ tests on trial. *New Scientist,* **80:**337–339.

SLATER, E., and V. COWIE. 1971. *The Genetics of Mental Disorders.* London: Oxford University Press.

WALLACE, B. 1975. Genetics and the great IQ controversy. *American Biology Teacher,* **37:**12–19.

WILSON, R. S. 1972. Twins: Early mental development. *Science,* **175:** 914–917.

WILSON, R. S. 1978. Synchronies in mental development: An epigenetic perspective. *Science,* **202:**939–947.

WINICK, M., K. K. MEYER, and R. C. HARRIS. 1975. Malnutrition and environmental enrichment by early adoption. *Science,* **190:**1173–1175.

ZAJONC, R. B. 1976. Family configuration and intelligence. *Science,* **192:**227–236.

Questions

1. Do genetic factors ever affect behavioral differences? If your answer to this question is positive, does it follow that the behavior of a person is determined by her or his genotype?
2. What is the evidence that an extra Y chromosome (either in XYY or in XXY) may modify behavior?
3. Is there any evidence of an inherited basis for manic-depressive psychosis or schizophrenia?
4. Who were the Jukes and Kallikaks? What contribution did they make to our understanding of the biological basis of behavioral traits?
5. One kind of consanguineous union is the double first-cousin marriage. Draw a pedigree showing such marriages. Do tests of the progeny from such marriages suggest the existence of recessive genes affecting intelligence?
6. Some of the strongest arguments for a genetic basis of behavioral characteristics come from a study of monozygotic twins. What is the strongest argument against the use of such data?
7. What is the history of the development of intelligence testing?
8. What is meant by *IQ?*
9. Most biological characteristics that show continuous variation from one extreme to another, like height, are distributed in a population roughly according to a normal curve. Can the same be said of intelligence?

10. Is it possible to design tests that cannot be criticized on the basis that a child's family and cultural background may have an influence on his or her performance?
11. Is there a simple explanation for the relatively poor performance of children in large sibships on IQ tests?
12. How can data from adopted children provide us with an argument for a genetic component in intelligence?
13. You have your own view of the relative importance of any genetic basis of intelligence. What kind of evidence would you require to be convinced that your point of view, whatever it is, is incorrect?
14. In 1927 Oliver Wendell Holmes, Justice of the Supreme Court, upheld sterilization of the mentally retarded. After considering one affected family, he stated that "Three generations of imbeciles are enough." Do you agree with him? If so, how would you define mental retardation or imbecility for this purpose?
15. In 1973 two black girls in Alabama, ages twelve and fourteen, were involuntarily sterilized. The argument put forth to justify this action was that they were mentally incompetent. The protests that followed led to reforms that prohibit sterilization without the individual's consent. These reforms, however, raise the question of how one obtains truly informed consent from someone who is mentally retarded. What are your thoughts on this dilemma?

Genes in Populations

Use of Population Analyses

The beginning student ordinarily thinks of the study of human genetics primarily as an analysis of human pedigrees. It may come as something of a surprise, then, to learn that one of the most powerful tools for genetic analysis is an algebraic test of the frequencies of genotypes and phenotypes in the overall population, quite independent of any family relationships.

One would have thought, for instance, that the multiple allelic basis for the ABO blood groups depended on the observation that certain combinations of parents produce some kinds of progeny but not others, that an AB parent and an O parent could produce offspring either A or B but never AB or O. Actually, the one-locus, three-allele hypothesis was arrived at in a completely different way. Originally it was believed that there were two loci involved, an *A* locus and a *B* locus, with a dominant allele at each responsible for the production of the corresponding antigen on the cell surface. Our conclusion that these blood groups are determined by multiple alleles was suggested in 1925, after an algebraic analysis of the frequencies of the four different types in the overall population. We shall make a detailed examination of this analysis later in the chapter. First, we must understand the theoretical basis of population analyses.

The production of progeny representing a new generation is subject to an almost infinite number of complications: the dictates of social customs and taboos; limitations of choice by the stratification of society by class, wealth, ethnic, religious, or other distinctions; and the unpredictability of personal decisions, to say nothing of the biological unknowns in fertilization and development. At first sight it might appear that any attempt to predict frequencies of homozygotes or heterozygotes from one generation to the next would be certain to fail. This is not the case; the mathematical expressions for the distributions of most alleles in populations are extraordinarily simple and equally useful.

The Concept of Random Associations of Alleles

As an illustration, assume that a blood group exists with two alleles, A and A', with no discernible effect and recognizable only after testing with rare antisera. Only a few persons would ever be aware of their genetic composition—whether AA, AA', or $A'A'$—and they would have no reason to let this information enter into their decision to have (or not to have) offspring; their spouses would be equally unconcerned about their genetic composition at this locus. For this reason we can bypass all the complications mentioned above and look at the production of a new individual (as far as this locus is concerned) as the combination of one randomly selected allele present in the sperm with a second random allele present in the egg. If the frequency of the allele A' is 10 percent, the other 90 percent being A, then the chance that the sperm will carry A' is 10 percent, or one tenth. The same would be true of the egg.

Another way of visualizing the random distribution of alleles in the gametes, and the consequences after zygote formation, is to consider the following analogy. Imagine that there are two containers, one with a large number of sperm, the other with an equally large number of eggs. The proportions of sperm with the allele A and sperm with A' in the first container are the same as those in the general population; the same is true of the two alleles in the eggs in the second. Zygotes are then formed by the combination of randomly selected sperm and eggs. If we know the relative proportions of gametes with A and A', we can predict the frequencies expected for the genotypes of the zygotes.

The Hardy-Weinberg Rule

Mendel showed that the ratio of the three genotypes in the F_2 after a cross of AA with $A'A'$ is $1AA : 2AA' : 1A'A'$. The question of the relative proportions of the three genotypes AA, AA', and $A'A'$ from the population standpoint came up only a year or two after the rediscovery of Mendel's paper, when a number of researchers pointed out that the Mendelian ratio of $1 : 2 : 1$, derived from experimental matings, did not describe the genotype frequencies of alleles in the general population, where the proportions of parental genotypes were unknown. In 1908 an English mathematician named G. H. Hardy and a German physician named W. Weinberg independently suggested the algebraic basis for this distribution of genotypes. As we have shown in Chapter 7, the argument is quite simple and straightforward, requiring nothing more than a knowledge of high school algebra. Unfortunately the generally accepted name for this formula is the *Hardy-Weinberg law*, which sounds more forbidding than it really is. We shall refer to it as the *Hardy-Weinberg rule*.

This rule is based on the simple principle of probability that the chance that two independent events will happen simultaneously is the product of their individual probabilities, times the number of different possible arrangements. Take a pair of alleles, A and A'. First, an egg is randomly selected from a large pool containing the proportion p of allele A, and q (or $1 - p$) of allele A'. In other words, the chance of reaching into the pool and getting an egg with allele A is p, and the chance of getting one with A' is q. The same applies with a pool of sperm. The probability that the zygote formed by the two chosen gametes will have genotype AA is $p \times p$, or p^2. The chance of getting an egg and a sperm both with A' is $q \times q$, or q^2. For the heterozygote AA', the chance that the egg has A and the sperm A' is pq. Or the sperm might be A and the egg A', which also has the probability of pq. So the total expected frequency of heterozygotes is $pq + pq$, or $2pq$.

Therefore the algebraic binomial distribution $p^2 + 2pq + q^2$ corresponds to the expected distribution of the genotypes AA, AA', and $A'A'$. Hence it is clear that if we know the frequencies of the alleles at any locus, we can easily calculate the expected frequencies of the different genotypes and their accompanying phenotypes. As an example, if the allele S' for sickle-cell anemia has a frequency of 0.1, and the normal allele S a frequency of 0.9, then the genotypes $S'S'$, SS', and SS should be found in the relative proportions $(0.1)^2 : 2(0.1)(0.9) : (0.9)^2$, or $0.01 : 0.18 : 0.81$.

RESTRICTIONS ON RANDOMNESS. In some cases our assumption of complete randomness in the selection of eggs and sperm is obviously not valid. If, for instance, brunettes have a predisposition to marry brunettes, and blonds to marry blonds, then zygote formation with respect to hair color will not be random in the population. This kind of preference in mating based on phenotype is referred to as *assortative mating*. Similarly, if a group of people form an inbreeding unit and do not freely exchange genes with the population at large, as is the case in many religious and ethnic groups throughout the world, they represent an *isolate* and cannot be included without qualification with the rest of the population. In the same way, it would be unreasonable to consider two sibs as two randomly selected people from the population: we know that they have a specific nonrandom genetic relationship to each other.

It must be true of the vast majority of inherited characteristics, however, that the alleles at the loci responsible are combined essentially at random. For example, most people do not know their own blood group, with respect either to ABO or to any of the other dozen or more cell-surface antigen types. Even if this were general knowledge, it seems unlikely that an appreciable number of impending marriages would be cancelled because the bride's and groom's cell-surface antigens did not seem to match suitably. A population involving random mating with respect to a set of alleles or loci is said to be *panmictic* (noun; *panmixis*) for that set.

Algebra of the MN Blood Groups

The best way to show that this algebraic rule actually holds for human populations is to examine a simple allelic case in which the available data give the frequencies of all three genotypes. The MN blood groups serve as an excellent example. At a single locus there are two major alleles, *M* and *N*. The alleles are *codominant*; that is, the heterozygote MN has both the M and N antigens. Because fewer than one couple in a million are aware of their MN types, assortative mating cannot possibly affect the randomness in this case. With test sera it is simple to determine whether individuals are phenotypically of group M *(MM)*, group N *(NN)* or group MN *(MN)*. This determination has been done for several populations (Table 19-1). For each population studied, a very good estimate of the allelic frequencies within that population can easily be made. Because all of the loci of homozygous M individuals and half of the MN heterozygotes carry *M*, we need simply to take the frequency of M homozygotes and add to it half the frequency of the heterozygotes to get a direct measure of the frequency, *p*, of the *M* allele. Similarly the frequency of the *N* allele, *q*, can be computed as the frequency of N homozygotes plus half that of the heterozygotes. (Alternatively, as there are only two alleles that must add up to 100 percent, $q = 1 - p$.)

Given those two frequencies, it is now possible to use the binomial distribution to arrive at the expected frequencies of both types of homozygotes as well as the frequencies of the heterozygotes. This operation has been performed in Table 19-1 on the line labeled "Calculated." It can readily be seen that there is astonishingly close agreement between the observed and the expected frequencies. The skeptic might wonder if there is not an element of circular reasoning in the performance of this calculation, as the figures on which the agreement is based come from the data themselves. It need only be pointed out that a wide variety of populations might have allele frequencies of .6 and .4—including the extreme case in which 60 percent are *MM* and 40 percent are NN—but only the single distribution of 36 percent *MM*, 48 percent *MN*, and 16 percent *NN* precisely fits the binomial expectation for those allele frequencies.

Table 19-1   The percentages of the MN groups in three populations selected because they represent populations in which the two alleles are found with quite different frequencies.

Population	Number		M	MN	N	p_M	q_N
German	16,000	Observed	30.6	49.1	20.3	.552	.448
		Calculated	30.4	49.5	20.1		
Melanesian	1,148	Observed	1.0	9.6	89.4	.058	.942
(New Guinea)		Calculated	.3	11.0	88.6		
Eskimo	739	Observed	87.4	12.2	0.4	.935	.065
(Greenland)		Calculated	87.4	12.2	0.4		

The preceding illustration in Table 19-1 shows that for the MN blood-group alleles, panmixis can be assumed to exist in the selected populations. Such a demonstration gives us confidence in applying this simple algebraic formulation to many other genetic characteristics in the human population, for otherwise we might be led astray by considerations such as social taboos or stigmas, laws governing marriage, or family relationships that might be imagined to have a significant effect in modifying genotypic frequencies in the population.

Calculation of Gene Frequencies from the Homozygote Frequency

Let us apply the algebra to the case of the simple recessive condition. Cystic fibrosis is a simple autosomal recessive disease that affects about 1 in every 1,600 whites (it is found much less commonly in blacks and Orientals), usually causing death before age eighteen. In total numbers of persons affected, it is the most prevalent genetic disease in the United States. It is therefore important to know how frequently individuals in the population carry this disease in the heterozygous state.

If the frequency of the homozygous recessive, q^2, equals 1/1,600, q is equal to its square root, or 1/40, and p then equals $1 - 1/40$, or 39/40. $2pq$, the frequency of heterozygotes, equals $2 \times 39/40 \times 1/40$, or about 1/20. Thus, from knowing the frequency of the homozygous, affected individuals, we can very easily compute the expected proportion of heterozygotes in the population.

The chance that two individuals (such as a married couple) taken at random from the population are both heterozygous is $1/20 \times 1/20$, or 1/400. The chance that any child produced by such a couple will be homozygous recessive is 1/4, so the overall chance of the production of an affected child is $1/400 \times 1/4 = 1/1,600$, in agreement with the original statement of the frequency of occurrence. (That is, in fact, a circular calculation, because it amounts to saying that $2pq \times 2pq \times 1/4 = q^2$, if p is approximately equal to 1.)

FREQUENCY OF RH INCOMPATIBILITY. In some cases it is important to estimate the frequency with which certain kinds of children are produced. For instance, it is interesting to compare the actual frequency of erythroblastosis fetalis (less than 1 percent) with a theoretically expected frequency based on the calculated number of Rh-positive children from Rh-negative mothers.

If the frequency of women who are Rh$^-$ is 16 percent, then it follows that all eggs of these women—16 percent of the eggs in the population—are potentially capable of producing Rh-incompatible children. Rh$^+$ children of Rh$^-$ mothers, however, must have an R allele from the father. The chance that an allele is R is quickly calculated. If rr homozygotes are found with a frequency of 16 percent, then $q^2 = .16$ and $q = .4$. Because the frequency of the r allele is .4, the frequency of R must be $1 - .4 = .6$. Therefore the chance that an r egg from an rr mother (.16) will be fertilized by an R sperm

(.6) is about 10 percent (.16 × .6 = .096). Because erythroblastosis actually occurs with a frequency of less than 1 percent, we can say that erythroblastosis is found less than one tenth as often as theoretically expected. It was, in fact, the pursuit of the factors that prevented Rh incompatibility from occurring with the theoretically expected frequency of 10 percent that led to the therapy of Rh immunization.

The ABO Blood Groups

The classic example of the usefulness of the algebraic approach in solving genetic problems is the analysis of the ABO blood groups. The earliest genetic data suggested that these groups were determined by two independent loci, an *A* locus and a *B* locus (Table 19-2). Although this was a reasonable hypothesis at the time, it has since been replaced by the generally accepted hypothesis of one locus with three alleles. There appears to be one unambiguous way of distinguishing the two hypotheses, as shown in Table 19-3. An individual of group AB, according to the early theory, having married someone of group O, could produce progeny of groups A, B, AB, or O. According to the three-allele hypothesis, however, their progeny can be only A or B, never AB or O. The distinction seems clearly demonstrable. Curiously, if the progeny of AB × O matings are tabulated (Table 19-4), we see that prior to 1925 the data favored the first hypothesis, because both AB and O individuals are found among the progeny of such matings.

DISPROOF OF THE TWO-LOCUS HYPOTHESIS. We shall now show that the two-locus hypothesis is untenable, using only population data. The algebraic argument is simple. If A and B are independent, it should be possible to treat them as independent algebraic quantities. To do this, we apply the elementary principle of probability that the chance that two independent events will occur simultaneously is the product of their individual probabilities. That means that the frequency of persons in the population of group AB (having both antigens simultaneously) should equal the product of the

Genotype	Phenotype
AABB	AB
AaBB	
AABb	
AaBb	
AAbb	A
Aabb	
aaBB	B
aaBb	
aabb	O

Table 19-2 A listing of the genotypes that would be responsible for the four ABO blood-group phenotypes on the assumption that there are two loci, one responsible for the A antigen and the other for B, with one dominant allele at each locus being sufficient for the expression of its antigen.

Table 19-3 The most obvious genetic test of two different hypotheses of the genetic determination of the ABO blood groups. According to the two-locus hypothesis, the progeny of a marriage of AB × O could give all four phenotypes (if the AB parent happened to be doubly heterozygous), whereas according to the one-locus three-allele hypothesis the progeny could be only A and B.

	Two-locus Hypothesis	*One-locus Three-allele Hypothesis*
Phenotypes of parents	AB × O	AB × O
Possible genotypes of parents	*AaBb × aabb*	*AB × OO*
Possible genotypes of progeny	*AaBb, Aabb, aaBb, aabb*	*AO, BO*
Possible phenotypes of progeny	AB, A, B, O	A, B

frequencies of all those with the A antigen times the frequencies of those with the B antigen. This frequency is not, however, simply the frequency of those in group A times the frequency of those in group B, because AB individuals also have the A antigen and the B antigen, so this class must be added to both.

$$\text{Frequency of AB} = (\text{frequency of all A}) (\text{frequency of all B})$$
$$= (A + AB) (B + AB)$$

This equation means that for any population, irrespective of the frequencies of the blood groups in it, we should get good agreement by making this simple multiplication. This calculation has been carried out in Table 19-5. There is clearly a wide discrepancy: for the French population, the expected frequency of AB, .059, is almost twice as large as the observed, .034, and the discrepancy is also very great for the calculations using the Japanese data.

Clearly something is wrong. We might suspect the data of being unreliable, possibly because of statistical variability in the number used, or because of errors in blood group typing. Perhaps the populations tested were not panmictic within themselves with respect to the ABO blood groups, or there may have been some unknown factor, such as low viability or lethality, causing the observed frequencies to deviate from the expected. Actually the real reason is that the basic genetic assumption on which our algebra is based (i.e., that there are two independent loci involved) is faulty.

Table 19-4 The phenotypes of offspring recorded from matings between O and AB over a period of time, showing how the observations changed as the theory of the genetic determination changed. (A. H. Sturtevant, *A History of Genetics*, Harper & Row, New York, 1965.)

Published Records	O	A	B	AB	Number of Papers
Up to 1910	2	2	2	3	1
Up to 1925	27	80	59	24	18
During 1927–1929	2	228	234	1	6

Group	Number	Frequencies 0	A	B	AB
French	103,242	.441	.435	.090	.034
Japanese	240,204	.323	.364	.228	.085

Table 19-5 Blood group frequencies for several populations, showing that the test for independence of the A and B groups fails.

Calculations To See Whether AB = (A + AB) (B + AB)

Group	A + AB	B + AB	Calculated AB: (A + AB) (B + AB)	Observed AB
French	.469	.124	.058	.034
Japanese	.449	.313	.141	.085

Let us look at the data in still another way. Suppose that there are four kinds of gametes in the population pool: those with the AB factors, those with A, those with B, and those with neither. We may represent all combinations of these gametes in a checkerboard (Figure 19-1) to produce all possible zygotic types. Because we know that the A and B factors are inherited as dominants, we can assume that any zygote receiving one A factor (or more) will have the A antigen, those receiving one B or more will have the B antigen, and those receiving at least one of each will have both A and B antigens. Furthermore we can specify the expected frequency for each zygotic combination by multiplying the marginal frequencies, which we represent algebraically by p, q, r, and s. Because we have four observed zygotic types in the human population, we should find it possible to work back from the observed population zygotic frequencies to the theoretical frequencies of gametes that best fit the data. Examination of the phenotypes in the sixteen cells in Figure 19-1 shows that the following relationships hold:

1. Cell 16 = group O = s^2

2. Cells 11 + 12 + 15 + 16 = B + O

$$= r^2 + 2rs + s^2$$
$$= (r + s)^2$$

Gametes from Male Parent \ Gametes from Female Parent	AB p	A q	B r	O s
AB p	p^2 AB 1	pq AB 2	pr AB 3	ps AB 4
A q	pq AB 5	q^2 A 6	qr AB 7	qs A 8
B r	pr AB 9	qr AB 10	r^2 B 11	rs B 12
O s	ps AB 13	qs A 14	rs B 15	s^2 O 16

Figure 19-1 A checkerboard showing the expected population frequencies of the ABO blood groups based on the assumption that two independent factors, A and B, determine those groups. This formulation leads to a logical absurdity and forces us to assume instead that there are three alleles at a single locus. The squares are numbered for convenience in referring to the expressions given in the text; the expected frequencies are given in small letters and the genetic constitutions in capital letters.

3. Cells $6 + 8 + 14 + 16 = A + O$
$$= q^2 + 2qs + s^2$$
$$= (q + s)^2$$

From the extensive data for the French population in Table 19-5 we see that the frequencies are

$$O = .441 \qquad A = .435 \qquad B = .090 \qquad AB = .034$$

Using the relations shown in Figure 19-1 giving the relative frequencies expected for the O, A, and B groups, and substituting the numerical values for these groups, we can rewrite the equations as follows:

1. Frequency of group O, $.441 = s^2$

2. Frequencies of groups B + O, $.090 + .441 = .531 = (r + s)^2$

3. Frequencies of groups A + O, $.435 + .441 = .876 = (q + s)^2$
From 1, $s = \sqrt{.441} = .664$
From 2, $r + s = \sqrt{.531} = .728 \qquad r = .728 - s = .064$
From 3, $q + s = \sqrt{.876} = .936 \qquad q = .936 - s = .272$

We have now calculated the frequencies of the gametes carrying the A, B, and O factors. There is one class left, that carrying both A and B, occurring with frequency p. However,

$$p + q + r + s \text{ must equal one}$$

therefore $p = 1 - q - r - s$
$$p = 1 - .272 - .064 - .664$$
$$p = 0(!)$$

In other words, the AB gamete type has a zero frequency. Could this be a mistake? We can make a simple check, as we have not yet used any information provided by the AB group itself. If $p = 0$, then the first row and the first column of cells also equal zero. The only source of AB zygotes is from cells 7 and 10, and each of these has an expected frequency of qr, for a total of $2qr$. We have just calculated q to be .272 and r to be .064. Therefore

$$2qr = 2(.272)(.064) = .035$$

This result agrees very well with the observed population value of .034; we may feel reassured that our calculations, and reasoning, are not in error.

(The student is invited to verify that this startling result is true for ABO frequencies from other large populations as well.)

Why does the gamete type carrying both A and B together have a zero frequency, according to our calculations? It is possible to think of reasons. Perhaps gametes with A and B together are lethal or do not produce viable zygotes. This, however, is not an acceptable explanation because such a loss of alleles would have led to the gradual reduction of the A and B alleles in the population over many centuries, so that they should now be eliminated. Other complicated genetic explanations can be similarly devised, and similarly refuted.

The correct explanation is also the simplest: Gametes with both A and B simultaneously are nonexistent because in AB individuals who theoretically might produce such gametes, the A and B factors are always found on homologous chromosomes and so they separate from each other during meiosis. Furthermore, as they are not found together on the same chromosome strand by crossing over to give a single chromosome of constitution AB, we can infer that they are at the same locus, that is, they are allelic. Finally, because there is a third property, O, which is an alternative to A and B, we further deduce that there is a third allele, O, necessary to complete the genetic explanation.

In this way, from population data alone and without recourse to pedigree studies, we can conclude that the early two-locus hypothesis is incorrect and that three alleles at one locus fit the data better. It was a simple calculation like this that caused F. Bernstein in 1925 to challenge the two-locus hypothesis and propose instead the one-locus, three-allele hypothesis.

VALIDITY OF THE THREE-ALLELE HYPOTHESIS. Subsequent work has proved Bernstein to be correct; published data after 1927 show a precipitous drop in the number of AB and O offspring from AB × O parents (Table 19-4). Just why the early results included so many impossible progeny is not at all clear; perhaps the antibodies available for testing in those days were not as discriminating as those available now, or perhaps the technicians responsible for typing the erythrocytes regularly misclassified individuals of group AB as being either A or B, or groups A and B as being O. In any case, this shift in the data after 1927 to agree with the one-locus, three-allele hypothesis is a very nice example of the closer agreement of experimental results with theory when the observer "knows" that certain results are impossible and that other results are consistent with an accepted theory. In retrospect it appears that some of the early workers made calculations on the expected frequency of AB from the total frequency of A (A + AB) multiplied by the total frequency of B (B + AB), just as we have done here, and demonstrated the internal inconsistency of the data. Such calculations, however, remained a puzzling curiosity until Bernstein provided the alternate hypothesis of three alleles at one locus, which proved to be consistent with the known population frequencies of the four blood groups.

At the present time the algebraic test for consistency is considered a necessary step in the formulation of any new genetic hypothesis, when such a test is feasible. Any genetic interpretation, however well it might agree with pedigree data, cannot be considered acceptable if it involves algebraic inconsistencies in a population analysis, as the two-locus hypothesis of the ABO blood groups does.

Equilibrium of Allele Frequencies

TIME REQUIRED FOR EQUILIBRIUM TO BE ESTABLISHED. If one population of humans with one set of allele frequencies meets another set and complete panmixis occurs, the genotypes of the next generation will be found in the proportions $p^2 AA : 2pq AA' : q^2 A'A'$, where p and q are the allele frequencies in the combined population. Therefore, in this case, it has taken only one generation to reach Hardy-Weinberg equilibrium.

In actual practice the stratifications of populations prevent complete panmixis. Data from European countries with superficially homogeneous populations show different blood group frequencies from one section of a country to another, with each of the sets of data internally consistent with the Hardy-Weinberg equilibrium, a relic no doubt of the historical immobility of populations. Adding together data simply because they are taken from the same national or ethnic group can lead to disagreement with the theoretical expectations of the frequencies of the genotypes. For this reason the populations used for calculations in this chapter have been chosen from a specific province (Nord) in France and one city (Miyasi) in Japan. On the other hand, it is surprising that studies of gene frequencies in highly heterogeneous populations, such as the whites of New York City, show reasonable approximations to the Hardy-Weinberg expectations.

There are few special circumstances in which the equilibrium is delayed for one or more generations, even with panmixis. These cases are found when our simple analogy of gametes randomly selected from pools of each sex having the same allele frequencies breaks down. For instance, if a group of males all of type M should mate with a group of females all of type N, then the F_1 would necessarily all be of group MN; the alleles would not be in equilibrium in the F_1, but would be in equilibrium in the following generation.

EQUILIBRIUM FOR LINKED LOCI. It is a logical extension of the Hardy-Weinberg rule that, with panmixis, equilibrium will readily be reached for more than two alleles at the same locus, as in the ABO blood groups, and even for sets of alleles at independent loci, equilibrium is reached rather rapidly. It may not be obvious that similar considerations apply to two different loci located on the same chromosome.

As an example, some individuals have light colored hair and blue eyes and most of the rest have dark hair with brown eyes. The other combinations, light hair with brown eyes and dark hair with blue eyes are much less common. It is a common mistake to attribute this association of hair and eye pigmentation to linkage, that is, to postulate that there is a hair pigmentation locus and an eye coloration locus, both located on the same chromosome. According to this idea, most people would have the alleles for light hair and blue eyes or for dark hair and brown eyes because the chromosomes carrying those loci would generally have the "light" set or the "dark" set.

People who are blue-eyed brunettes or brown-eyed blondes would therefore carry the less common recombinant chromosomes. Were this in fact the case, after a number of generations (the exact number depending on the closeness of the two loci on the chromosomes and the randomness of mating), crossing over in heterozygous individuals would give rise to recombinant chromosomes, some carrying the brunette and blue alleles together, others the blond and brown, and eventually the alleles at the two loci would be randomly associated in any population study. Any apparent association in the population at that point would have to be attributed to some other cause—in this case, to a single locus responsible for a generalized pigment reduction affecting both hair and eyes. The test provided by the Hardy-

Weinberg rule is a very powerful tool for many genetic hypotheses, but it is useless by itself in determining linkage between two loci. For that type of analysis, specific pedigree data are necessary.

LINKAGE DISEQUILIBRIUM.   Let us take the case where there are two (or more) alleles at one locus and another syntenic set of two (or more) alleles at a second locus. Whatever the relative frequencies of the combinations were initially, crossing over occurring regularly over a period of time will randomize all alleles at both loci with respect to each other, or, in other words, the frequencies of the allelic combinations will reach equilibrium. If it is not reached, the alleles are said to be in *linkage disequilibrium* (LD).

More specifically, if an allele A at one locus occurs with a frequency *p* (and all other alleles at that locus together total *1-p*) and an allele B at a different locus occurs with a frequency *q* (once again, all other alleles at the second locus totaling *1-q*), then the combination AB should be found with the frequency *pq*. Linkage disequilibrium exists when the observed frequencies of the allelic combinations do not agree with the simple algebraic expectation.

We can illustrate this with an actual example. Over 600 persons in northern Germany were categorized for the HLA antigens A, B and C and, by family studies, the actual haplotypes were determined. It was found that the frequency of the A1 allele was 16 percent and that the B8 allele was 10 percent. One would expect, then, that they would be found together on the same chromosome in 1.6 percent of all cases. In fact, by family studies, it was found that 8 percent of all chromosomes in that population carried the A1B8 haplotype.

Such deviations from the algebraic expectations have two ready explanations. The two specific alleles and their products may interact in such a way as to produce a disproportionately beneficial result. Selection would therefore favor the two alleles simultaneously, and their combination in a haplotype would occur more frequently than expected, even though crossing over would tend to randomize the alleles in each generation.

Another explanation is that the population under study is of relatively recent origin, having been formed by the amalgamation of two populations with grossly different haplotype frequencies, and that not enough time has passed for equilibrium frequencies to have been reached.

In the German study, 101 cases of apparent LD between the A, B and C loci were examined. Those loci most closely linked (B and C, see Figure 12-4) showed LD in 70 percent of all combinations, whereas those most distant (A and B) were in LD only 23 percent of the time. The workers considered that the reasonable explanation for most of the cases was that the population being studied had been formed from two smaller populations in the not-too-distant past. According to this hypothesis, the two initial populations had different haplotype frequencies and the alleles were now slowly reaching equilibrium in the combined population with the most distant loci approaching it more rapidly than the more closely linked ones. The A1B8 haplotype, however, deviated unusually from expectation and is suspected of being held at a high level by some as yet undetermined positive selection pressure.

CONSTANCY OF EQUILIBRIUM FREQUENCIES. A corollary to the principle that equilibrium is usually arrived at within one or at most a few generations is the idea that such an equilibrium, once attained, will be repeated in each subsequent generation with the same frequencies of genotypes. The English mathematician Hardy developed the simple algebraic expectations after a statistician asked why, if brown eyes are dominant to blue, they did not take over the entire allelic population, resulting in the inevitable elimination of blue eyes. That query clearly resulted from confusing the meaning of *dominant,* which simply describes the phenotype of the heterozygote, with some erroneously imagined genetic advantage of that particular allele in the course of evolution.

Factors Changing Allele Frequencies

SELECTION. In a panmictic population the genotypic frequencies shift when the relative frequencies of the alleles change. By far the most important cause of changes in allele frequency is *selection,* operating differently on different genotypes. If one of the genotypes is "superior" (i.e., produces more offspring), then the alleles in that genotype will increase at the expense of those in other genotypes. The converse argument holds for an "inferior" genotype. In this way *natural selection* will increase the frequency of advantageous alleles and decrease that of deleterious ones.

MUTATION. *Mutation* from one allele to another will, theoretically, change gene frequencies, but mutation rates are so low (approximately 1 per 1 million) that the effect of mutation every generation will not be as great as that of selection. The great significance of mutations in a population is that they must provide the genetic variability on which selection can act.

GENETIC DRIFT. When a population is very small, the gene frequencies in one generation may not coincide with those in the next because random selection of a small number of gametes from the population of gametes may just happen to result in a higher or lower percentage of an allele than in the previous generation. This is basically a statistical, or chance, phenomenon and is known as *genetic drift*. Because genetic drift is relatively insignificant in populations as large as, say, a thousand people, this chance variation can have played an important role only when the human population consisted more of smaller isolates than it does now, although, to be sure, genetic drift may still be effective in certain small groups—religious or ethnic isolates, for instance.

These factors affect gene frequencies in a population over long periods of time. It should be noted, however, that except for a few cases such as sickle-cell anemia, in which one of the genotypes has a much reduced viability, they do not produce detectable deviations in the Hardy-Weinberg equilibrium at any one time. Our calculations of expected frequencies of genotypes—of the blood groups, for instance—are based on data too limited to show any effects of these additional disturbances.

Additional Reading

HARDY, G. H. 1908. Mendelian proportions in a mixed population. *Science,* **28:**49–50.

MOURANT, A. E., A. C. KOPÉC, and K. DOMANIEWSKE-SOBCZAK. 1976. *The Distribution of the Human Blood Groups and Other Polymorphisms,* 2nd ed. London: Oxford University Press.

NOVITSKI, E. 1976. ABO blood groups and the Hardy-Weinberg equilibrium. *Science,* **191:**478.

STERN, C. 1973. *Principles of Human Genetics,* 3rd edition. San Francisco: Freeman.

STURTEVANT, A. H. 1965. *A History of Genetics.* New York: Harper & Row.

Questions

1. The expressions showing the expected frequencies of the three genotypes *AA, Aa,* and *aa* are p^2, $2pq$ and q^2, where p is the frequency of *A*, q the frequency of *a* and where $p + q = 1$.
   (a) If the frequency of cystic fibrosis in a certain population is 1 in 2,500, what is the frequency of the recessive allele?
   (b) About how frequent are heterozygotes?
   (c) What proportion of all children are produced from matings of two heterozygotes? If one quarter of their children are homozygotes, what is the expected frequency of affected children?
   (d) The figure arrived at in (c) is about equal to the value in (a). Is this outcome reasonable?

2. Suppose you read in a text that about 1 black out of 10 is heterozygous for the recessive allele for sickle-cell anemia, and that 1 in 4,000 is homozygous. Are these two statements consistent with each other? What is wrong?

3. If 36 percent of a population is of group M, how many would you expect to be of group N? Of group MN?

4. Make a list of characters for which there is probably some degree of assortative mating. Make another list for which assortative mating probably does not occur.

5. Can you suggest several reasons that large populations might divide themselves into isolates?

6. Is the population of the United States panmictic?

7. How is it possible to show that populations of humans found in different parts of the world are panmictic?

8. Would populations that consisted of one third of each of the two homozygote and one third of the heterozygote classes be panmictic?

9. Using the analogies of gametes found in a common pool from which specific alleles are selected at random, we can readily see that irrespective of the frequencies of genotypes in any generation, the Hardy-Weinberg equilibrium will be set up after one generation of panmixis. Is it necessary to show that this equilibrium will continue for the next generation?

10. If the frequency of the recessive allele responsible for a specific type of albinism is 1 in 20,000, what is the frequency of heterozygotes in a population? Would you use the same calculation if the question referred to the frequency of red or green color blindness in a population?
11. Can you show very simply, without the use of complicated mathematics, that the population data for the distribution of the ABO phenotypes do not agree with the hypothesis of two independent loci, an *A* and a *B* locus?
12. Why was it necessary to await the mathematical analysis of population data to refute the two-locus hypothesis of genetic determination of the ABO blood groups, although its inaccuracy should have been immediately apparent from the kinds of children produced in marriages of AB × O?
13. What are the factors responsible for altering the frequencies of alleles from one generation to the next? Why are such changes of any significance?
14. What is meant by linkage disequilibrium? How can it be demonstrated?

20

Genetic Variation in Human Populations

The most superficial examination of people from different parts of the world forces one to conclude that there is considerable diversity from one group to the next. Furthermore these differences must be largely genetic, rather than environmentally caused, because the migration of one group from its native land to another part of the world, as Orientals to the Americas, does not alter the essential physical characteristics of that group. Our interest here is both in establishing that these genetic differences in human populations involve a vast array of allelic variation and, also in considering the various mechanisms by which such variation must have arisen.

Human Population Differences

THE MEANING OF RACE. It is convenient to use the word *race* to refer to human population groups occurring in different parts of the world. Traditionally this word has referred to the physical differences by which one group of people distinguish themselves from another group, independent of the extent of the genetic basis for these differences. We shall use the more precise definition of race provided by Curt Stern: a race is a group, more-or-less isolated geographically or culturally, who share a common gene pool and who, statistically speaking, are somewhat different at some loci from other populations.

413

Of the total number of loci possessed by the human species, numbering somewhere about 100,000, all but a trivial number of their alleles must be common to all groups. There are a small number of alleles found primarily in one population and rarely in another. The skin color differences among African, Oriental, and European populations are obvious examples, but a list of such allelic differences would represent a minute fraction of the total loci. There is no case known in which any allele of significance is present in one racial group and not in others.

The physical differences that anthropologists use to separate the human species seem unambiguous and suggest that racially distinct groups may carry substantially different sets of alleles at a number of loci. Thus a simple question can be asked: Are two persons taken at random from the same ethnic group likely to have the same alleles more often than two persons from two different ethnic groups? A study was made involving eighteen polymorphic loci (ten blood groups, three blood serum proteins, and five enzymes), all loci showing polymorphisms, that is, loci with two or more alleles occuring with frequencies of a few percent or more. From this study it could be shown that the genetic differences between races are small compared with the genetic differences occurring within individual races.

The word *race* has a distinct usefulness in the discussion of human populations; it is unfortunate that over the years it has developed connotations of prejudice and intolerance to the extent that some workers (particularly in the social sciences) are reluctant to use the word at all. However, avoiding the use of the word would require either the substitution of a synonym or the omission of any discussion of the genetic variability that characterizes different segments of humanity.

Darwinism and Neo-Darwinism

When Darwin proposed his theory of natural selection by the survival of the fittest in the 1860s and showed how this theory could account for the process of evolution, one essential step was missing from his otherwise convincing argument. Darwin knew nothing about the nature of the hereditary material or how it might be transmitted from one generation to the next. This was a serious conceptual gap, because it was not at all clear how a new beneficial change could maintain itself in the population. On the contrary, it seemed reasonable to imagine that it would be diluted by half in the immediate progeny, by another half in the next generation, and so on, disappearing rather quickly. Darwin's ideas about the mechanism of inheritance, based on a biological communication system, whereby the different cells of the body send messages to the germ cells, was essentially Lamarckian; that is, it was fundamentally based on the assumption of the inheritance of acquired characters. Although this idea may appeal to the biologically unsophisticated who are willing to accept it as a valid explanation for such peculiarities as the length of the neck of the giraffe (who must reach to eat the leaves of a tree), it is a hypothesis for which there is no scientific support despite serious attempts by many conscientious scientists to confirm it.

DE VRIES'S "MUTATION THEORY." It was not until about 1900 that Hugo de Vries, working in the Netherlands, pointed out the possibility that the sharp, clear-cut morphological changes he saw in the evening primrose might provide the variations on which natural selection could act. In retrospect it appears that de Vries was quite incorrect in his interpretation of the details of this phenomenon. In fact, almost all of the apparent mutations he found were gross chromosomal abnormalities—trisomies, duplications, and deficiencies—but he did introduce the word *mutation* and correctly interpret its importance in evolution.

The work with *Drosophila* started by T. H. Morgan in 1910 led to the accumulation of large numbers of new spontaneous mutant types. Although these new mutations affected virtually every aspect of the flies' morphology, in some respects they were similar. In the first place, these new mutations were almost without exception recessive, and, second, from the developmental standpoint the mutational change had, in fact, a relatively slight effect on the total animal. No mutational change caused the sudden conversion of one species into a new one or, for that matter, from one type in a species into another very dissimilar type. These two characteristics—recessivity and relatively slight developmental effects—were fashioned into a modified theory of evolution known as *neo-Darwinism,* according to which the developmentally small, usually recessive mutational changes arising spontaneously and affecting parts of the animal at random were transmitted in a Mendelian fashion and constituted the variability on which natural selection could then act. Both of these properties of mutation were consistent with the observation that evolution must be a very slow process, for although evidence for it could be unambiguously derived from a large body of biological and geological data, the instances in which speciation appeared to have been noted within recorded time were so few as to appear to be exceptional textbook cases rather than illustrations of a universal phenomenon.

Factors Affecting Allele Frequencies

MUTATION. Mutation, broadly defined to include chromosomal changes such as inversions and duplications, is the ultimate source of all genetic variability in a population. Genes and chromosomes are quite stable and remarkably effective in maintaining their precise individuality over a large number of replications. Thus, in humans, where there may be from thirty to fifty successive mitotic divisions in the germ cells in each generation, only 1 gene in 100,000 to 10 million, roughly, suffers an obvious change after that many replications. Of course, the vast majority of such random mutational changes are either neutral or deleterious in their immediate effect on the organisms.

What is the frequency of potentially beneficial mutations? Most workers in the field make a rough guess of something like one out of every thousand mutational changes, but this is pure speculation. It may be argued that such mutations must be rare because mutation as a recurrent phenomenon would have produced such beneficial alleles many times in the previous history of the species, allowing for their earlier incorporation into the genotype of that species. Of course, mutations that at one time may have been deleterious or neutral may be beneficial at a later time, when superimposed on a new genetic background, or under new environmental conditions. Even so, the overall frequency of beneficial mutations must be very low.

The time of action of new mutational changes may be critical. In a single-celled organism with no differentiation of cellular structure, a new mutant need function only under a limited set of cellular conditions. However, in multicellular organisms with a vast variety of tissue types, any beneficial mutational change would have to operate in precisely the right tissues. In other words, in addition to a potentially beneficial mutational change, which is an event of low frequency in itself, that change must also become operational in the right place at the right time. For instance, consider the case of a uniquely new mutational change that results in the synthesis of a hitherto unknown digestive enzyme in an animal. This enzyme would have to be produced in cells specialized in the digestive tract, where it would confer an advantage to the animal. However, if its action were not limited to these cells and also extended to other cells in the system, the net result could be disastrous to the animal, which might suffer from having its own cells digested away. Thus the problem of location and timing in highly organized multicellular systems adds an additional probability of low value that decreases still further the chance of incorporation of a potentially beneficial change.

SELECTION. The all-powerful force of evolution is *natural selection,* to which every individual of every species is exposed every second of its lifetime and by which, operating through the phenotypes of the individuals, specific genotypes are either discriminated against or allowed to produce more than their fair share of progeny. This process leads to a change in gene frequencies in favor of those that are "best fitted." Although the force of selection is directly effective at the level of the organism and of population groups and acts indirectly on the genotype and even more indirectly on specific genes, our understanding of how selection operates is sharpened if we assume that selection is operating directly on specific loci and overlook momentarily the obvious complication that an organism is more than the sum of its individual genes. By far the best example of the operation of selection on a single locus in humans is the high frequency of the alleles responsible for those abnormal hemoglobins that confer resistance to malaria.

SELECTION AGAINST A RECESSIVE. The most common form of mutational change is that of the deleterious recessive. That all diploids normally carry such deleterious recessives has been shown repeatedly in experimental organisms. For humans these recessives can be demonstrated by a comparison of the viability of progeny of first-cousin marriages with the viability of progeny in the general population. It is very important to understand that all people of all ethnic groups carry their fair share of potentially damaging genes. These are commonly referred to as the *genetic load,* that is, the burden placed on future generations by our deleterious genes.

Because most recessive alleles of low frequency are found in heterozygotes, selection against homozygotes reduces allele frequencies slowly, the speed decreasing as the allele drops in frequency (Figure 20-1). This slowness serves as a strong argument against any program, involving any expense or effort, that would attempt to eliminate recessive alleles in a population by curtailing the reproductivity of the homozygotes.

Natural selection operates constantly to reduce the frequency of deleterious alleles. They would be eventually eliminated completely except that mutation replaces the lost ones. When the rate of loss of recessives by homozygosity is equal to the rate of gain by mutation, an equilibrium condition for that allele is reached, whereupon the frequency remains stable. This relationship is most dramatically illustrated by the sex-linked recessives with a severe phenotypic effect where the frequency of affected males approximately equals the mutation rate.

SELECTION FOR A RECESSIVE.   We have already noted that selection in favor of beneficial recessives is relatively ineffective because selection for a recessive depends on the production of the homozygote, which appears only very rarely for alleles of low frequency. Figure 20-2 shows the extremely low rate at which a gene with a hypothetical advantage of 50 percent would be incorporated into the population. It may be wondered why, if such selection is going on at a large number of loci, we do not see evidences of it in the population at the present time. Possibly our analytic tools are not adequate to identify cases in which a specific allele confers some definite advantage to the homozygous individuals.

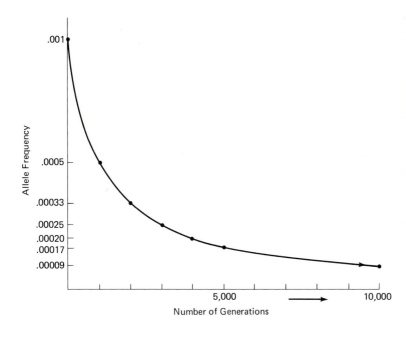

Figure 20-1 Graph showing the extremely slow decrease in frequency of a deleterious recessive that, when homozygous, completely wipes out the reproductivity of the affected individual. Starting with an initial allele frequency of one per thousand, it will take 1,000 generations to reduce it by 50 percent and 10,000 generations to reduce it to about 10 percent of its original frequency.

Figure 20-2 Selection for a recessive that endows the homozygote with 50 percent greater reproductivity than normal. Even with this high positive selective value, it will take approximately 2,000 generations to increase the allele frequency from .001 to .002.

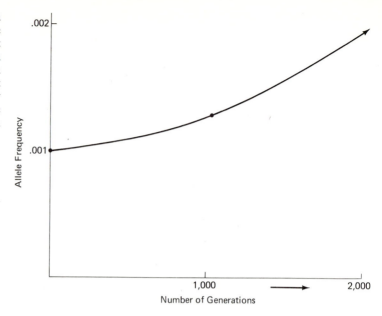

SELECTION AGAINST A DOMINANT. Selection against a deleterious dominant is immediate and decisive. If its effect is drastic, then selection eliminates the dominant allele immediately, either by the loss of the individual carrying it or by his or her failure to produce progeny carrying it. Some unknown proportion of early abortions must be caused by new, dominant, lethal mutations.

SELECTION IN FAVOR OF A DOMINANT. On rare occasions a beneficial dominant mutation may arise. When it does, its frequency in the population will rise rapidly, depending on the extent of the greater reproductivity of the individuals carrying it (Figure 20-3). Incorporation into the population will be even swifter if the homozygote for the new dominant allele has a higher fitness than the heterozygote. It should come as no great surprise that the only case of an apparently beneficial dominant is one in which the homozygote is actually less fit (i.e., sickle-cell anemia), this latter detrimental feature having prevented the complete incorporation of that allele into those populations.

MUTATION PRESSURE. The fact that most changes appear to be recessive created a dilemma in evolutionary thought. If virtually all mutations of evolutionary significance were recessive, then natural selection would have to act on the homozygotes, a class of very low frequency. Thus, if one mutational event occurred in a diploid population of size 500,000, the gametic frequency of the allele would be 1 in 1 million, or 10^{-6}; the expected frequency of homozygotes for that specific allele would, in later generations, be the square of that allele frequency, or 10^{-12}, that is, once in every million million individuals, or, if each generation of that species consisted of 500,000 individuals, once every 2 million generations, as $5 \times 10^5 \times 2 \times$

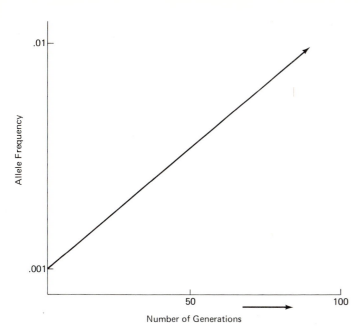

Figure 20-3 The immediate and rapid rise in frequency of a beneficial dominant. In the case chosen, the dominant allele is assumed to have a reproductivity only 10 percent greater than normal. In a few (about 100) generations, it will increase in frequency in the population tenfold.

$10^6 = 10^{12}$. It should be noted that for most species, other than large mammals, 500,000 must be a gross underestimate of population size, and for each increase in population size by a factor of 10, the incidence of homozygotes for a specific mutated allele must decrease by a factor of 100. Further, regardless of how much better this new hypothetical homozygote should be, it would disappear into the gene pool, to reappear some thousands or more generations later with an imperceptibly increased frequency of the homozygote.

Several extensions to the general theory took care of these points. For one thing, allele frequencies might gradually change by repeated mutation of one allele to another. Thus, if the predominant allele at a locus, A^1, mutates to A^2 regularly, over a period of time the frequency of A^2 will increase at the expense of A^1. This change in frequency caused by mutation is called *mutation pressure*.

GENETIC DRIFT. Although the total number of any species may number well into the many millions or billions, the population is subdivided into smaller interbreeding groups, with some migration between them. These subpopulations gain or lose some alleles or some genetic combinations, so that each is slightly different from the next. Migration of a few individuals from one group to the next allows for a testing of different alleles in different genetic backgrounds. At the same time, those groups that have acquired some beneficial alleles may expand at the expense of other groups, which may then be at a slight disadvantage. This gain and loss of genetic variability may be promoted by a small population size because when the breeding population is small, alleles are more likely to change frequency just by chance. The tendency of small populations to have shifting allele frequencies is known as *genetic drift*.

REGULATORY GENES IN EVOLUTION.   Comparative studies on the similarities between humans and other higher primates have shown that there are relatively few differences in protein structure between these groups. In fact, more often than not, quite distantly related species even carry the same array of alleles, such as those for the ABO blood groups, that humans do. On the average, only about 1 percent of the DNA of a human is different from that of a chimpanzee; the rest is identical.

How, then, can we account for the extensive differences between these two groups? One current explanation is that many of these morphological differences result from mutations that modify the timing and action of regulatory genes. By a simple shift in the time at which a gene, or a set of genes, is triggered into action, an entirely new set of developmental reactions result that can have profound effects on the final product. This hypothesis has an additional point in its favor. From our studies of genetic effects in humans we know that alleles that modify development are likely to be dominant (as opposed to alleles that modify enzymatic activity, which tend to be recessive). As we saw earlier, selection for a dominant allele with only slight beneficial effects is immediate and rapid, unlike that for new recessives, even if they are highly beneficial to the homozygote.

In summary, we can be sure that the genetic structure of the human species is changing, although at a rate not obviously perceptible to us. Although it is certain that the gene pool is not static, that some alleles are increasing in frequency and others are decreasing, for all practical purposes it may be regarded as being in equilibrium. The major factor in altering gene frequencies is *selection*. *Mutation* is constantly introducing new alleles into the population in each generation. In some populations that are small or consist of very small isolates, *genetic drift* may be important because chance may determine whether certain alleles will persist or be lost.

Other Aspects of Evolutionary Change

HETEROSIS.   When two inbred lines of plants or animals are crossed with each other, the F_1 often exceeds either of the two parents in physical vigor. This is known as *hybrid vigor* or *heterosis*. It is best known in its commercial application in corn, where by trial and error with various lines of highly inbred corn it was found that certain crosses could produce hybrids with very high yields. (As a matter of practical procedure, four inbred lines may be used to get two hybrids, which then produce further hybrids.) It has been suggested from time to time that hybrid vigor might play a role in human crosses. A few studies comparing physical characteristics (such as height) of previous generations with the present one suggest an increase not entirely attributable to improved nutrition and therefore possibly the result of the recent breakdown of isolated breeding groups. Some students of history interested in breeding patterns in different societies relate the blossoming and subsequent decline of civilizations to a change in heterozygosity, but there is little to support the hypothesis that humankind might now be improved, in physique, mentality, or any other way, if outbreeding increased, with a consequent increase in heterosis. Neither is there any reason to believe that an increase in inbreeding that would decrease heterozygosity and reveal more recessive traits would benefit humankind. Perhaps the amount of genetic variability within each human population group is so great that whatever advantage might be promoted by heterosis is already found within those populations.

POLYMORPHISMS. Variation with respect to blood group and histocompatibility antigens has been mentioned earlier. Polymorphisms exist for other gene products and are well known in those cases in which enzymes have been subjected to structural analysis, usually by electrophoretic mobility. More often than not, it can be shown that within a class of active enzymes there are some with demonstrably different mobilities; such equivalent enzymes are referred to as *isozymes (iso meaning equal)*. Isozyme polymorphism has been found in virtually every organism investigated for its presence and has provided new insights into the relationship of organisms to each other and of different population groups within species. It is tempting to speculate that these polymorphisms represent one kind of genetic variability involved in evolutionary progress. Another point of view is that these polymorphisms are neutral changes that do not modify the activity of the enzymes in question to the extent that they would be of any great evolutionary importance. In any case these allelic polymorphisms do vary in frequency from one population group to another, and because related groups are usually more similar to each other than unrelated ones, they provide a powerful tool for the anthropologist interested in determining the interrelations of various groups of people.

POLYPLOIDY AND GENETIC REDUNDANCY. In the course of evolution there must occur from time to time an event that increases the number of loci in the genome; humans, for example, must have more genes than bacteria have. A simple and effective means of accomplishing such an increase is found in plants, where a doubling of the diploid chromosome number may produce a viable and fertile tetraploid individual. The establishment of a stable tetraploid line in plants is aided by three conditions not found in the higher animals. First, a single chromosome-doubling event in the premeiotic cells of a plant can give rise to both pollen and eggs with that doubled chromosome number, and these gametes, being located on the same plant, can produce tetraploid individuals, which, again producing gametes of both sexes, may perpetuate the polyploid line indefinitely. Second, a sex chromosome mechanism such as the XY combination in humans is usually absent. When sex chromosomes are present in varying numbers in polyploids, they cause the development of intersexes and other sterile or poorly fertile types that make it more difficult for an isolated chance incidence of polyploidy to establish itself as a viable vigorous polyploid line. Third, vegetative reproduction in plants by runners, tubers, and so forth allows a single individual with an unusual chromosome number to proliferate without running the hazards either of meiosis or of the reproduction of deleterious sterile types produced when a gamete from a polyploid combines with one from a diploid.

In animals such as humans, in which the two sexes are represented in different individuals and in which sex is genetically determined, polyploidy

Figure 20-4 The value of duplications in evolution. *A.* Under ordinary circumstances, if a gene, *R*, which is essential to the organism, mutates to another form, *X*, with a completely different function, the new mutant cannot be incorporated because the homozygote, now deficient for *R*, will be lethal. *B.* However, if the locus is duplicated elsewhere, alleles at one of the two loci may mutate to a novel allele, *X*, with a new function because of the presence of another set of *R* alleles.

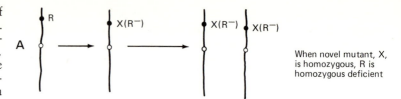

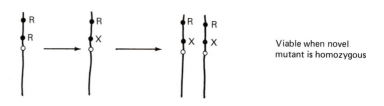

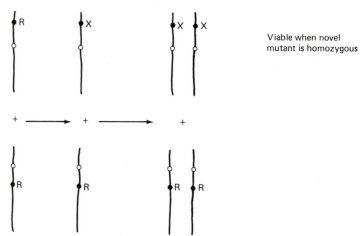

is not a suitable means of adding to the gene pool. In any case, as we saw earlier, polyploid fetuses in humans are aborted early. Additions to the total number of loci must come primarily from the occasional accident of duplication of loci on single chromosomes. Such duplication makes it possible for one set of loci to mutate in order to undertake a completely new function, at the same time that the second set retains the original function, which itself may be essential to normal development (Figure 20-4).

Primate Variability

TASTE SENSITIVITY. It is interesting that primates other than humans show gene diversity not unlike that of humans. About 70 percent of American whites have a strong revulsion to the bitter taste of a substance called *phenylthiocarbamide* (PTC); the rest find it relatively innocuous. The latter appear to be homozygous for a simple recessive allele found with a frequency of about .55. A large number of different primates have been tested for their ability to taste PTC. Chimpanzees appear to show clear bimodal distribution of the ability to taste the substance, just as humans do, and in one pedigree two nontaster parents produced only nontaster offspring, similarly to humans. In a small sample of spider monkeys (fewer than twelve), no tasters were found, whereas all marmosets were tasters. It is not known why this polymorphism should be present in the primates, but it has been suggested that it is connected with their diet. Certain plants, relatives of cabbage and cauliflower, have a particular taste related to that of PTC and are known to affect thyroid disease in humans. One suggestion is that the effect of certain nutritional factors on the thyroid is the important element in maintaining this taster polymorphism in various members of the primates.

BLOOD GROUPS. Large-scale tests have been made on the primates for their blood group antigens. In most cases, the common human antigens A, B, M, and N can be found in some if not all of the different primates investigated. It should be kept in mind, of course, that the investigations involve a test for agglutination of the erythrocytes of the lower primates when tested by the specific antibody active in humans, and this test is no proof of the absolute identity of the antigens in both cases. However, for purposes of argument, we will assume that they are the same.

A large number of primates have been tested for the blood groups; the results are summarized in Table 20-1. Different primate groups are quite different, and no single antigen of the ABO group is present in all primates. The gorillas are rather interesting in that their erythrocytes have neither the A, the B, nor the H antigens, although the saliva shows a presence of B and H group substance, but not A. The MN antigens appear to be present in most of the primates (including chimpanzees and gorillas) but absent in others (some gibbons and orangutans), and one allele is present and the other absent in still other forms.

Table 20-1 The presence of the ABO antigens among primates other than humans. The numbers tested are given for the first three listed; only the presence or absence is indicated for the remaining five.

Primate Tested	Number Tested (when available)	O	A	B	AB
Chimpanzee	250	35	215	0	0
Gibbon	52	0	10	20	22
Orangutan	26	0	22	1	3
Gorilla	—	+	−	−	−
Baboon	—	+	+	+	+
Macaque monkey	—	+	+	+	+
Rhesus monkey	—	−	−	+	−
Patas monkey	—	−	+	−	−

TRANSPLANTATION ANTIGENS.   Both the rhesus monkey and the chimpanzee have been shown to have a transplantation locus of much the same sort as humans; in fact, chimpanzee serum contains antibodies with specificity against human HLA antigens, and two of the antigens present in the chimpanzee appear to be alternative alleles just as they are in humans. In one study involving sixty-two chimpanzees, there was evidence to indicate the likely presence of seven of the known human antigens and the possible presence of six others.

COMPARISON OF PRIMATE KARYOTYPES.   Chromosome preparations of some of the primates show very strong resemblances to those of the human, and banding patterns make it possible to homologize the different chromosomes. As an example, it is clear that there are few differences between the chromosomes of chimpanzees and those of humans. There appear to be at least six pericentric inversion differences, and one translocation, the latter probably being involved in the reduction of the chromosome number from 48 to 46. In any case, except for the difference in total numbers, the chromosomes appear to be so similar that it would be difficult for anyone but an expert cytologist to tell the difference (Figure 20-5).

The similarities extend beyond the gross physical appearances of the individual chromosomes. In the chimpanzee, the rhesus monkey, and the gorilla, the major histocompatibility complex—the loci responsible for the antigens important in transplantation—is found on chromosome 6, as it is in humans. Similarly three other independent loci producing distinctive enzymes are found together on chromosome 6 in all four primates mentioned above. That genetic content of chromosome 21 of humans and the chimpanzee are much the same is supported by a case of apparent trisomy-21 in the chimpanzee, where that animal appeared to have some of the morphological features, as well as the extra chromosome, of typical human Down syndrome.

DNA HOMOLOGIES.   DNA similarities (or lack thereof) may also be deduced by the mixing of purified DNA of various species together. When there exist considerable lengths of DNA with complementary DNA sequences, two single strands of DNA of different origin will reassociate. Thus, if single-stranded DNA from a human male is mixed with an excess of similar DNA from a human female, the autosomal DNA of the two sexes will reassociate, but the Y chromosome DNA will not. The human Y DNA so isolated may then be mixed with Y DNA from other primates as a test for reassociation, which would indicate similarity in their DNA sequences. When Y DNA from humans was checked against Y DNA from gorillas, they failed to reassociate; therefore it would seem that the human Y nucleotide sequences are mostly different from those of a gorilla.

Despite the pronounced morphological differences among the upper primates, apparently there are not very many genetic differences. We have noted that the human and the chimpanzee differ by no more than about 1 percent of all loci. This estimate suggests that much of the recent divergence of these groups from some common ancestor has resulted largely from a limited number of mutational events that affect morphological differen-

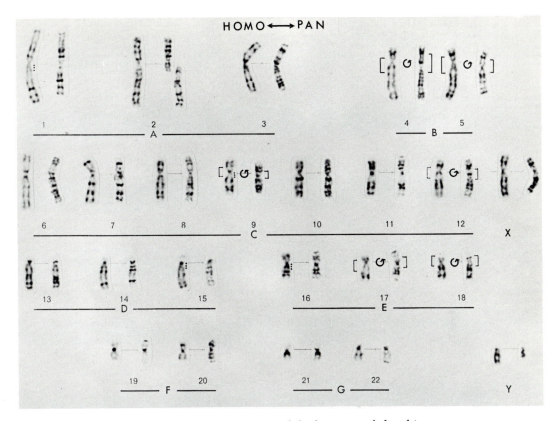

Figure 20-5 A comparison of the chromosomes of the human and the chimpanzee, those of the human being the left member of each pair, and those of the chimpanzee being on the right. The only striking difference between the chromosomes is that the chimpanzee has two smaller chromosomes that correspond to the larger number 2 in humans. The circular arrows indicate hypothetical inversion differences that have occurred as the human and the chimpanzee both evolved from some more primitive primate. (From C. Turleau and J. de Grouchy. New observations on the Human and Chimpanzee Karyotypes. *Humangenetik,* **20:**151–57, 1973.)

tiation rather than a large number that affect the pathways of protein synthesis.

Detailed analyses of the amino acid sequences of proteins found in the higher primates show relatively few differences. The hemoglobin of the chimpanzee is identical in all 287 amino acids to that of humans, and the hemoglobin of the gorilla has only two amino acids different from that of humans. Studies of a dozen enzymes show similar close relationships.

Human Variability

THE FOUNDER EFFECT.   One aspect of genetic drift is seen in human populations when a small group of immigrants leave a larger population, establish a colony, and then expand. The final result is a new population with allele frequencies that reflect the alleles carried by the few migrants. They may carry some alleles with a higher (or a lower) frequency than that of the main group, and these may then appear later in the migrant population with a frequency higher (or lower) than that of the original population. This chance shift in gene frequencies found when a small group detaches itself from the larger population is known as the *founder effect.*

Evidences of the founder effect can be seen in a number of isolated groups in the United States. The Amish, a religious sect found in a few eastern states, have unusually high allele frequencies for several diseases. (To be sure, the detection of these is facilitated by the high degree of inbreeding, which leads to more homozygosity.) Studies in Finland show about twelve recessive disorders, rare in general but of a high frequency in that region, all presumably the result of the founder effect.

BLOOD GROUP DIFFERENCES.   All population groups show some allelic diversity, and the alleles they carry may have quite different frequencies. Tables 20-2, 20-3, and 20-4 give the frequencies in various populations of the ABO, MN, and Rh blood groups. Some of these differences are quite interesting. The Basques, an isolated population living in the Pyrenees, the mountain range between France and Spain, have the *Rh*⁻ allele of the Rh blood-group system with a high frequency. It is the only population known to have this allele with a frequency over 50 percent, whereas the frequency of that same allele in the American Indians and the Australian Aborigines is zero. On the other hand, a blood group allele known as *Di*ᵃ is found only in Asians and American Indians. A wide array of other differences can be seen from an examination of the blood-group frequency tables.

Table 20-2   The frequencies of the O, A, B, and AB blood groups in a selected sample of populations, some with greatly different frequencies.

Population Tested	Total	Phenotypic Frequencies				Allele Frequencies		
		O	A	B	AB	A	B	O
Iowa City, United States	49,979	22,392	21,144	4,695	1,748	.264	.067	.669
Nord, France	103,242	45,524	44,878	9,303	3,537	.271	.064	.664
Dublin, Ireland	36,879	19,981	11,919	3,926	1,053	.195	.070	.735
Armenians, U.S.S.R.	44,632	12,913	22,272	5,883	3,564	.351	.112	.537
Moscow, U.S.S.R.	63,549	21,598	23,932	13,105	4,914	.261	.154	.585
Uzbekistan, Asiatic U.S.S.R.	2,400	735	770	643	252	.242	.208	.550
Ibadan, Nigeria	26,027	13,411	5,559	6,038	1,019	.136	.146	.718
Bihar, India	9,257	2,909	1,995	3,551	802	.165	.273	.562
Eastern Islanders, Oceania	754	247	492	8	6	.418	.009	.573
Pure Cherokee Indians, North Carolina	166	157	6	3	0	.018	.009	.973
Miyagi, Japan	240,204	32.3%	36.4%	22.8%	8.5%	Allele frequencies not calculated		

Table 20-3   The frequencies of the MN blood types in a sample of populations, some with widely different allele frequencies.

Population Tested	Total	Phenotypic Frequencies			Allele Frequencies	
		M	MN	N	M	N
New York City, U.S.A.	3,268	1,037	1,623	608	.566	.434
Danes, Copenhagen	4,319	1,270	2,147	902	.543	.457
Armenians, U.S.S.R.	5,728	1,743	3,222	763	.585	.415
Eskimos, Greenland	569	475	89	5	.913	.087
Papuans, Oceania	240	5	43	192	.110	.889

DISTRIBUTION OF GENETIC DISEASE.   Some genetic diseases are very highly limited in range, such as Tay-Sachs disease, found among the Ashkenazic Jews from Central Europe and virtually not found in other groups of Jews or Caucasoids. Tuberculosis is less frequent among Jews than among other peoples, and scarlet fever is relatively rare among Asians and blacks; this finding is evidence, along with the concordance rates in twin studies, that there exist genetic factors conferring resistance to these diseases.

By far the best known of the genetic diseases with an unusual distribution is sickle-cell anemia, caused by homozygosity for an allele producing abnormal hemoglobin. Although it occurs primarily in the black population from Africa, it is also found in Italy and Greece and, with a lower frequency, in other parts of the world (Figure 20-6). In fact, the heterozygote for these alleles has increased resistance to malaria (Table 20-5), thereby conferring an advantage on populations in malaria-infested regions of the world, despite the serious effects on the homozygotes (Figure 20-7).

CHROMOSOME POLYMORPHISMS.   The discovery of the bands on mammalian chromosomes has made it possible to compare, in detail, the chromosomes of one form with another. The karyotypes of some of the other pri-

Table 20-4   The frequencies of the Rh phenotypes in various populations.

Population Tested	Total	Phenotypic Frequencies		Allele Frequencies	
		Rh+	Rh−	R	h
Massachusetts, U.S.	91,817	76,415	15,402	.591	.409
Bohemians, Czechoslovakia	80,518	66,811	13,707	.587	.413
Spanish Basques, Pyrenees	480	343	137	.466	.534
Eskimos, Western Alaska	2,522	2,521	1	.980	.020
Javanese, Jakarta	48,964	48,964	0	1.00	0
Aborigines, Australia	281	281	0	1.00	0

Figure 20-6 The distribution of the allele for hemoglobin S, responsible in the homozygote for sickle-cell anemia, found in Africa mostly south of the Sahara and north of the River Zambezi, in all the major Mediterranean countries, in the Middle East, and in South India and Assam. (E. M. Edington and H. Lehmann. The Distribution of Hemoglobin C in West Africa, *Man* **36**:34–36, 1956.)

Assam

mates are quite similar to that of humans (Figure 20-5). If the variability of chromosome structure among primates is quite limited, one might guess that there would be little difference in chromosome structure among different races of humans, and this is precisely the case. Some chromosome polymorphisms are found with a low frequency, the level of which may be some-

Table 20-5 The apparent correlation of the presence of the sickle-cell allele, present in heterozygotes, with the severity of malaria in that region. (From A. C. Allison, in *Genetic Polymorphisms and Geographic Variations in Disease*, ed. B. S. Blumberg, Grune & Stratton, New York, 1961. By permission.)

Tribe	Locality	Number Tested	Percentage Heterozygotes	Malaria Severity
Ganda	Kampala	334	19.5	+
Amba	Bundibugyo	220	39.1	+
Chiga	Kabale	206	1.0	−
Suba	Rusinga Island	173	27.7	+
Simbiti	Musoma, Kanesi	126	40.5	+
Sukuma	Mwanza	175	26.9	+
Kituyu	Nairobi	227	.4	−
Kamba	Machakos	213	0	−

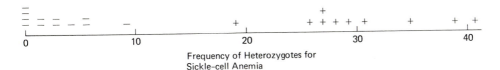

Figure 20-7 The frequencies of heterozygotes for the sickle-cell alleles, which give a direct measure of the allele frequency, compared with the relative severity of malaria, against which the allele may provide some protection. Twelve populations have low frequencies of heterozygotes, below 10 percent, and are found in regions of low (−) malaria incidence. Ten populations have frequencies of 19 percent or higher and are all found in regions with severe malaria problems. (Data from A. C. Allison, in *Genetic Polymorphisms and Geographic Variations in Disease*, 1961, ed., B. S. Blumberg. By permission of Grune & Stratton, Inc., New York.)

what different from one group to the next. One study has shown that the Y chromosome may be somewhat larger in the Japanese than in other groups, but it should be kept in mind that this chromosome must be largely inert genetically. It is possible, even quite likely, that as cytological techniques improve, some minor chromosome differences may be discovered between different races, but the failure to uncover any now, with the methods presently available, denies the existence of differences of any great extent.

OTHER INTERESTING DIFFERENCES.   Japanese, Taiwanese, and Koreans have a low threshold for the effect of alcohol; after drinking a small amount that has no detectable effect on whites, they may show marked flushing of the face and evidences of minor intoxication. Biochemical studies have shown that the Japanese have a high frequency of an unusual allele for the production of alcohol dehydrogenase, the enzyme that breaks ethyl alcohol down to acetaldehyde, as well as a high incidence of certain alleles responsible for a delayed oxidation of acetaldehyde, one of the main contributors to the intoxicating symptoms of alcohol. It is possible that the increased sensitivity to alcohol is the result of the presence of these alleles in this ethnic group, and this may be a genetic characteristic common to all populations of Oriental origin. The susceptibility of the American Indian (also of Oriental origin) to "firewater" has become legendary through innumerable written and filmed versions of the Wild West. There is another interesting genetic similarity between Asians and American Indians: a study of 3,000 Indians from sixteen tribal groups demonstrates that the dry cerumen (ear wax) that is characteristic of Orientals is also found among the American Indians.

Caucasians appear to have two different loci (i.e., four genes) responsible for the manufacture of the alpha chain of the hemoglobin molecule; Melanesians, on the other hand, have only one locus, or two genes. American blacks may have a variable number; any one person may have two, three or four alpha chain genes. If only two are present, the disease known as *thalassemia minor* results.

Human erythrocytes have an enzyme called pyridoxine kinase. In one test for this enzyme all black subjects had an average activity of almost exactly half that of the white subjects, a finding suggesting that the two different groups carry different alleles.

LACTOSE INTOLERANCE. Most humans, as well as other mammals, have an enzyme, lactase, that breaks down the major sugar in milk, lactose. This enzyme usually disappears a few months after birth, at which time lactose cannot be broken down; diarrhea and intestinal problems result. Thus most adults cannot drink milk without suffering ill effects. However, in a few groups, particularly Europeans and those of European extraction, lactase persists throughout adult life, making it possible for them to drink milk freely. Other groups,—Africans, American Indians, blacks, and Orientals, for example—are often homozygous for the recessive allele responsible for the early disappearance of lactase. This genetic difference is not fully appreciated by those benevolent agencies that send milk (dried) to Africa or the Orient or by those school authorities who urge all children, regardless of ethnic background, to "drink plenty of milk."

Pigmentation Differences

Differences in skin color in the various races undoubtedly have an explanation in the different selection pressures that are imposed on the various phenotypes and therefore change the frequency of the genes responsible for the production of the pigment melanin. The most generally accepted theory is that the lighter skin colors represent an adaptation to conditions of scarce ultraviolet light, where greater efficiency in converting and synthesizing vitamin D is made possible by the greater absorption of that light. The darker skins are more common in the tropical zones, serving as protection against the harmful effects of ultraviolet light (i.e., carcinogenesis) while still permitting the necessary level of vitamin D synthesis. Another theory suggests that the pigmentation provides a protective mechanism for the eye in those regions where solar radiation is at a maximum, still another that the presence of the pigments is correlated with changes in the immune system, providing greater protection against infection.

Different studies on the skin color of progeny of matings of individuals of different color have led to estimates of the number of loci involved at from two to half a dozen, with the figure of three to four as the preferred number. If we assume that there are three loci, each with two alleles, the black genotype would be represented by AABBCC and the white by A'A'B'B'C'C'. These alleles are postulated to be additive and equal in effect on pigmentation, so that an F₁ of composition A'AB'BC'C would be intermediate in color. Genotypes with any combination of three alleles from one race and three from the other, such as A'A'B'BCC or AAB'B'C'C, possible from subsequent intermarriages, would have a similar phenotype.

Nothing is known of the genetic basis for the color differences between the Oriental and other races.

Miscegenation, or Racial Mixture. Until several centuries ago, groups of humans maintained their individuality over long periods of time simply because of limitations of transportation and communication. At present, isolation is virtually impossible, and small isolates are becoming dispersed at a continually increasing rate.

There is no evidence that racial admixture, or *miscegenation (misce =* *"mixture")*, is accompanied by any particularly positive or negative effects from the standpoint of genetics. We do not know of any instances in which miscegenation has been either advantageous or disadvantageous except possibly for those rare cases of the crossing of groups that are grossly different in size. A Pygmy woman, for instance, might have great difficulty bearing a child whose father is from a neighboring tribe with larger stature, because of the inadequacy of the maternal birth canal to allow the passage of the large fetus; such a serious medical problem might limit the degree of race crossing. This represents an exceptional case, however, and in general it appears that race crossing is without serious consequences.

At first glance it might seem that with sufficient crossing between the races, over a long period of time, racial distinctions would disappear and the human population would be relatively homogeneous. This might be the case if inheritance were of a blending nature, but it is not. For instance, this would almost certainly not be true of differences in pigmentation.

Consider the consequences if two populations of equal size, one AABBCC and the other A'A'B'B'C'C', married completely at random for several generations. Using our simple rules of probability, we can calculate that the chance that a person will have all six alleles of a specific type is $(1/2)^6$, or 1/64. Thus slightly less than 2 percent of the population would be of each of the original genotypes. Any assortative mating of similar phenotypes would increase the proportion of types at the extremes of the distribution.

These simple considerations make it clear that humans cannot be molded into one phenotypically similar conglomerate; genetic diversity is a permanent feature of the human species.

The Degree of White–Black Admixture in the United States. In this connection it is of interest to consider the amount of racial mixing that has occurred historically between whites and blacks in the United States. We can examine this case by comparing the blood group frequencies of these two groups along with the frequencies of alleles in the region of Africa from which the ancestors of most U.S. blacks came during the days of slave trading in the New World. The method is simple. Suppose that one group has a frequency of an allele L of 100 percent and the other a frequency of 0 percent. Then, if a hybrid population has a frequency of 25 percent, it can be hypothesized that the population was made up of one quarter of the population of 100 percent and three quarters of the population of 0 percent. No other set of values will give the right proportion.

It is clear that we can directly calculate the proportions of two populations that must go into a hybrid population in order to give the gene frequencies found in it. An allele for a variant of the immunoglobulin molecule has a frequency of 100 percent in the African and 0 percent in Caucasian groups.

In one black population in the United States it has a frequency of 73 percent. This finding must mean that 27 percent of the allelic content of that population comes from whites. The alleles in the Duffy blood group can be used similarly. Here the frequency is 43 percent in whites and 0 percent in blacks in Africa. The American black population has an allele frequency of 9.4 percent, suggesting a 22 percent contribution from the white population to the black. These figures may vary, of course, depending on the population being studied. A group of blacks from South Carolina, studied by this method, showed only 6 or 7 percent white alleles. Information from several studies of this type may be combined graphically to illustrate the relative "distance" between different population groups (Figure 20-8).

It is not possible to make any corresponding calculations with any great validity on the proportion of black genes in the white population of the United States because of several complicating factors. Much of the miscegenation occurred during the early days of slavery, whereas the population of the United States is derived to a great extent from the massive immigration from Europe that occurred in the nineteenth century and the early part of the twentieth century. The fact of pervasive racial discrimination in the United States ensured that the extent of introduction of genes from the black population into the white would depend to some extent on the frequency with which individuals with black ancestors crossed the color line to become part of the white population.

Figure 20-8 A diagrammatic scheme showing the relationship of four populations to each other, based on morphological data (*A*) and on allele frequencies for blood groups and hemoglobin types (*B*). (W. S. Pollitzer. The Negroes of Charleston [S.C.]; A Study of Hemoglobin Types, Serology and Morphology, *Am. J. Phys. Anthropol.* **16**:241, 1958.)

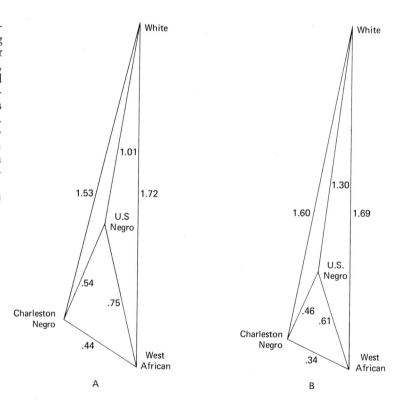

The Hemoglobin Diseases

We have previously discussed many aspects of the hemoglobin molecules, from the standpoint of simple inheritance, genetic counseling, and mutation.

HISTORY. After sickling was first noted in a black student in 1910 and was shown to be an unusually common disease in the black population, it was shown to be a hereditary disease (1923). In 1949 L. S. Pauling and his co-workers compared the electrical charge on the hemoglobin of normal red cells and that of a person with sickle-cell anemia electrophoretically, that is, by putting the hemoglobin in solution and measuring their rates of migration in an electrical field. The two kinds of hemoglobin were distinctly different. Further, the hemoglobin of a heterozygote gave a result roughly equivalent to a mixture of the two types. This finding paralleled the kind of result expected if the hemoglobin difference depended on a simple allelic difference.

Subsequently, V. Ingram at Cambridge University used the enzyme trypsin to cleave the polypeptide chains of hemoglobin, which it does at points where lysine or arginine is found. The polypeptide is broken up into smaller fragments, peptides, which separate if they are allowed to migrate at their own characteristic rates in a suitable medium. An electrical field is applied at right angles to give a two-dimensional separation. Figure 20-9 shows a peptide present in normal adult hemoglobin, but not in sickle-cell hemoglobin, and one present in sickle-cell and not in normal adult hemoglobin.

NORMAL HEMOGLOBIN STRUCTURE. The typical hemoglobin molecule is composed of four chains, two of a type called *alpha* (α) and two *beta* (β), and each of these four chains has an iron-containing segment called a *heme*

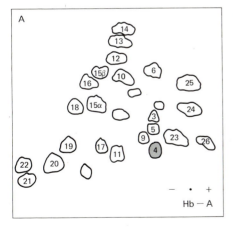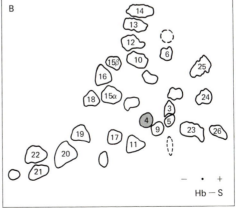

Figure 20-9 The separation of the components of normal hemoglobin (*A*) and hemoglobin S (*B*), showing that they differ by only a single group. (C. Baglioni. An Improved Method for the Fingerprinting of Human Hemoglobin, *Biochim. Biophys. Acta* **48**:392, 1961.)

Figure 20-10 Four polypeptide chains, two alpha and two beta, are associated to form normal adult hemoglobin. A heme group containing iron, which gives hemoglobin its oxygen-binding capacity, is indicated by a disc in each of the four chains. (From *The Structure and Action of Proteins* by R. E. Dickerson and I. Geis. Benjamin and Cummings, Publishers, Menlo Park, Calif. 1969.)

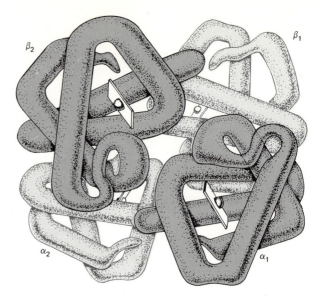

(Figure 20-10). The heme is constant in all forms of hemoglobin; the variants are found in the polypeptide, or globin, chains. Normal *adult hemoglobin, HbA,* consists of two identical alpha and two identical beta chains, with a structural formula $\alpha_2\beta_2$. The length of the alpha polypeptides is 141 amino acids, the beta 146. A different form of hemoglobin, *fetal hemoglobin (HbF)*, is found in the fetus and is replaced by adult hemoglobin shortly after birth. Its alpha chains are like those of HbA, but the beta chains are different and are designated as gamma (γ), hence they are given the formula $\alpha_2\gamma_2$. The gamma chain of the fetal hemoglobin differs from the beta chain of adult hemoglobin by thirty-nine different amino acids. It can be readily surmised from our knowledge of the production of polypeptides by codon sequences in the chromosome that the alpha, beta and gamma polypeptide chains are synthesized by distinctly different gene loci.

The peptide difference between normal hemoglobin, HbA, and sickle-cell hemoglobin, HbS, is found on the beta chain of the hemoglobin molecule and can be shown to result from the substitution of valine for glutamic acid at the sixth amino-acid position in that chain. This is the only difference between the two hemoglobins, and all of the medical symptoms that come from the presence of sickle-cell hemoglobin are the result of what appears to be at first sight a trivial substitution of one amino acid for another. Under conditions of a low oxygen pressure, the HbS molecules aggregate and form long, rodlike conglomerations that distort the normal spherical shape of the erythrocytes and cause them to assume elongated or sickle shapes. An individual who is heterozygous for the gene for sickling produces two kinds of hemoglobin, HbA and HbS. Those homozygous for the sickle allele produce only HbS.

For a reason that is not clear, the vast majority of the abnormal hemoglobins have their amino acid substitutions in the beta sequence. However, alpha chain variants are known, and in one case a single family showed both an alpha chain (Hb-Hopkins-2) and a beta chain (HbS) abnormality. The independent segregation of these two abnormalities in the sibship showed that the genes responsible for the alpha and the beta chains must be at separate loci, and either some distance from each other on the same chromosome or on separate chromosomes. This has been confirmed independently by the allocation, by other methods, of the locus of alpha to chromosome 16 and of beta to chromosome 11.

PRENATAL DIAGNOSIS OF SICKLE-CELL ANEMIA. Until recently the only way of detecting sickle-cell anemia was to test fetal blood. Obtaining it, however, was a rather risky operation, which resulted in about 5 percent fetal deaths.

The prenatal diagnosis of this disease has since been accomplished by the detection of the defective gene in cells obtained by amniocentesis. Although the fibroblasts obtained in this way do not manufacture hemoglobin, these cells have the alleles for hemoglobin synthesis, and the DNA on both sides of these alleles can be analyzed. There exists an enzyme, Hpal, that cuts DNA whenever the sequence GTTAAC occurs. Such a sequence is found on both sides of the gene that controls the synthesis of the beta chain of hemoglobin. When DNA with the normal beta-chain allele is cut, the result is a DNA fragment with about 7,600 bases; when DNA with the abnormal allele for the defective beta chain is cut, the length of the fragment is about 13,000 bases (in more than half the cases). By an analysis of the DNA of the parents, it can be determined whether they belong to those cases where the allele for sickling is detectable, and, if so, the DNA of fetal cells can be checked so that it can be determined whether the fetus is homozygous normal, heterozygous, or homozygous for the sickle allele.

Another approach is to obtain rare fetal cells from the maternal circulation. Tests sensitive enough to accomplish this are being developed; they have the great advantage of bypassing the necessity of amniocentesis.

The Thalassemias

The thalassemias affect more persons worldwide than any other hereditary defect, with a total of 6 million cases running from the area of the Mediterranean through the Middle East to Malaysia. Like sickle-cell anemia, this defect may owe its widespread occurrence to the greater resistance of the heterozygote to malaria.

There are various types of thalassemia; in some the production of alpha chains is reduced (*α-thalassemia*), and in others the beta chains (*β-thalassemia*). There are as many basic causes of thalassemia as there are ways in which a normal amount of gene product may be interfered with. The series of steps, as well as five major points at which normal gene function may be altered, is shown in Figure 20-11.

ALPHA-THALASSEMIA. There are four different types of alpha-thalassemia, two rather mild, a third serious, and the fourth fatal to the newborn. The existence of these four types led initially to the suggestion that there are two loci, rather than one, responsible for the alpha chain, a guess that has subse-

Figure 20-11 Scheme showing five major points at which hemoglobin synthesis may be interfered with to produce thalassemia. A regulatory gene located some distance away (*1*) or adjacent to (*2*) the structural gene producing the polypeptide chains may be abnormal; the structural gene (*3*) itself may be altered; transcription of the nuclear gene into messenger RNA (*4*) may be interfered with, or the translation of the mRNA into protein (*5*) may be defective. (R. Williamson, 1977, *New Scientist,* **75**:405.)

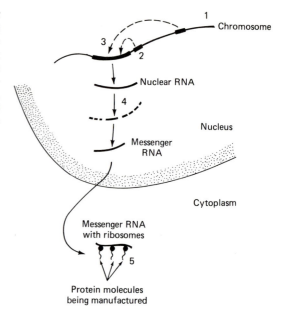

quently been confirmed. The actual existence of two loci has been shown by isolating mRNA for the alpha chain from red cells and manufacturing complementary DNA (cDNA) by the use of a special enzyme, *reverse transcriptase.* Radioactive compounds can be used in this synthesis to "mark" the cDNA. When this cDNA is added to a culture of cells, it hybridizes with the normal homologous DNA, and the spots at which hybridization has occurred are marked by the concentration of radioactivity at that region. Because there are two regions at which this hybridization occurs in a normal cell, it can be deduced that there are two loci in the cell responsible for the alpha chain.

In those seriously affected with alpha-thalassemia, there are no alpha genes at all, and the fetus dies. In the milder forms of alpha-thalassemia, from one to three alpha genes may be present, and the seriousness of the defect is rougly proportional to the number of alpha genes absent. Because the alpha genes are present in all cells, including the fetal cells in the amniotic fluid, it is now possible to detect such seriously affected fetuses by amniocentesis.

BETA-THALASSEMIA. In some cases of beta-thalassemia, a gene for the beta chain appears to be deleted. In other cases, the genes for the chain appear to be normal, but either the mRNA ordinarily found in the cytoplasm is absent or the nuclear mRNA is absent.

The defective production of these chains is complicated by the fact that their function may be taken over to some extent by the delta and gamma chains, which are usually present only in fetal life but which may be produced in greater quantities by beta-thalassemics. Nevertheless, despite the reduced severity of beta-thalassemia, in general, compared with alpha-thalassemia, the homozygotes for these defects usually suffer severe anemia, particularly after a few years of postnatal life, as the amount of fetal hemoglobin gradually diminishes.

Additional Reading

CHIARELLI, A. B., ed. 1971. Comparative genetics in monkeys, apes and man. *Proceedings of a Symposium of Comparative Genetics in Primates and Human Heredity, Ernice, Sicily.* Longon: Academic.

CORLISS, C. 1974. Review: Masatoshi Mei and Arun K. Raychoudhury. 1972. Gene differences between Caucasians, American Negroes, and Japanese populations. *Science,* **177**:434–439.

GOLDSBY, R. A. 1971. *Race and Races.* Springfield, Ill.: Thomas.

HIRADA, S., et al. 1980. Liver alcohol dehydrogenase and aldehyde dehydrogenase in the Japanese: Isozyme variation and its possible role in alcohol intoxication. *American Journal of Human Genetics,* **32**:8–15.

KAN, Y. W., and A. M. DOZY. 1980. Evolution of the hemoglobin S and C genes in world populations. *Science,* **209**:388–391.

KING, J. C. 1971. *The Biology of Race.* New York: Harcourt Brace Jovanovich.

KING, M.-C., and A. C. WILSON. 1975. Our close cousin, the chimpanzee. *New Scientist,* **65**:16–18.

KOLATA, G. B. 1980. Thalassemias: Models of genetic diseases. *Science,* **210**:300–302.

MARX, J. L. 1978. Restriction enzymes: Prenatal diagnosis of genetic disease. *Science,* **202**:1068–1069.

MENOZZI, P., A. PIAZZA, and L. CAVALLI-SFORZA. 1978. Synthetic maps of human gene frequencies in Europeans. *Science,* **201**:786–792.

MORTON, N. E. 1969. Birth defects in racial crosses. In *Proceedings of the Third International Conference on Congenital Malformations,* pp. 264–274.

REED. T. E. 1969. Caucasian genes in American Negroes. *Science,* **165**:762–768.

ROGERS, J. 1977. Diagnosing blood diseases in the womb. *New Scientist,* **73**:718–719.

Questions

1. How much of the genetic variability in the human species is found within races and how much between races? In consequence, may the word *race* be meaningless?
2. What was Darwin's view of the nature of inheritance? Why was he not aware of Mendel's paper on plant hybridization?
3. What contributions did Mendel, de Vries, and Morgan make to clarify the mechanism of genetic change over a long number of generations of species?
4. Why are new mutations more likely to be deleterious than beneficial?
5. Explain why selection for (and against) a new recessive allele is relatively ineffective, particularly when the allele has a low frequency. How does this selection differ from selection for or against a dominant allele?
6. What is meant by *mutation pressure? Genetic drift?*
7. Why may mutations in regulatory genes be, on the average, more impor-

tant to evolutionary change than mutations in structural genes? Do we have any reason for believing that this is the case?

8. Is there any reason to believe that heterosis has ever played a significant role in human evolution?
9. Would the speed of evolution in a polyploid species differ, depending on whether the effective mutations are dominant or recessive?
10. Do humans and other primates show similar genetic variability?
11. In what sorts of populations are we likely to see evidences of the "founder effect"?
12. Do you think that you would be able to distinguish a metaphase cell of a human from that of a chimpanzee? What are the differences?
13. Which populations of humans are characterized by possessing quite distinctive alleles, or allele frequencies?
14. What is the enzymatic basis for lactose intolerance?
15. Is there any reason to believe that miscegenation has a deleterious effect on the species?
16. If the human population should be completely panmictic over several centuries, would racial differences disappear?
17. How is it possible to calculate the extent to which the U.S. black population has genes from the white population?
18. How can different hemoglobins be isolated and analyzed? What is the difference between normal adult hemoglobin, fetal hemoglobin, and sickle-cell hemoglobin?
19. How can sickle-cell anemia be diagnosed prenatally?
20. What is the difference between alpha- and beta-thalassemia? Are the thalassemias serious diseases?

21

Manipulation of the Human Genetic System

Biological Manipulation in General

The selective breeding of desirable traits in plants and other animals to improve crops and domesticated animals is as old as civilization. Now, with our expanding horizons in human genetics, it is becoming more possible to apply some of this new knowledge to the benefit of humans. In some cases the procedures for doing this seem unacceptable; in others, there is room for rational discussion.

Furthermore, it is now feasible to reach within the cell and, in one way or another, to alter the genetic mechanism so that it is able to fulfill some function not previously possible. In popular terms the deliberate alteration of the genetic composition of a species is referred to as *genetic engineering*.

ANIMAL AND PLANT BREEDING IN THE PAST. Genetic engineering has been going on since prehistoric time. The domestication and selection of breeds of animals have been carried on for thousands of years. Babylonian records show that artificial pollination of the date palm occurred several thousand years B.C., although it may not have been immediately obvious to the primitive agriculturist that the production of the new generation depends on a contribution from each of the two parents.

It seems intuitively obvious that two individuals of the same species will

439

produce progeny more similar to themselves than to other individuals taken at random from the population. Simple selection for specific traits is the means by which large numbers of varieties of cultivated plants and animals came into existence. It is only within the last century that our knowledge of genetics has advanced to the point at which more efficient mating schemes have been developed. Professional plant and animal breeders have perfected elegant and intricate matings systems for manipulating the genetic constitution of living things to produce what is, in their opinions, some desired end results. Perhaps the best-known case is that of hybrid corn, in which highly inbred strains, relatively weak and poor-yielding plants, are mated to each other to produce hybrids that far surpass the parental types in yield. The point here is not so much that this has been and is being done, with great success, but that this biological experimentation is socially acceptable and, in fact, highly valued by society. No one has seriously criticized the production of high-yielding grain crops, for instance, on the grounds that such genetic manipulation interferes with God's or Nature's original design of living things.

THE SPECIAL CASE OF *HOMO SAPIENS.* However, these comments do not apply to humans, who occupy a special place in their own concept of the universe. There are many reasons that it would be difficult, even impossible, to apply simple genetic principles to the breeding of people, but even if it were possible, there are restraints of many kinds (including persuasive ethical considerations) that limit the extent to which we may feel free to tinker with our own constitution. Nevertheless, with the present high level of sophistication in the biological sciences, some possibilities for genetic engineering to benefit the human population are being seriously considered. In this chapter we discuss some of these possibilities, as well as some of the many problems that must be faced before any of these procedures could be considered practicable.

One popular conception of the ultimate goal of the geneticist is that of drastically altering the human species, a view promoted by hundreds of low-cost Hollywood films. It is supposed on the one hand, that the scientist's ideal is to make all individuals absolutely identical and on the other, that it is to create multitudinous human freaks.

The fact is that geneticists, being human themselves, feel that the race as it is now composed would not be improved much by a tampering with its phenotype. Little would be gained by adding another eye, a couple of antennae, or a pair of wings. Technological advances have made many of the improvements of science fiction obsolete. No pair of wings could compete with a jet aircraft; telepathy is accomplished far better by the selective communication made possible by the telephone, the radio, and the television than by ESP. We can survive at temperature extremes, at the bottom of the ocean, and in outer space; changes in the genome that might promote adaptations to such conditions would be superfluous. At the present time, in fact, most thoughts about the alteration of human genetic makeup are directed toward the alleviation or elimination of genetic defects that disturb our well-being.

PRESENT PROBLEMS AND FUTURE PROSPECTS. A list of the new approaches that are either with us now or will undoubtedly be available to us in the near future—almost certainly within the next century—would include the control of the sex of offspring, the use of sperm banks to provide for fertilization of eggs in special cases, the transplantation of organs from one person to another (already discussed in Chapter 12), the implantation of fertilized eggs in a foster mother, the modification of prenatal and postnatal development to provide for "better" individuals *(euphenics)*, the production of new individuals with the same genetic composition as that of a currently existing one *(cloning)*, and the manipulation of the genetic material either to repair defective genes or to improve the performance of currently existing ones.

As we consider each of these aspects of possible future applications, we will be faced constantly with a number of questions. Who will decide what is a desirable or an undesirable trait? How is it possible to guard against an "incorrect" determination when there is no hard-and-fast basis for that judgment in the first place? When does the life of a human start? Is it permissible to destroy a sperm and an egg but not a fertilized egg, an embryo, or a fetus? Where does one draw the line? Would a program meant to improve the state of humanity be likely to diminish our sense of dignity as human beings by putting us at the same level as economically useful plants or animals? Would any plan for the "improvement" of the human race simply reflect the fears, biases, and prejudices of the individuals making the decisions? If a program of therapy must be limited to a fraction of those who would probably benefit from it, who will decide which individuals will receive the treatment and which will not? Are there any forseeable long-range problems that might develop as a consequence of a certain line of research that is so potentially hazardous that it should be prohibited? Might the proposed procedure conflict with the religious or ethical beliefs of a sizable fraction of the population, and if so, to what extent would society have the right to curtail the freedoms of the individual for the purpose of "improving" the race?

These, and many other questions like them, are neither trivial nor rhetorical, nor do they have a specific "right" answer. The response to any of them will be an individual one, determined, on the one hand, by the prevailing religion and philosophy of society at the time and, on the other, by the age, background, and convictions of each person.

Selection at the Level of the Gamete

CONTROL OF SEX. In principle, choosing an offspring of a desired sex is possible at the present time, as the sex of a fetus may be ascertained as early as amniocentesis can be performed. However, few persons would agree that the sex of a fetus is, in itself, a matter of such importance as to justify amniocentesis, and even fewer would agree that a fetus's having the "wrong" sex would be sufficient cause for a "therapeutic" abortion.

With astonishing regularity, reports are published both in the daily press and in reputable scientific journals suggesting that the sex of an offspring may be predetermined by alkaline or acid treatments of sperm or by physical separation of X- and Y-bearing sperm prior to the fertilization of the egg. Most of these reports result from a relatively limited number of trials in humans, with the work being done by protagonists of the point of view that sex predetermination is now possible. Because these experiments with humans can be readily accomplished as well with laboratory animals, such as mice, where large numbers of progeny can be produced, ultimately the convincing data will no doubt come from animal laboratories rather than hospitals. Furthermore, when such data is presented by a worker who had previously been skeptical of, rather than disposed toward, the thesis that such modification of sex at fertilization is now feasible, reports of positive results will be more credible.

In any case, even if these reports should not be verified by more extensive and strictly controlled experiments, it is likely that a procedure will be developed in the near future that will change the present approximately equal probabilities of the two sexes at fertilization. Whereas it was necessary previously to undertake actual breeding experiments to test schemes for the differential selection of one kind of sperm over the other, the discovery of the fluorescence of the Y chromosome within the sperm after quinacrine staining makes it possible to evaluate new techniques for the separation of X and Y sperm much more effectively.

There is one forseeable problem. If the separation depends on physical differences between the X and the Y sperm, such as the size or the volume of the sperm head, then aneuploid sperm, those with an extra autosome, or with one missing, would necessarily be selected at the same time that the X or Y sperm were being selected, as they would have similar size and volume differences. For this reason, we might expect that such a separation procedure used to increase the probability of fertilization by either an X- or a Y-bearing sperm would, at the same time, increase the likelihood of a grossly abnormal embryo.

It is interesting that a procedure like the choice of the sex of offspring, which was considered a bold, spectacular advance not very long ago, is now theoretically possible by amniocentesis, but it raises little more than subdued yawns from the population at large, reminiscent of the reaction of the world to the fifth walk on the moon. Perhaps there is a lesson to be learned here: that some of the anticipated advances will turn out to be not quite as exciting as we now imagine they will be.

Artificial Insemination by Donor. More than 10,000 women in the United States each year produce progeny who have been conceived by artificial insemination by donor (AID), and a quarter of a million persons now alive were conceived in this way. The sperm are obtained, for the most part, from doctors and medical students, who may in this way be the fathers of a large number of children. (One male has been the father of more than fifty children.) Records are usually not kept of the names of the donors, or of the mothers and children. The absence of documentation may serve to prevent any future stigmatization of the children and to preserve the anonymity of the fathers. It may also help avoid legal entanglements in an area that has been poorly defined by law.

Doctors performing this operation usually do not test the males selected as donors, either for chromosome abnormalities or for common genetic diseases. In at least one case of AID, the child was born with Tay-Sachs disease. This tragedy could very easily have been avoided, as both the mother and the donor must have been heterozygous for the recessive; a simple, readily available test would have revealed their carrier status.

Another problem that may arise is that when such children grow up, they may marry each other and themselves produce offspring, unaware that they in fact had the same father and therefore are half sibs. This outcome may be a hazard in smaller communities that have a limited number of donors.

Finally, there are the legal problems still to be resolved, of whether every child has a right to know the identity of the father and whether that father has certain responsibilities, such as financial support, toward his offspring. On the other hand, does the sperm donor have certain rights, too? Can he, for instance, lay claim to paternal rights, such as weekly visits to his child, or even custody of the child? Is he, or the husband of the mother, the legal father?

SPERM BANKS.   In the 1930's H. J. Muller suggested that some benefit to society could result from the creation of sperm banks where the sperm of distinguished males could be preserved for later use. (In theory a similar system might be set up for the preservation of the eggs of distinguished women, but the techniques for doing this have not yet been developed.) Would it not be wise, asked Muller, to select as sperm donors individuals who have been shown to be free of genetic defects and who, furthermore, are not run-of-the-mill, as most of us are, but are outstanding in some respect that might be attributable in part to an unusual genotype? Thus one might preferentially choose the sperm of individuals who are distinguished as composers, humanitarians, or authors, rather than the mentally disturbed, social outcasts, dissidents, and so on, although, to be sure, these two lists may overlap.

The common example advanced in favor of the propagation of such genes is given by the case of Albert Einstein: How many women, faced with the necessity of having a child by AID, might prefer that he be the father rather than some unknown and impecunious medical student? Or might a woman prefer to have an offspring by such a distinguished male rather than by her relatively untalented husband? Or if she is unmarried, might she particularly consider having an offspring, if such a distinguished "father" were available? These possibilities do not take into consideration a negative aspect, that any person, however well endowed genetically, also has his fair share of deleterious alleles, and the excessive reproductivity of any one person runs the risk of increasing the frequency of defective alleles in the population.

It would, of course, be simpleminded to imagine that such germinal selection would produce the characteristics of the individual from whom the sperm were obtained. In the first place, the sperm would be a random haploid gene complement of the diploid condition that made up the individual. In the second place, virtually all of the characteristics that make one human seem superior to another are multifactorial in nature, to whatever extent that "superiority" is genetically determined. Nevertheless it would seem a simple matter of common sense to suggest that if artificial insemination is to be a widespread practice, the donor sperm should come from individuals who have been checked for deleterious alleles, insofar as that is possible. And for those who believe that there is a substantial genetic component in ability and accomplishment, the selection of extraordinary donors holds out the additional hope that perhaps sometime in the future a new, unusual genotype, including some genes perpetuated by the selection of donors, will appear and make some outstanding contribution to civilization.

Asexual Reproduction

PARTHENOGENESIS. The production of a new individual from an egg, without any genetic contribution from a male, is known as *parthenogenesis*. There are about a thousand animal species, mostly insects and other invertebrates, that reproduce exclusively by parthenogenesis. It has never been observed as a natural form of reproduction in any mammalian species.

It might, in theory, happen in humans in a number of different ways. An unfertilized egg might be induced to initiate cleavage divisions. An embryo with this origin would be haploid. We know, however, that just about every human diploid cell carries one or more seriously defective genes, which are tolerated because they are recessive and because the homologous chromosome carries a good functional allele. We would therefore expect just about every haploid cell to be defective or lethal because the deleterious recessives would be uncovered. In fact, no haploid humans, or even mosaics partly haploid, have ever been seen.

In another type of parthenogenesis known in lower organisms, unfertilized eggs fuse with other meiotic products (i.e., a polar body or another egg) to restore the diploid condition. Such a zygote would have a greater chance of being viable, depending on the specific meiotic products so fused. Still another possibility would involve the induction of an oogonial (and therefore diploid) cell to behave like a fertilized egg and to commence cleavage. If these events were induced in a test tube outside the mother's body, the cells so induced to imitate development could then be returned to the mother, or to a foster mother, to continue embryonic and fetal development.

Some experiments with mice, although not completely successful, have given positive results. Two unfertilized mouse eggs can be fused by the use of Sendai virus. The resultant diploid product then may divide until it reaches the sixty-four-cell stage, before it succumbs. In a more successful effort, one of the two nuclei is removed from a mouse egg, immediately after fertilization and before the nuclei of the two gametes have had a chance to fuse, and the remaining nucleus is chemically induced to start mitosis. The resulting viable embryos are all females. Finally, it has been possible to fuse cells of parthenogenetic origin (which would not ordinarily survive) with normal embryonic cells to produce adult mice. In this experiment, because the normal cells were albinos and the parthenogenetic cells had an allele for dark coat color, it was easy to spot the dark fur of parthenogenetic origin on the white background.

Any embryo developing from these events could have only maternal genes, and any resulting individual would probably bear an unusually strong physical resemblance to the female providing the egg. In 1955 an English newspaper ran an article on parthenogenesis, asking to hear from any woman who thought she had a parthenogenetic child. It was stipulated at the outset that the child would obviously have to be female. Of the nineteen mother-child pairs who answered the challenge, eleven were disqualified after an interview. Six of the remaining eight were eliminated because the child possessed a blood group antigen not carried by the mother. One pair had different eye colors. Skin grafting between the remaining pair showed histoincompatibility between the two.

CLONING. A *clone* may be defined as a group of genetically identical individuals descended from a single cell. Although this term has been used by plant breeders for years to designate many varieties that are propogated asexually, as by cuttings or runners, it is now popularly used in two ways. One is the isolation of a specific segment of DNA and its propagation in a microorganism; this we will call *DNA cloning*. It is discussed in detail near the end of this chapter. Here we will consider the second common use, the production of a number of higher organisms, such as humans, with identical genotypes. This is *organism cloning*.

In the cloning of organisms, the nucleus is removed from an ordinary somatic cell and is implanted in an egg, which develops to produce an individual with precisely the same genes as those in the cell from which the nucleus was taken (Figure 21-1). The individual produced is then the "identical twin" of the animal from which the cell was removed. This procedure was first accomplished in the frog, where nuclei taken from cells of a developing blastula have replaced the nucleus of an egg, and the egg has then developed to produce a normal mature animal. This kind of cloning is relatively simple in the frog because its eggs are very large, as is obvious to anyone who has looked at the gelatinous masses, dotted with frog eggs, in brooks and pools during the spring of the year. The operation of removing the nucleus of the frog egg is technically difficult but not impossible. Furthermore frog eggs ordinarily develop to maturity in a pond, where there are wide fluctuations in conditions, such as temperature and surrounding media, so that there is no problem in persuading the eggs to develop in the laboratory.

On the other hand, the mammalian egg, as we have noted before, is quite small, demands very special conditions, and must be implanted in a female who is hormonally receptive to the development of an embryo. Progress has been reported in the first steps toward successful cloning in mammals. Nuclei from blastocyst cells of one mouse strain have been injected into the freshly fertilized egg of another after the original nucleus and the sperm nucleus, not yet fused, were removed. Normal mice have been born having only the genes of the foreign nucleus.

From the technical standpoint, cloning, although difficult, is now possible. However, the limited success with mammalian cells so far has depended on the use of nuclei from early embryonic cells, not those from adult tissue. It is possible that the cells of the mature mammalian organism are incapable of forming a new organism, as a zygote does. If this proves to be

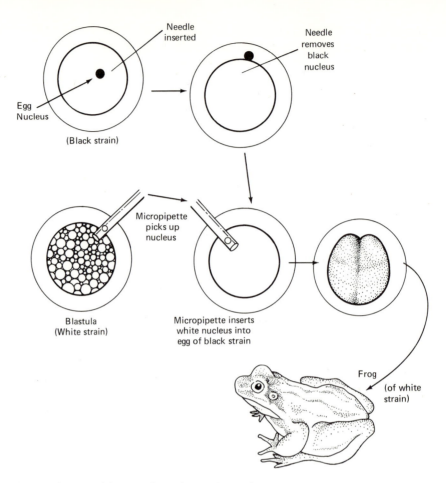

Figure 21-1 Cloning of frog nuclei. The nucleus of an egg is removed with a needle, and a nucleus from a blastula is inserted into the egg. The egg then undergoes cleavage and eventually becomes an adult frog.

the case, then cloning humans would seem pointless since the procedure would depend on the use of cells obtained from an embryo with unknown characteristics.

Apart from any philosophical, ethical, or religious objections that might arise from the possibility that humans could be produced by unorthodox methods, some apprehensions stem from the thought that it might become possible to populate the world with specific genotypes. Thus one unpleasant picture is that of aggressive nations creating multitudes of physically capable, somewhat unintelligent, and psychologically abnormal individuals as professional armies. Equally repelling is the thought that powerful egotists might decide to create many individuals patterned exactly after themselves. Because of such potential problems, many concerned scientists have suggested that measures should be taken now to prevent any further development of cloning techniques.

Probably some of the distaste for the application of techniques like cloning is unwarranted, at least to the extent that it is imagined that a group of persons with identical genotypes would be exact carbon copies of each other. There is also the implication that such a group would somehow form a psychic bond, not just among themselves, but particularly with their donor, whose motive might be to achieve some measure of immortality. From the study of data from identical twins, even those raised together, the concordance of psychological traits is imperfect at best. It is quite likely that if a group of cloned individuals were ever to be produced, they would, as they diverged psychologically during early life, consist of a wide assortment of types and would not be simple replicas of each other.

At first sight it might appear that techniques that would increase the population size should be discouraged. Why should science and society make an effort to develop procedures that would inevitably accentuate population problems, to the detriment of humanity, when child production is already carried out very efficiently and without great expense? The best answer to this question is that techniques like cloning have a great usefulness in other ways. For instance, one of the greatest puzzles facing the biologist at the present time is that of differentiation: why certain cells become muscle and others nerve, for instance, and also why some cells become cancerous whereas others fail to differentiate at all. The understanding (and possible cure) of a large number of diseases depends on mastering such basic biological phenomena. It seems certain that the experimentation on nuclear transplantation, and the knowledge about differentiation that will come from it, will eventually be responsible for saving human life and eliminating untold suffering. The potential problems presented by the misuse of these techniques seem inconsequential in comparison.

Modifications of Normal Prenatal Development

Since the publication of Aldous Huxley's *Brave New World,* the specter of the mass production of embryos and fetuses, carefully designed for specific purposes, has caught popular fancy. This area of biological engineering has several different aspects.

EXTRAUTERINE FERTILIZATION. In England in 1978 the birth of Louise Brown astounded the world; she was the first child born from a human egg fertilized outside the body. The egg was allowed to go through several cleavage divisions and, two and one half days later, was returned to its natural abode in the uterus of the mother. Drs. Robert Edwards and Patrick Steptoe had finally succeeded in this difficult task, which makes it possible for women who would otherwise be sterile, because of blocked or degenerative Fallopian tubes, to have their own children. This first successful effort was shortly followed by a number of others throughout the world, particularly Australia, and it is to be expected that this method of circumventing sterility problems caused by defective Fallopian tubes will become more routine. Although the initial case was the first success in a large number of trials, the chances of success increase as the techniques improve.

It should be noted that this procedure has been in use for years in cattle and sheep, where fertilized eggs have been transported long distances and then implanted in surrogate mothers. The extension of this technique to humans was clearly only a matter of time.

Desirable as this procedure may seem to be to sterile couples, it presents new possibilities with respect to reproduction. For instance, it is now within the realm of possibility for a woman to have her egg fertilized by her husband's sperm and then to have it transferred to the womb of another woman, a surrogate mother. The legal questions raised by such a procedure seem endless. Which of the two would be the legal mother of the child? If the surrogate mother were paid a fee for carrying the fetus and for renouncing any rights to the newborn, would she also have no right to ask for a therapeutic abortion if the fetus proved to be grossly defective? And suppose that the surrogate mother becomes emotionally attached to the fetus during pregnancy and decides that she cannot give it up. Even if she has signed a legal document to surrender the child, who will force her to do so?

Finally, this kind of medical advance opens up other possibilities. The scientific curiosity that motivates work with fertilization and embryonic development in the higher primates may lead to other types of experiments. Who will prevent the ardent student of anthropology, in the study of the biological relationships among the higher primates, from fertilizing the egg of a gorilla with the sperm of a human to see if a hybrid can develop? Such experiments, should they ever be carried out, would, in all probability be fruitless because the wide taxonomic differences between these groups cause development to stop at an early embryonic stage. Nevertheless some hybrids have appeared as the natural offspring of species of the higher primates: the gibbon and the simiang, as well as the baboon and the rhesus monkey. The thought of such experimentation makes one wonder about the wide variety of human, legal, and ethical problems that experiments with fertilization will lead us into in future years.

EXTRAUTERINE DEVELOPMENT. Strickly speaking, the phrase *test-tube baby* should be applied to development outside of the mother from the fertilized egg to the newborn. The contributions of the mother to the embryo and fetus during development are so substantial and complex that it may very well be impossible to duplicate these conditions outside the womb. The mother not only provides simple nutrients, which might be easily supplied *in vitro*; but also supplies more complex proteins, such as antibodies, which may not be easily supplied *in vitro*, as well as an environment for development of the fetus that may be unique to the womb. It is not just a question of the warmth and the oxygen supply, which could be duplicated easily, but also of the precise cellular structure of the uterus, which is responsible for specific physical and chemical signals during development. If the embryologist should reach the level of sophistication that would make it possible to reproduce these exactly, a question would immediately arise about whether the effort was worthwhile, as the entire procedure is being accomplished quite efficiently and naturally at relatively little expense. It is difficult to imagine circumstances, in civilization as we know it or as we project its course in the near future, that would justify developing this line of research as an end in itself.

EUPHENICS.   Whereas *eugenics* is aimed at improving the genotype, *euphenics* is the improvement of the phenotype. This might be accomplished, for instance, by a modification of the course of development of the embryo *in utero* to produce somewhat "better" individuals. It has been suggested, for example, that the cells giving rise to the neurons of the brain might somehow be induced to undergo (by some as-yet-unknown treatment) just one more mitosis during early development. The number of brain cells could then be doubled, and it might be anticipated that the effect on intelligence would be striking. Some people might object to a simple treatment of the embryo or fetus *in utero* that would promote the development of a much higher intelligence in the newborn. Others, however, would welcome it.

Treatment of the fetus *in utero* to thwart a genetic disease has been accomplished in several cases. Fetuses diagnosed to have an inherited defect in vitamin B_{12} synthesis in one case and a biotin disorder in another developed normally after the mothers were given large doses of those vitamins.

Selection at the Level of the Organism

EUGENICS.   The improvement of the overall genetic composition of humans by an alteration in the frequency with which different persons, or groups of persons, have children, was termed *eugenics* by Francis Galton in the latter part of the nineteenth century. Historically this concept has had a very dismal record. The "upper" social classes in England and then in America used it to justify their privileged position by maintaining that they possessed superior genes. The 1924 immigration laws of the United States were biased in favor of Northern Europeans and against others—Asians and Southern Europeans, for instance—on the implicit assumption that the former had "better" genes than the latter. It has played a role in the discrimination against blacks in the United States. And Nazi Germany put to death millions of humans on the grounds that they were genetically inferior.

For these reasons society is reluctant to admit any validity to eugenic arguments. Nevertheless we must concede that any genetic information that a couple may obtain that modifies their attitude toward having offspring is, in itself, eugenic. Further, such action taken by potential parents would , with few exceptions, meet with the approval of society. Thus, if one of a couple has Huntington disease and the couple decides not to have children because they do not wish to take the 50 percent chance of condemning their offspring to this dread disease, few would find their decision objectionable.

There are two essential differences between this example and the earlier ones. In the first place, the curtailment of reproduction in this example is voluntary, which is acceptable, and is not imposed by outside authority which may not be acceptable. Second, the defect in question is a well-established medical entity and not some diffuse characteristic with an ill-defined genetic basis, such as innate abilities.

Programs of negative eugenics that would alter the genetic composition of humans by interfering with the right of individuals to choose mates or to have offspring, or with the right of every offspring to be allowed to live, are generally regarded as unacceptable because they conflict with fundamental human freedoms. To be sure, these rights are limited at the present time. Congenitally feeble-minded individuals are generally put into institutions, where they are prevented from marrying and producing offspring. However, such institutionalization may be caused by concern for the welfare of the children born under such conditions rather than by a wish to prevent the perpetuation of any deleterious genes.

Although it might seem self-evident at first glance that negative eugenics might reasonably be practiced only in extreme cases, there is still cause for concern. That concern centers on the possibility that once criteria are set up that allow individuals in certain groups to be deprived of their reproductivity, the criteria can then be adjusted to fit the prejudices of those who happen to hold power in that group.

Positive eugenics, which involves the promotion of more than average numbers of offspring from individuals considered to be superior, is faced with an equal criticism, as the definition of *superior* can be adjusted to suit the fancies of those who happen to be making the rules at the time. A good example of this sort of bias is the experience of the Oneida Community of New York, a Utopian society founded in the nineteenth century to improve the lot of humankind. The children were to be bred by matings determined by committee decision. It turned out that a disproportionate share of the offspring were fathered by one man, who was also the founder of the community. This fact did not go unnoticed by the other men, and along with other disagreements, it led to the dissolution of the community.

CARRIER DETECTION.   The detection of homozygotes *in utero* provides a means whereby grossly defective embryos may be aborted, if abortion is acceptable to the parents. Unfortunately heterozygous parents are usually not identified as such until after an abnormal offspring has been produced, and if an aborted embryo is later replaced by a normal child, that child has a two-thirds chance of being heterozygous. Thus, although monitoring for homozygotes by amniocentesis may decrease the frequency of homozygotes born, it will *increase* the frequency of the defective allele in the population because in two thirds of the cases the poorly reproductive homozygotes may be replaced by normally reproductive heterozygotes.

Clearly, if all heterozygotes could be identified, and if they either chose not to have any children or chose not to have heterozygous children, any defective gene could be eliminated in one generation (except for rare new mutations). In theory this elimination is possible; in practice it is not. Even if all heterozygotes (e.g., for Tay-Sachs disease) could be identified, it would be difficult to persuade all of them that they should not have children because half of their offspring, who would be quite normal, would be heterozygous for this defective gene. It would be even more difficult to persuade such heterozygotes of the advisability of a therapeutic abortion of a phenotypically normal embryo because it was heterozygous for a defective allele. The destruction of phenotypically normal embryos heterozygous for an allele that, in the homozygous state, produces a serious defect would probably be considered an unacceptable procedure at the present time. Large sums are spent looking for cures for many genetic diseases—which, when found, will almost certainly turn out to be incomplete solutions—although a theoretical solution is now at hand in the detection of heterozygotes, along with a program to educate them to the undesirability of perpetuating the defective alleles.

Such an approach may seem inconsistent with the knowledge that every person carries quite a few deleterious recessive alleles. If we were to suggest that all of those with defective alleles not produce offspring, would not this suggestion logically include every member of the human race? Not quite. Some of the defective alleles are recessive lethals that act early in the development of the homozygote. Because they cause no deaths of newborns and no stillbirths, and because they may cause embryonic loss so early that it is undetected by the mother, they create no social or personal trauma. At the other end of the spectrum are those alleles with nonspecific effects, detectable as an overall loss of vitality only, when progeny of consanguineous marriage are compared with progeny at random in the population. These alleles also are unmanageable in the sense that we cannot identify persons with these alleles. What are left are the genetic diseases with serious debilitating effects in the homozygotes. If the human could be manipulated genetically as are plant-breeding stocks, there would be no problem in removing the heterozygotes for the most serious medical diseases from the breeding population, without undue complications to its total reproductive capacity.

Intervention in the Operation of the Somatic Cell

ENZYME THERAPY.   In some cases the question arises as to whether or not a simple genetic defect caused by an enzyme deficiency might be ameliorated by injection of the missing enzyme. This possibility has great appeal and is very often used as an argument against the application of any measures that might be used to decrease the frequency of presumably deleterious genes in the population. The classic example in such discussions is diabetes and its treatment by insulin. How unnecessary it would have been, goes the argument, to have tried to eliminate any alleles predisposing a person to diabetes when subsequent work has shown that it can be cured by injections of insulin. This is probably the best example of enzyme therapy, and therefore it illustrates some of the problems associated with this approach. Many cases of diabetes are not manageable by insulin treatments, and when they are, the treatment leaves something to be desired. To be sure, the health of diabetics has been improved enormously by insulin treatment (and the new kinds of human insulin being produced by cloning techniques will lead to greater improvements), but the fact remains that on the average they are not as well as normal people, nor do they have as long a life expectancy.

Similarly, antihemophilic serum is of great help to hemophiliacs, but it does not actually allow them to lead completely normal lives. Although such therapy is of inestimable value to the affected person, there is no doubt that, given a choice, that person would vastly prefer not to have the disease. Similarly, when we consider a "cure" for sickle-cell anemia, it should be kept in mind that if our experiences with other comparable diseases can serve as a guide, any cure that is found in the near future is likely only to relieve the condition to some extent and not really to cure it.

Furthermore such therapy allows for increased reproductivity of the affected individuals, so that the defective allele may no longer be subject to the naturally-occurring negative selection that is ordinarily thought to hold it at a low level in the population. This example illustrates a dilemma that plagues medicine at the present time: The effective treatment of the symptoms of individuals with genetic diseases may actually lead to a slow increase in frequency in the population as a burden to future generations.

There are other problems associated with enzyme therapy as a potential cure-all for human ailments. One is that in the vast majority of genetic disabilities an enzyme defect has not been found—or has been only tentatively identified, if at all. Purification of the enzyme, if possible, may be costly, and whether society will be willing to underwrite the expense involved may depend on its perception of the benefits to be gained in relation to the sums expended. Furthermore enzymes very often perform their normal function in specific cellular environments and not in the bloodstream, where they would appear after injection. And finally, in many cases the recipient reacts to these foreign proteins by developing antibodies against them and thus becoming sensitized to them. This sensitization prevents their further use. So what seems like a simple, straightforward way of handling a genetic problem may, in fact, be difficult—and in many cases impossible.

GENE REPAIR BY BASE MANIPULATION.   Gene repair has very often been suggested as a means of using the modern advances of molecular biology for the benefit of the human race. The vision of altering a defective gene in such a fashion as to enable it to carry on its normal function is one that has caught the attention of the public. However, this goal is very far from realization. At the present time we do not know the precise location of the genes responsible for the most serious genetic diseases, and when this is accomplished, it will be a challenging problem to alter the base pairs in the DNA of defective genes in a germ cell in such a way as to make them normal.

Insertion of Foreign Genes

There is no reason that genes not normally found in human cells (or not found in some persons) cannot, in principle, be introduced into humans from other organisms, with potentially beneficial results.

As an example, humans lack a single enzyme needed to manufacture ascorbic acid (commonly referred to as *vitamin C*). Although the race manages to survive on the small amounts present in the diet, some workers have presented strong arguments that the optimal amounts for well-being and disease resistance are very much larger than those now in the diet of the average person. If this is the case, the insertion of the necessary gene into our genetic complement might have immediate beneficial effects. This insertion of genetic material might be accomplished in any one of a number of ways.

GENE ACTIVITY PROVIDED BY A MICROORGANISM. A virus may be capable of performing some function that a mammalian cell cannot, and its activity may substitute effectively for the missing mammalian one. There are some indications that variations in blood enzyme levels may depend on the presence of viruses. Levels of the enzyme thymidine kinase, found in the blood, has been increased by infection by herpes simplex, the common virus that causes cold sores. Another virus, which causes the Shope papilloma, has decreased the amino acid arginine in the blood of persons exposed, possibly by providing the enzyme that breaks it down.

In a more direct experimental approach, investigators have injected into newly fertilized mouse eggs viral genes that confer resistance to a specific drug. The eggs were transferred to the wombs of foster mothers. Of 150 newborn mice, 2 showed evidences of the viral genes. It was not determined whether these transplanted genes were functional, whether they were free in the cell or incorporated into chromosomes, or whether they could be transmitted to progeny along with the rest of the genome. It remains to be shown also, before such experiments can be extended to humans, that such foreign genes will not have any adverse effect, such as carcinogenesis. Nevertheless this procedure appears to be a first step in the ultimate "cure" of genetic defects by the incorporation of a normal gene into a cell containing a defective one.

Cloning Recombinant DNA

It is now possible to take segments of DNA from a human (or any other organism), to recombine it with the DNA of a virus, to allow the virus to infect the cells of some other organism, and so to transfer the genes of one species into the cells of another. The procedure is illustrated in Figure 21-2.

USE OF RESTRICTION ENZYMES. To clone a gene from a human in cells of *Escherichia coli,* for instance, the procedure would be as follows: The DNA is chemically isolated from the nuclei of a person's cells. This DNA is treated with a *restriction enzyme,* which has the unusual characteristic of cutting the DNA strand at regions of specific base sequences. Several dozen different restriction enzymes are known that cut DNA at different series of base sequences. In our example the enzyme cuts open the helix wherever it sees the sequence AATT in one direction and the complementary TTAA in the other.

The DNA of the nucleus is thus chopped up into many smaller pieces. Some of these cuts may occur within a gene when that gene has the AATT sequence somewhere along its length. Such genes are ruined. On the other hand, when the cuts occur in the DNA on either side of the gene, the pieces may contain the entire gene.

INCORPORATION INTO PLASMID DNA. At the same time, the restriction enzyme is added to a smaller DNA segment of a virus or a *plasmid.* A plasmid is a self-replicating fragment of circular DNA that, although not an integral part of a microorganism, can invade one and, once inside, can reproduce in great quantities. This particular restriction enzyme also cuts the

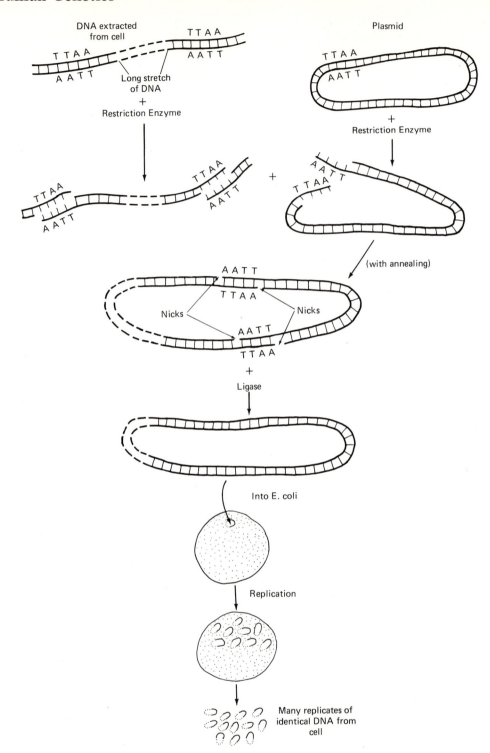

Figure 21-2 Cloning DNA segments in bacteria. DNA from any source (*upper left*) and DNA from a plasmid or other carrier (*upper right*) are treated with a restriction enzyme which cuts the DNA at specific base sequences. The DNA from the two sources then join to form a larger circle of DNA, recombinant DNA, which can be introduced into a microorganism and grown in quantity.

plasmid DNA at the sequence AATT. At this point the two kinds of DNA segments, from humans and from the plasmids, are mixed together. In a small fraction of the cases, the small, single-stranded sections of AATT pair, or "anneal," with the TTAA sections of the DNA from the other species. In this way a new piece of DNA, recombinant DNA, is manufactured. When AATT anneals with TTAA at both ends, a new, larger circle of DNA may be formed in which the gene from the human is incorporated into the plasmid.

CLOSING THE CIRCLE. After the annealing process, there are still points at which the DNA backbone is discontinuous. These "nicks" may be closed by the addition of a special enzyme, a *ligase*, which performs this function in normal cells.

TRANSFER OF GENES FOR DRUG RESISTANCE. How can it be shown that gene insertion has, in fact, occurred? Obviously, because these steps are completed in only a minute fraction of the total DNA mixture, some sort of selection system is needed to isolate the desired cases. One scheme is to incorporate a gene that confers resistance to some chemical. In the first successful experiments in 1973, a gene from the bacterium *Salmonella* that renders it resistant to the antibiotic streptomycin was incorporated into a plasmid and was added to the intestinal bacterium *E. coli*. The cells of *E. coli* that incorporated the plasmid with the gene for streptomycin resistance could easily be isolated, because treatment of all of the cells with the drug killed off the normal ones and only the few with the plasmid survived.

EXONS AND INTRONS. Inserting *functional* mammalian genes into plasmids presents greater difficulties. In the first place, some of the genes of higher organisms contain intervening sequences that interrupt the code for manufacturing polypeptides. These sequences are ordinarily removed from the messenger RNA by special enzymes in the mammalian cell, enzymes that are absent in the bacterial cell. Synthesis in the bacterial cell would be unable to differentiate between the exons (which code for the polypeptide) and the introns (which are nonsense as far as coding is concerned). Polypeptide synthesis of a mammalian gene with intervening sequences in a bacterium could result in a long polypeptide that has no resemblance to the one produced by the mammalian cell.

Such a problem is not insurmountable. It is possible to circumvent it by manipulating not the gene but the messenger RNA, after it has had the intervening sequences removed. An enzyme, reverse transcriptase, promotes the production of DNA complementary to that RNA; that is, a copy of the gene is manufactured with the intervening sequences removed. This DNA can then be inserted into a plasmid, which, when put into a microorganism, may then manufacture large quantities of the polypeptide coded for by the gene.

On the other hand, it has been found that yeast genes for transfer RNA which have intervening sequences can be introduced into frog cells, which then manufacture yeast transfer RNA. The sequence of nucleotides that separate introns from exons may be essentially the same, and recognized by the same enzymes which remove the introns over a wide range of living forms.

By the use of such extraordinarily clever tricks, biologists and chemists have succeeded in a large number of gene transfers. Insulin, used in the treatment of diabetics, is one of the first medically useful compounds to be manufactured by recombinant DNA. Interferon, an antiviral agent found in human leukocytes and connective tissue, that may attack cancer cells, is found ordinarily only in extremely minute quantities. This substance is being produced in larger quantities by bacterial cells.

INCORPORATION OF GENES INTO CHROMOSOMES.   When such plasmids are introduced into mammalian cells, the plasmids may function and replicate independently of the chromosomal DNA; whether this replication will give rise to any problems to the organism must be determined for each case. There is, however, some chance that the plasmid may become incorporated into one of the chromosomes of the complement. This process may be readily visualized in the following way (Figure 21-3). A sequence of bases on the plasmid—say, GCCTAG—may find the complementary set, CGGATC, on the chromosome. A crossoverlike event can then incorporate that segment into the chromosome. If the linear sequence of bases is not interrupted by this event—as it would be if the "crossover" occurred with the gene—then the gene could be completely functional. This may be one way of approaching the problem of "repairing" genetic damage. A defective gene (for Tay-Sachs, for instance) might still be present, but there could be, in addition, a normal gene that would carry out the necessary syntheses to prevent the disastrous disease.

HAZARDS OF RECOMBINANT DNA EXPERIMENTS.   The possibility of moving genes from one organism to another has aroused apprehensions, as well as

Figure 21-3   One scheme for explaining how circular segments of recombinant DNA could become incorporated into the DNA or chromosomes of some unrelated organism. If there exists a segment of homology on both the recombinant DNA and the recipient, then an exchangelike event occurring between the two in the region of homology inserts the recombinant DNA.

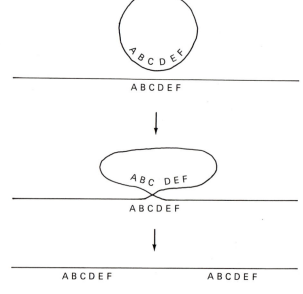

Figure 21-4 The anticipated consequences to the human morphology after introduction of the gene for wings from *Drosophila*.

hopes and excitement, in the scientific world (Figure 21-4). A number of workers in the field have wondered whether transformed bacteria with the capacity to synthesize new polypeptides might prove to be unusually virulent or otherwise unmanageable. Guidelines have been set up for the control of recombinant DNA experiments. The insertion of an antibiotic-resistant gene into a pathogenic microorganism, for instance, is not to be carried out except under the most secure conditions. Similarly the transfer of a gene that manufactures a potent poison or of the DNA from a virus that may produce tumors may be considered an operation of high risk to be carried out only under conditions that safeguard against the escape of the bacteria. On the other hand, strains of bacteria have been developed with complex nutritional requirements that can be satisfied only under very special laboratory conditions. Such strains carrying recombinant DNA would have great difficulty surviving outside the laboratory and so, it is hoped, would not present any great hazard.

The early fears about possible problems with recombinant DNA have now largely subsided. It is assumed that the benefits to be gained from this type of research far outweigh any potential hazards. Progress in this area may be followed in the daily newspaper.

A Look Forward

Our knowledge of the genetic constitution of the human is expanding at an exponential rate. Most of the information included in this book was unknown as recently as thirty years ago, and as is clear from this section, we may expect to see even more spectacular developments in the immediate years to come. Those advances that unequivocally improve the condition of the species will be welcomed; others that transgress on individual rights, religious beliefs, or preconceived ethical standards will be viewed with suspicion, or even hostility. It appears that the scientific and intellectual progress of the species has far outdistanced our innate capacity to deal with such completely new circumstances. One good example of such inadequacy is the failure to control the ever-expanding population, largely attributable to advances in medicine, technology, and agriculture. The innate urge of the human species to procreate is a basic biological attribute and appears to be very difficult to control because most persons agree that this is a fundamental human right.

However, much as we may be the servants of our biological makeup, we are not its complete slaves. Over the past several centuries new cultural factors have supplemented the genetic transmission of information from one generation to the next. Thus, although the information in the DNA of our chromosomes is enormous, the equivalent of several thousand volumes of print, this is a puny fraction of the knowledge stored in libraries, available to anyone who can read. The human brain is capable of profound deductions, but we can use various forms of communication to integrate the

Table 21-1   Comparison of biological and cultural evolution. Note that the evolutionary destiny of all species was dependent upon interaction of genetic constitution and environment beyond its control. Humans alone have the means to control both their environment and, to some extent, their genetic constitution. (A. G. Motulsky. 1968. *Journal of Heredity,* **59**:329.)

	Biologic Evolution	*Cultural Evolution*
Mediated by	Genes	Ideas
Rate of change	Slow	Rapid and exponential
Agents of change	Random variation (mutations) and selection	Usually purposeful. Directional variation and selection
Nature of new variant	Often harmful	Often beneficial
Transmission	Parents to offspring	Wide dissemination by many means
Nature of transmission	Simple	May be highly complex
Distribution in nature	All forms of life	Unique to humans
Interaction	Human's biology requires cultural evolution	Human culture required biological evolution to achieve the human brain
Complexity achieved by	Rare formation of new genes by chromosomal duplication	Frequent formation of new ideas and technologies

thoughts of many thousands of persons. The biological components of human evolutionary progress are now being dwarfed by these factors in our cultural evolution (Table 21-1).

This is not to imply, however, that natural human biological evolution has stopped. The forces of natural selection are always present and their impact cannot be denied. The human species, if it is to survive, will, a million years from now, surely be different from its present form and psyche. One part of the change will be directly related to cultural evolution, another part to the pressures that have always acted on the human being as another creature in a complex environment. But for the first time, it is now possible to predict with certainty that some of the change will result from the deliberate application of knowledge that will enable the human to change its genome in predictable ways.

Additional Reading

BLATTNER, F. R., et al. 1978. Cloning human fetal γ globin and mouse α-type globin DNA: Preparation and screening of shotgun collections. *Science,* **202:**1279–1284.

CLARK, M., S. BEGLEY, and M. HAGER. 1980. The miracles of spliced genes. *Newsweek,* March **17:**62–71.

Closing in on cloning. 1981. *Time,* January 19:75.

COHEN, S. N. 1975. The manipulation of genes. *Scientific American,* **233:**25–32.

KORNBLUTH, J. 1980. Souls on ice. *Esquire,* April:25–32.

LEWIN, R. 1981. Biggest challenge since the double helix. *Science,* **212:**28–32.

MARX, J. L. 1981. Three mice "cloned" in Switzerland. *Science,* **211:** 375–376.

McGINTY, L. 1980. Biotechnology and the scientific ethic. *New Scientist,* **85:**147–148.

NATHANS, D. 1979. Restriction endonucleases, simian virus 40, and the new genetics. *Science,* **206:**903–909.

On the way to a clone. 1979. *Science News,* **116:**68.

OSBORN, F. 1973. The emergence of a valid eugenics. *American Scientist,* **61:**425–429.

SINSHEIMER, R. L. 1976. Recombinant DNA—on our own. *BioScience,* **26:**599.

SINSHEIMER, R. L. 1977. An evolutionary perspective for genetic engineering. *New Scientist,* **73:**150–152.

SMITHIES, O., et al. 1978. Cloning human fetal γ globin and mouse α-type globin DNA: Characterization and partial sequencing. *Science,* **202:** 1284–1288.

SNOWDEN, R., and D. MITCHELL. 1980. Anonymous AID for the childless couple. *New Scientist,* **85:**828–829.

WETZEL, R. 1980. Applications of recombinant DNA technology. *American Scientist,* **68:**664–675.

What future for science? 1978. *New Scientist,* **77:**71.

Questions

1. How old is genetic engineering (when it is broadly defined)? Can you give some examples that go back to antiquity?
2. What are some of the ethical questions that are being raised at the present time by proposals to modify the human genome?
3. Would the possibility of predetermining the sex of offspring prior to birth represent a great advance in the control of human destiny? Is it, in fact, now possible to do so? Can the sex of offspring be predetermined prior to the fertilization of the egg?
4. What is AID? For what reasons has it been criticized? Does a child resulting from AID have a legal or moral right to know who its true father is?
5. Discuss the idea of sperm banks, with the pros and cons of their application.
6. In an apparently bona fide case of parthenogenesis in humans, what tests would you suggest to ascertain the likelihood of the phenomenon? List the tests in the order of their simplicity or feasibility.
7. In what two different senses has the word *clone* been used in this chapter? Why has the prospect of cloning created considerable apprehension?
8. Has fertilization of the human egg outside the human body ever been accomplished? What is the scientific term for this procedure? Is it appropriate to refer to the resulting infants as *test-tube babies?*
9. Consider the following experiment, which the successes with extrauterine fertilization and reimplantation make possible. Imagine that ten or twenty eggs from different women are fertilized by sperm from an equal number of males. As the eggs begin to cleave, a single cell is taken from each and the sex of the zygote is determined. The cells of all like-sexed morulas are then combined to form a single embryo, and the resulting chimera is implanted to complete development. The resulting person would have at least one positive attribute: it would be able to accept transplants from a vastly increased number of potential donors. Can you think of any other positive or negative features of such a procedure? Do you think that such experimentation with human cells should be permitted? Suppose that some of the cells added to the mixture came from a primate other than a human.
10. What is the difference between eugenics and euphenics? Which of the two is more feasible at the present time? Why has eugenics been in disfavor for many years? Can you think of cases in which eugenic measures are, in fact, regularly taken, with public approval?
11. If amniocentesis leads to the detection of a defective fetus, which is then not allowed to come to term, would the resulting decrease in the number of homozygotes born with the recessive in question lead to a decrease in the frequency of that defective allele in the population?
12. Because every person carries at least five to ten defective genes, is it reasonable to try to decrease the frequency of defective genes in the population? Would this effort logically require that no person produce offspring?

13. If a person is unable to synthesize an essential enzyme, why are injections of the missing enzyme not an immediate and effective solution?
14. Is it now possible to repair defective genes by changing the base pairs of the abnormal DNA?
15. Describe the principle of cloning recombinant DNA: use the terms *E. coli, restriction enzyme, plasmid, annealing, nicks.*
16. What problems do introns present in making mammalian genes functional in bacteria? How might reverse transcriptase be used to take care of this difficulty?
17. What future hazards in pursuing genetic engineering are foreseen by some as perilous or even potentially disastrous?
18. Discuss the contribution that cultural evolution and genetic engineering will possibly have made to the human condition within several thousand generations from now.
19. Was the work of Drs. Edwards and Steptoe in arranging for the *in vitro* fertilization of the egg that gave rise to Louise Brown moral? Are there any potential dangers in such experimentation? What if the egg had been fertilized by sperm from a man not the husband of Mrs. Brown? Or what if the sperm had come from one of the higher primates, like the gorilla?
20. Who is the legal father of a child born by artificial insemination? Does a child so produced have a right to know who the biological father is?

Glossary

These abbreviated and simplified definitions are designed to aid in understanding this text; they may, therefore, be more limited in scope than definitions found elsewhere. Many technical terms that have been left out of this glossary are defined in detail in the text itself.

acentric a chromosome or a chromatid lacking a centromere.

acrocentric a chromosome or chromatid with a nearly terminal centromere.

acute exposure a radiation dose applied in a short period of time (as opposed to chronic).

agglutination the clumping of cells by attachment to antibodies (not to be confused with clotting).

allele one of an array of possible mutational forms of a gene at a specific locus.

allergen a substance which induces hypersensitivity.

allergy unusual sensitivity to certain substances which have no effect on most people.

allograft a graft of tissue from a donor of one genotype to a host of different genotypes but of the same species.

alpha particle a helium nucleus which has two protons and two neutrons.

amniocentesis the puncturing of the amniotic cavity to obtain amniotic fluid, usually for prenatal diagnosis of fetal disorders.

amnion the inner of three walls of the amniotic fluid-filled sac in which embryos develop.

anaphase that phase of nuclear division during which the chromatids are converted to independent chromosomes, separate and move to opposite poles.

anaphase bridge a chromosome with two centromeres which proceed to opposite poles at anaphase, resulting in a "bridge" connecting the two groups of separating chromosomes.

anaphylaxis an acute immune response of hypersensitive individuals.

androgen any substance with male sex hormone activity; a male sex hormone.

anencephaly a congenital tube defect in which the brain and head are severely reduced; usually fatal.

aneuploid an organism with one or a few chromosomes present more or less often than normal.

ankylosing spondylitis a hereditary inflammatory arthritis, mostly affecting young men, characterized by low back pain and stiffness.

antibody a specific complex protein produced in response to the introduction of a specific antigen into an animal.

antigen a substance which may stimulate the production of specific antibodies against it.

antihemophilic globulin a component of the blood plasma essential for the normal clotting (and deficient in the blood of hemophiliacs).

antihistamine an agent which counteracts the effects of histamine; used in treatment of allergies.

autoimmunity the production of antibodies by a person's own antigens.

autosome a chromosome other than a sex chromosome.

autosurveillance hypothesis the idea that an autoimmune reaction may limit the growth of cancers.

base pair a pair of hydrogen-bonded nitrogenous bases (one purine and one pyrimidine) that join the component strands of the DNA double helix.

bilirubin a yellow or orange compound produced by the breakdown of hemoglobin.

bivalent a chromosome pair resulting from synapsis of homologous chromosomes.

blastocyst the primitive embryo at the time of implantation.

blastula the hollow sphere of cells formed during very early development.

bridge a chromosome with two centromeres which proceed to opposite poles at anaphase, resulting in a "bridge" connecting the two groups of separating chromosomes.

carcinogen a physical or chemical agent that induces cancer.

carcinoma a form of cancer consisting of a malignant growth originating in epithelial tissue found in the linings of the respiratory, digestive and other systems, in skin and various other organs.

carrier a heterozygote.

centromere the specialized organelle on the chromosome to which the spindle fiber is attached during nuclear division.

chiasma (plural, chiasmata) the cross-shaped point of junction between chromatids seen at prophase, commonly thought to result from crossing over between nonsister chromatids.

chimera an individual composed of genetically different cells derived of different origin.

chorion the middle layer of the amniotic fluid-filled sac in which embryos develop.

chromatids the daughter strands of duplicated chromosomes, both joined by a single centromere, which become separate chromosomes upon division of the centromere.

chromatin the nucleoprotein fibers of which chromosomes are composed.

chromomere a densely coiled region of chromatin on a chromosome, giving the extended chromosome a beaded appearance.

chromosome the darkly staining bodies within the cell made up of a large number of genes and a centromere region.

chromosome map representation of the linear arrangement of genes on a chromosome, as deduced from genetic and cytological observations.

clastogen a physical or chemical agent capable of producing chromosome breaks.

cleavage division an early mitotic division of a fertilized egg.

clone a group of genetically identical cells or organisms descended from a single common ancestral cell or organism.

coagulation time the time required for blood to clot.

code the correspondence of the 64 possible base triplets with the twenty different amino acids.

codominance the expression of both of two different alleles in a heterozygote.

codon a triplet of three bases in an RNA molecule specifying a single amino acid.

colchicine a drug that inhibits the formation of the spindle and delays the division of centromeres.

color blindness defective color vision involving one or more of the three visual pigments that absorb red, green or blue light.

compound chromosome a chromosome made up by the rearrangement of entire arms of other chromosomes; a Robertsonian translocation.

concordance the presence of a certain trait in both members of a twin pair.

congenital defect a defect that exists at the time of birth.

consanguinity relationship by descent from a common ancestor.

crossing over exchange of genetic material between members of a chromosome pair.

cytology the branch of biology dealing with the structure, function and life history of the cell.

cytomegalovirus a virus capable of inducing congenital malformations; is undetectable in the mother but can be fatal or cause severe mental retardation to the infant.

deficiency abnormal chromosome resulting from loss of some loci.

deletion see deficiency.

deoxyribonucleic acid the nucleic acid of the chromosome (DNA).

dicentric a structurally abnormal chromosome or chromatid with two centromeres.

dictyate the stage of the first meiotic division in which the oocyte remains from late fetal life until ovulation.

dictyotene see dictyate.

differentiation the complex of changes involved in the progressive diversification of the structure and functioning of the cells of a developing organism.

dimer the abnormal combination of two adjacent thymine bases on the DNA strand.

diploidy the presence of two of each chromosome (the sex chromosomes may represent an exception).

direct-acting mutagen a compound that is active without the intervention of any metabolic change.

discordance the presence of a certain trait in one member of a twin pair and not in the other.

dizygotic (fraternal) twins resulting from two different ova fertilized by different sperm.

DNA the nucleic acid of the chromosome, deoxyribonucleic acid whose sequence of base pairs carries the genetic information of the cell.

dominant an allele expressed phenotypically in the heterozygote.

dosage compensation the celllular mechanism that compensates for the activity of genes which, as a consequence of their location on the X chromosome, exist in two doses in the female but only in one in the male.

double helix a strand of DNA hydrogen-bonded to a complementary strand of DNA, with the combination assuming a long spiral configuration.

doubling dose the amount of radiation it would take to produce the same number of mutations as are produced spontaneously.

Drosophila a common fruit fly found world-wide and historically the most extensively studied experimental organism, from the standpoint of genetics and cytology.

duplication presence of part of a chromosome in duplicate.

dysmorphology a morphological developmental abnormality, as seen in many syndromes of genetic or environmental origin.

electrophoresis the differential movement of molecules through a liquid under the influence of an electric field, used in separating different proteins.

elliptocytosis a disease in which the red blood cells are elliptical in shape.

embryo an organism in early stages of development; in humans, from fertilization to the beginning of the third month of pregnancy, after which it is termed a fetus.

empiric risk estimate that a trait will occur or recur in a family based on past experience rather than knowledge of the mechanism responsible.

endosperm triploid nutritive cells surrounding and nourishing the embryo in seed plants.

enzyme a protein molecule that catalyzes a specific chemical reaction.

epicanthus a fold of skin extending over the inner corner of the eye, most commonly found in Orientals.

equilibrium a state of balance between opposing forces; in genetic equilibrium, the forces which tend to change allele frequencies, selection and mutation, have become counterbalanced so that there is no net change in allele frequencies from one generation to the next.

erythroblastosis a condition marked by an excess of erythroblasts (primitive red blood cells which possess hemoglobin) in the blood.

erythrocyte   red blood cell.

Escherichia coli   the colon bacillus, a favorite experimental organism in the study of molecular genetics.

estrogen   an ovarian hormone which stimulates uterine growth.

eugenics   improvement of the genetic constitution of the human species through selective breeding.

euphenics   the modification of prenatal development to improve well-being.

exchange transfusion   whole-body blood transfusion in which all of the blood of a fetus or newborn infant is replaced.

exon   a sequence of DNA that is translated into protein; the exons of a gene may be interrupted by introns, which are not translated.

falciparum malaria   most dangerous form of malaria, to which persons carrying a sickle cell allele show some resistance.

fallopian tube   duct extending from the vicinity of the ovary to the uterus.

familial   any trait which tends to occur in certain families.

fertilization   the union of two gametes to form a zygote.

fetus   in humans the postembryonic stage from beginning of the third month to birth.

fission   a method of asexual reproduction seen in bacteria, protozoa and other lower forms of life in which one cell divides into two or more parts, each of which then develops into a complete individual; also, the separation of one atom into the smaller ones, often accompanied by the production of radiation and other energy.

fluorescent   emitting light that results from a reaction going on within the emitter.

fluorescent polymorphism   chromosome variation detectable by the presence of fluorescence.

fluoroscopy   X-ray examination by means of a screen which flouresces when rays from an X-ray tube strike it.

founder effect   a population, started by colonizers, with a frequency of alleles different from that of the parent population.

fragile site   a point on a chromosome where breaks may occur spontaneously; a chromosome weakness that appears to be heritable.

frame shift mutation   an addition or loss of genetic nucleotides of the DNA, causing the translation of the corresponding mRNA to be shifted so that nucleotide sequences are read in a different register.

fraternal   (dizygotic) twins developing from two independently fertilized eggs.

freemartin   an intersex in cows arising from the masculinization of a female twin by its male sibling when their fetal circulations are continuous.

fruit fly   see *Drosophila.*

gamma globulin   the antibody-containing protein fraction of the blood.

gamma radiation   an electromagnetic radiation of short wavelength emitted from an atomic nucleus undergoing radioactive decay.

gene   the unit of inheritance, located at a specific region on the chromosome.

gene pool   the total genetic information possessed by the reproductive members of a population of sexually reproducing organisms.

genetic code   the correspondence of the 64 possible base triplets with the twenty different amino acids.

genetic drift   random fluctuation of gene frequencies in small populations from one generation to the next.

genetic load   the burden imposed upon a population by the presence of detrimental alleles.

genetically significant dose   the total gonadal dose of radiation from a given source with possible genetic effects, divided by the total number in the population, to give an average dose per person, of importance to succeeding generations.

genome   the totality of genes in the haploid set.

genotype   the genetic constitution of an organism (as distinguished from its physical appearance) usually expressed in terms of the one or few allelic pairs under consideration.

germ cells   the cells that produce gametes.

germinal mutation   mutations occurring in the germ line and eventually appearing in the gametes, or occuring in the gametes.

germ line   all the cells destined to become gametes.

gestation period   the time from conception to birth.

glucose-6-phosphate dehydrogenase   (G-6-PD), an enzyme important in the metabolism of sugar.

glycolipids   a lipid containing carbohydrate.

glycoprotein   a protein containing small amounts of carbohydrate.

gonial cells   cells that eventually produce gametes, and are found in the germ line in ovaries and testes.

gonad   an essential sexual gland; an ovary or a testis.

graft-versus-host (GVH) disease   a reaction arising when immune competent cells of a graft, as in the case of bone marrow, produce antibodies against the host tissues.

GVH disease   see graft-versus-host disease above.

H substance   the precursor to the A and B antigens.

Haemanthus katherenae   the African blood lily; a favorite species used for the time lapse photographic study of chromosome behavior during mitosis.

haploid   the chromosome set with only one member of each chromosome pair; in humans, the haploid number, n = 23.

haplotype   a group of alleles from closely linked loci (as the MLC complex) usually inherited as a unit.

HDN   see hemolytic disease of the newborn.

heme   an iron-binding molecule that forms the oxygen-binding portion of hemoglobin.

hemizygous   the condition of having only one set of genes, instead of two, as in the case of the genes of the X chromosome in a male; since males have only one X, they are hemizygous (not homozygous or heterozygous) with respect to X-linked genes.

hemoglobin   the oxygen-carrying pigment of red blood cells.

hemolytic disease of the newborn   several disorders in fetuses and infants in which the red cells are destroyed.

hemorrhage   abnormal discharge of blood, either internal or external.

hepatitis   inflammation of the liver of virus or toxic origin; usually manifest by jaundice and sometimes liver enlargement.

heritability   a statistical estimate of the degree to which variability in a trait is genetically determined among individuals in a population.

Herpes simplex   a filterable virus which causes inflammation of skin and all mucous membrames.

heterochromatin   chromatin that remains compacted and stains deeply in interphase.

heteromorphs   homologous chromosomes that differ morphologically.

heterosis   the greater vigor in terms of growth, survival or fertility found in hybrids and associated with increased heterozygosity; synonymous with hybrid vigor.

heterozygous   having dissimilar alleles at one or more loci.

hexachlorophene   bactericidal and bacteriostatic agent used in antiseptic (and surgical) soaps, various cosmetics and skin preparations.

hexosaminidase-A   the enzyme whose deficiency causes Tay-Sachs disease, in which large amounts of a substance called ganglioside accumulates in brain tissue.

histamine   a substance produced by the amino acid histidine thought to be responsible for the dilation and increased permeability of blood vessels which play a major role in allergic reactions.

histocompatibility   tolerance to transplanted tissue.

histoincompatibility   intolerance to transplanted tissue.

hnRNA   heterogeneous nuclear RNA; RNA in the nucleus still containing introns.

holandric   appearing only in males; said of characters determined by genes on the Y chromosome.

holandric inheritance   inheritance by genes on the Y chromosome.

homogentisic acid   a compound derived from the metabolic breakdown of the amino acid tyrosine (see *alkaptonuria*).

homologue   a single member of a chromosome pair.

homologous   the relationship of two chromosomes that are homologues of each other.

homozygosity   condition of having identical alleles at one or more loci under consideration.

humoral immunity   the presence of free antibodies in the plasma of the blood.

humoral response   the release of a specific antibody into the blood stream (as opposed to the whole-cell or cell-mediated response).

hybrid   a plant or animal resulting from a cross between parents that are genetically unlike.

hybrid vigor   heterosis.

hybridoma   a cell hybrid made by the fusion of a cancer cell with a lymphocyte that has been induced to produce antibodies.

hydatidiform mole   tumor produced by a grossly abnormal pregnancy without embryo or placenta; usually benign but can become malignant.

hydrocephaly   congenital condition involving increased accumulation of fluid in the brain.

hypophosphatemia   phosphates below normal concentration in the blood.

hypothyroidism   a diminished production of thyroid hormone.

hypotonia   a relaxed muscular condition; found in some persons with the Down syndrome.

hypotonic solution   one with a salt concentration lower than that found in normal cells.

immune-competent   a cell capable of producing antibody in response to an antigenic stimulus.

immune-deficiency disease   caused by a deficiency of cells capable of producing antibody.

immunoglobulin   the antibody molecule, composed of two pairs of polypeptides capable of binding specific antigens.

index case   the clinically affected family member through whom attention is first drawn to a pedigree of particular interest to human genetics; synonymous with proband and propositus (or proposita).

infectious mononucleosis   glandular fever with a great increase of atypical or abnormal mononuclear leukocytes in the blood.

interphase   the period between successive mitoses.

intersex   a class of individuals of a bisexual species which has sexual characteristics intermediate between the male and female.

intervening sequences   groups of excess DNA nucleotides that interrupt the sequence of essential nucleotides at irregular intervals within the gene; these are eliminated enzymatically from the mRNA transcript, leaving the nucleotides in uninterrupted sequence; synonymous with introns.

introns   see intervening sequences.

inversion   a chromosomal segment has been deleted, turned through 180 degrees and reinserted at the same position on a chromosome.

in vitro   in the test tube; designating biological processes made to occur experimentally outside of the living organism; opposite of in vivo.

in vivo   within the living organism; opposite of in vitro.

ionization   the loss or gain of an electron by an atom.

isochromosome   an abnormal chromosome formed by the attachment of two similar arms to the same centromere.

isogenic   genetically uniform, as isogenic lines of laboratory animals.

isograft   a tissue graft between two individuals of identical genotype.

isolate   a subpopulation in which mating takes place exclusively with other members of the same subpopulation.

isotope   one of the several forms of a chemical element, differing from its other forms in atomic weight but not in chemical properties.

isozymes   different molecular forms of an enzyme.

jaundice   a condition characterized by yellowness of the skin, of the whites of eyes, of mucous membranes and of body fluids, caused by deposition of bile pigment resulting from excess bilirubin in the blood.

karyotype   the chromosomes of an individual arranged in the standard classification.

kindred   the extended family including distant relatives.

kinship   blood relationship; consanguinity; relationship by marriage.

krypton   a colorless inert noble gas (does not usually combine chemically with any other element), found in release from nuclear reactors.

lactase an enzyme that breaks down milk sugar (lactose).

Lamarckian theory the idea that species evolve by the inheritance of acquired characteristics.

lectins substances found in animals and plants with antibody-like properties; when found in plants, called phytohemagglutinins.

lethal equivalent a gene carried in the heterozygous state which, if homozygous, would be lethal; or a combination of several genes in the heterozygous state whose total effect when homozygous as a group would cause one death.

leukemia a generally fatal disease characterized by an overproduction of white blood cells, or a relative overproduction of immature white cells.

leukopenia a decrease in the number of white blood cells.

linkage the association of alleles at different loci because of their location on the same chromosome, shown by deviations from independent assortment in genetic crosses.

lipid any of a group of fatty substances soluble in organic solvents but barely soluble in water.

locus the position that a gene occupies on a chromosome.

lymphocyte a cell with a round nucleus seen in the lymph nodes, spleen and blood.

Lyonization X-inactivation; random inactivation of all X chromosomes but one in somatic cells during early embryonic life, resulting in female mosaicism.

macrophage a large white blood cell which ingests foreign material in the blood stream; important in the immune response.

malignant describes a neoplasm (cancer) that metastasizes.

major histocompatibility complex (MHC) a group of closely linked loci involved in the production of antigens important in transplantation.

manic-depressive psychosis a psychosis in which there are alternating moods of depression and euphoria.

maple syrup urine disease a hereditary disease arising from an enzyme deficiency which is characterized by urine that smells like maple syrup.

marker chromosome altered chromosome conspicuously different from normal so that it is easy to trace in a pedigree.

meiosis the final two cell divisions in germ cells by which gametes containing the haploid chromosome number are produced from diploid cells.

meiotic nondisjunction the failure of two members of a chromosome pair to separate during meiotic cell division so that both pass to one daughter cell and none to the other.

melanin the dark pigment responsible for the coloration of skin, hair, and the iris of the eye.

messenger RNA the RNA molecules complementary to DNA which are formed by transcription and function in translation.

metacentric a chromosome with a centrally placed centromere.

metafemale a Drosophila female with three X chromosomes and a diploid set of autosomes, infertile and relatively inviable.

metamale a Drosophila male with one X chromosome and three sets of autosomes, infertile and relatively inviable.

metaphase   the stage of cell division in which the chromosomes have reached their maximum condensation and are lined up on the equatorial plane of the cell, attached to the spindle fibers.

microcephaly   congenital abnormal smallness of the head sometimes seen in idiocy.

microtubule   a submicroscopic hollow filament found in the cytoplasm of many motile cells; also as a component of the mitotic spindle.

miscarriage   loss of a human fetus before the twenty second week of gestation; a spontaneous abortion.

miscegenation   sexual relations or marriage between humans of different races.

mitosis   division of body cells resulting in the formation of two cells, each with the same chromosome complement as the parent cell.

mitotic nondisjunction   the failure of two members of a chromosome pair to separate during mitotic cell division so that both pass to one daughter cell and none to the other.

monoclonal   derived from the mitotic reproduction of a single cell.

monosomy   a condition in which one chromosome of one pair is missing.

monozygotic   twins derived from a single fertilized ovum; almost always genetically and phenotypically indistinguishable; opposite of dizygotic.

morula   an early embryo that consists of a cluster of cleaving cells, prior to the blastula stage.

mosaic   an individual or tissue with at least two cell lines differing in genotype or karyotype, as a result of mutation or nondisjunction occurring during development.

mrem   millirem; a thousandth of a rem.

mRNA   messenger RNA.

multiple alleles   the existence in a population of more than two alleles at a given locus.

mutagen   a physical or chemical agent that increases the frequency of mutation above the spontaneous rate.

mutant   a gene in which a mutation has occurred; an individual expressing such a gene.

mutation   the process of change of one allele into another; more loosely, a permanent heritable change in the genetic material.

mutation pressure   the tendency to an increased frequency of an allele in a population because of repeated mutation to that allele over a period of time.

myelomatosis   a type of blood cancer in which a single antibody-forming plasma cell starts uncontrolled growth.

nondisjunction   the failure of two members of a chromosome pair to disjoin during anaphase of cell division so that both pass to one daughter cell and none to the other; may be either mitotic or meiotic.

nontasters   individuals (30 percent of humans) not able to taste a bitter substance called phenylthiocarbamide (PTC).

nucleosome   a histone-DNA complex which is partly responsible for the compact form of the chromosome.

nucleotide   one of the bases, along with a sugar and phosphate group, that makes up a unit of DNA.

nucleus a structure inside the cell which contains the chromosomes at interphase.

oocyte the female germ cell during the stages of meiosis.

oogonium a cell in the female germ line which divides by mitosis and eventually gives rise to oocytes.

oubain a poisonous drug, historically used in arrow poisons, useful in certain genetic tests.

ovulation the release of a ripe egg from the mammalian follicle.

ovum an unfertilized egg cell.

panmixis random mating; opposite of assortative mating.

paracentric inversion an inversion which does not include the centromere.

parthenogenesis the production of a new individual from an egg without any genetic contribution from a male.

pedigree a diagrammatic representation of family relationships indicating the normal and affected individuals and other aspects of genetic interest.

penetrance the probability that an individual with a particular genotype will show the expected phenotype.

pericentric inversion an inversion with a break in each of the two arms of the chromosome.

phenocopy a developmental defect in which the phenotype imitates one produced by a specific gene.

phenotype the appearance of an organism resulting from the interaction of the genotype and the environment.

phocomelia absence of the proximal portion of a limb or limbs, the hands and feet being attached to the trunk by a single bone.

phytohemagglutinins antibody-like compounds found in plants.

plasma the fluid portion of blood including fiber.

plasma cell a lymphocyte that synthesizes and releases antibodies.

plasmid an extra-chromosomal hereditary element.

pleiotropy the phenomenon of a single gene (or pair of genes) being responsible for a number of distinct and seemingly unrelated phenotypic effects.

polar body the nonfunctional product of female meiosis.

polygenic character a quantitatively variable phenotype dependent on the interaction of numerous genes.

polymorphism the presence of two or more variants of a given gene or chromosome each with an appreciable frequency.

polypeptide a chain of amino acids linked together by peptide bonds.

polyploid an individual having more than two sets of chromosomes.

polytene chromosome a giant chromosome found primarily in flies consisting of many identical chromosomes closely associated along their lengths.

position effect the change in the expression of a gene when its position is changed with respect to the neighboring genes.

precursor a chemical compound from which another compound is formed.

proband the person through whom a pedigree is ascertained; same as propositus or index case.

progestin a hormone which prepares the lining of the uterus for the implantation of the fertilized ovum; synonym: progesterone.

promutagen a class of compounds which themselves are inactive but whose metabolic breakdown products are active.

prophase   the early phase of mitosis or meiosis from the time that the chromosomes first become visible to the time when they are maximally condensed at metaphase.

propositus   the family member who first draws attention to a pedigree of a given trait; same as index case and proband.

protein   a complex molecule of high molecular weight consisting of one or more polypeptide chains.

puberty   a period in life at which one of either sex becomes functionally capable of reproduction.

quinacrine   an antimalarial drug; a fluorescent dye used in chromosome cytology.

race   a group, more or less isolated geographically or culturally, who share a common gene pool and who, statistically speaking, are somewhat different at some loci from other populations.

rad   a unit of energy like the roentgen, but based on absorbed energy and applicable to both ionizing and nonionizing sources; in many cases it may be considered about the equivalent of a roentgen.

recessive   an allele whose main phenotypic effect is expressed only when present in two doses.

reciprocal cross   a cross identical to another with respect to phenotypes and genotypes involved but with sexes of the parents interchanged.

reciprocal translocation   mutual exchange of segments between nonhomologous chromosomes.

recombinant DNA   artificially synthesized DNA in which a DNA segment from one organism is inserted into the genome of another.

recombination   the occurrence of combinations of genes different from those in the parents, due to independent assortment or crossing over.

recurrence risk   the probability that some characteristic present in one or more relatives will be found in another member of the same or a subsequent generation.

reduction division   first meiotic division in which the number of chromosomes is reduced by one half.

regulator gene   a gene whose primary function is to control the rate of synthesis of the products of other genes.

rejection   refusal of an individual's body to accept transplanted cells, tissue or organs.

rem   roentgen equivalent for man, a roentgen adjusted for the atomic makeup of the human body.

rep   roentgen equivalent physical; a unit of absorbed radiation approximately equal to a roentgen.

restitution   the spontaneous rejoining of broken chromosomes to produce the original configuration.

restriction enzyme   any of the enzymes that cut DNA at specific nucleotide sequences.

reverse transcriptase   an enzyme that catalyzes the formation of DNA using RNA as a template.

ribonucleic acid (RNA)   a nucleic acid similar to DNA except that it contains the sugar ribose (instead of deoxyribose) and the base uracil instead of thymine.

ribosome   cytoplasmic organelle composed of RNA and protein, on which polypeptide synthesis from messenger RNA occurs.

ring chromosome   a structurally abnormal chromosome in which the end of each arm has been lost and the broken arms have reunited in ring formation.

RNA   see ribonucleic acid.

Robertsonian translocation   a translocation involving all of the essential genetic material of the long arms of two acrocentric chromosomes.

roentgen   the quantity of ionizing radiation that liberates 2×10^9 ion-pairs per cm^3 of air or 1.6×10^{12} ion-pairs per cm^3 of water.

satellite   a chromosome segment attached to the main part of the chromosome arm by means of a thin strand.

serum   the watery portion of blood remaining after fibrinogen has been removed from the plasma by coagulation.

sex chromosomes   chromosomes responsible for sex determination, the X and Y chromosomes.

sibship   a group of brothers and sisters.

sib   brother or sister.

somatic mutation   a mutation occurring in body cells (as opposed to germ cells).

sperm   the mature male gamete.

spermatocyte   a male germ cell during the period of meiosis.

spermatogenesis   male meiosis.

spermatogonium   a mitotically active cell in the testes which produces primary spermatocytes.

spina bifida   congenital defect in walls of spinal canal as a result of which the membranes of the spinal cord are pushed through the opening.

stillbirth   the birth of a dead fetus.

structural gene   one that codes for the amino acid sequence of a polypeptide chain.

submetacentric   describes a chromosome with a centromere between the center and one end.

superfecundation   successive fertilization of two or more ova from the same ovulation.

synapsis   pairing of homologous chromosomes in meiosis.

syndrome   a group of symptoms that occur together, characterizing a disease.

syntenic   located on the same chromosome.

T-cell   small lymphocytes committed by the influence of the thymus gland to be responsible for cell-mediated response to antigens.

taster   an individual able to taste a bitter substance called phenylthiocarbamide (PTC).

telocentric   a chromosome with a terminal centromere.

telophase   the stage of cell division that begins when the daughter chromosomes reach the poles of the dividing cell and lasts until the two daughter cells take on the appearance of interphase cells.

teratogenic effects malformations caused during embryonic development.

teratogen an agent that produces or raises the incidence of congenital malformations.

teratoma a severely deformed fetus containing embryonic elements of the primary germ layers (hair, teeth, etc.).

thalassemia a human anemia caused by defective hemoglobin synthesis.

titer the strength of a solution, or concentration of a substance in a solution.

transcription the formation of messenger RNA against a DNA template.

transfer RNA RNA of low molecular weight with an anticodon sequence specific for one amino acid.

transformation transfer of genetic information from one cell to another by means of free DNA.

translation the synthesis of a polypeptide chain using mRNA to direct the amino acid sequences.

transplantation transfer of cells, tissue or organ from one organism to another or to another part of the same organism; a graft.

triploid an organism having three haploid sets of chromosomes in each nucleus.

trisomy the state of having three of a given chromosome instead of the usual pair, as in trisomy-21, Down syndrome.

tritium radioactive isotope of hydrogen, with a half-life of 12.5 years; it emits particles of very low energy and is the most widely used radioisotope in biological studies.

trivalent in meiosis, three chromosomes paired in a single configuration.

tRNA see *transfer RNA*.

tumor promotors noncarcinogenic substances that enhance the effect of a small amount of a carcinogen that would otherwise be relatively ineffective.

tyrosinase an enzyme that converts tryosine to the intermediate steps in the formation of the pigment melanin; deficiency results in albinism.

ultrasound sound waves of frequency higher than audible range generated in equipment used to view internal tissue without radiation.

univalent a single unpaired chromosome during meiosis.

variable expressivity production of different abnormal phenotypes by the same genotype.

viability a measure of the proportion of surviving individuals of one type compared to another type.

X chromosome the sex chromosome found twice in normal human females and singly, along with a Y chromosome, in normal human males.

X-linked genes on the X chromosome, or traits determined by such genes.

Y-chromatin body a brightly fluorescing body in quinacrine-stained cells that distinguish them as Y chromosomes.

Y chromosome sex chromosome found in normal males, along with an X chromosome.

zygosity the type of twinning event that gave rise to a twin pair, either monozygotic (identical) or dizygotic (fraternal).

zygote the diploid cell formed from fusion of sperm and ovum.

Index

When there are several pages per entry, **boldface** is used to indicate a definition or major discussion, if any.